INSTRUMENTATION
BETWEEN SCIENCE, STATE AND INDUSTRY

Sociology of the Sciences

VOLUME XXII

The titles published in this series are listed at the end of this volume.

INSTRUMENTATION BETWEEN SCIENCE, STATE AND INDUSTRY

Edited by

BERNWARD JOERGES

Wissenschaftszentrum Berlin für Sozialforschung, (WZB) and
Technische Universität Berlin, Germany

and

TERRY SHINN

Sociology and History of Science,
CNRS/GEMAS, Paris, France

KLUWER ACADEMIC PUBLISHERS

DORDRECHT / BOSTON / LONDON

A C.I.P. Catalogue record for this book is available from the Library of Congress.

ISBN 0-7923-6736-7

Published by Kluwer Academic Publishers,
P.O. Box 17, 3300 AA Dordrecht, The Netherlands.

Sold and distributed in North, Central and South America
by Kluwer Academic Publishers,
101 Philip Drive, Norwell, MA 02061, U.S.A.

In all other countries, sold and distributed
by Kluwer Academic Publishers,
P.O. Box 322, 3300 AH Dordrecht, The Netherlands.

Printed on acid-free paper

TABLE OF CONTENTS

PART II   INTERSTITIAL WORLDS

PART III   PURVIEWS OF GENERIC INSTRUMENTS

IN CONCLUSION

CHAPTER 1

BERNWARD JOERGES[a] AND TERRY SHINN[b]

A FRESH LOOK AT INSTRUMENTATION
AN INTRODUCTION

In the 1930s, 40s, and 50s, the American Jesse Beams (1898–1977) developed the modern ultra-centrifuge (Elzen 1986; Gordy 1983). The device and the man do not fit neatly into any standard institutional, professional, or intellectual mold. Long-time chairman of the University of Virginia physics department, Beams also sponsored two firms, acted as a key consultant to four additional companies, participated in the Manhattan Project, worked for the military during the 1940s and 50s, and contributed to numerous NSF science programs. Beams was not the classical academic, engineer, entrepreneur, nor technical consultant. Although often located at or near the University of Virginia, his principal connection to that academic institution was the huge and well-equipped workshops that he developed there during decades of arduous endeavor (Brown 1967).

Beams' ultra-centrifuge had a parallel life. The ultra-centrifuge was a by-product of his 1924 doctoral dissertation which focused on rapidly rotating mechanical systems. Assigned by his thesis director to investigate the time interval of quantum absorption events, Beams developed a high-speed rotating technique for the accurate measurement of very short intervals of time. This device, and not the study of physical phenomena, was the centerpiece of his successful dissertation. An interest in multipurpose, multi-audience technical apparatus rather than a focus on the stuff of the physical world emerged as Beams' guiding logic. Yet this focus did not make Beams an engineer or technologist in the usual sense of the term.

His initial devices employed air-driven turbines. However, their performance was limited by mechanical factors as well as by air friction. He first augmented speed by introducing a flexible drive-shaft which allowed for adjustments in the center of gravity, thereby multiplying rotating capacity. He next placed the rotating vessel inside a vacuum, thereby eliminating air friction. But nonetheless shaft mechanics continued to restrict performance. To solve this, Beams employed magnets to spin his vessel. The vessel was suspended inside a vacuum, thanks to a magnet-based servomechanism. This constituted his consummate ultra-centrifuge which rotated at previously unheard-of rates.

1

B. Joerges and T. Shinn (eds.), Instrumentation: Between Science, State and Industry, 1–13.
© 2001 *Kluwer Academic Publishers.*

The ultra-centrifuge became an important element in bio-medical research on bacteria and viruses, and soon figured centrally in medical diagnosis and treatment. Beams engineered devices for radioactive isotope separation in the late 1930s which were effectively tested in the Manhattan Project and became commercially viable in the 1950s and 1960s. The Beams ultra-centrifuge served in early ram jet propulsion research, and it was also used to do physics and engineering research on the strength of thin films. A Beams device rotating at over three million revolutions per second was used by physicists to measure light pressure. A somewhat different instrument enabled enhanced precision in the measurement of the gravitational constant.

As an author Beams published abundantly, sometimes in disciplinary periodicals, but much more of his written output appeared in instrumentation journals, such as the American *Review of Scientific Instruments*. A high proportion of his writings took the form of unpublished technical reports and he co-sponsored half a dozen patents. Beams' written productions were equally divided between the public and private spheres: between articles and patents on the one hand (public), and confidential reports and consultancy on the other (private). Concurrent with these publications, he continued to build far-reaching artefacts.

Beams and his devices crossed innumerable boundaries. Beams circulated in and out of institutions and shifted from employer to employer. He belonged to many organizations, movements, and interests. He was neither a-institutional nor anti-institutional, but rather multi-institutional. He had no single home; his home lay everywhere. He explored and exploited the laws of nature as embedded in instruments and, like Beams himself, his ultra-centrifuges also crossed a multitude of boundaries. They were open-ended, general-purpose devices which came to perform a host of functions and found their way into a variety of non-academic publications and applications.

A special vocabulary and way of seeing events developed in conjunction with the Beams device. Light pressure and gravitation, isotope separation and thin films, microbes and viruses came to be spoken about in terms of rotational speeds and centrifugal pressures. "Rotation," and with it "specific density," emerged as a lingua franca for a disparate spread of fields and functions, extending from academia and research to industrial production and medical services. The rotational vocabulary and imagery of Beams' instrument percolated outward. Beams' approach and his artefacts thereby helped coalesce dispersed technical, professional, and institutional worlds.

RESEARCH-TECHNOLOGY

The Beams ultra-centrifuge is just one instance of what we label "research-technology." The term research-technology first arose in the early 1930s in an exchange of letters between the Dutch Nobel Laureate Pieter Zeeman (1865-1943) and the French physicist Aimé Cotton (1869-1951) – director of the laboratory that housed the Bellevue giant electromagnet (Shinn 1993). In the context of that correspondence, "research-technology" referred to multipurpose devices for detection, measurement and control that were conceived and developed by a community connected to both science and industry – yet at the same time also separate from each of these. In this book, we appropriate their conception of research-technology, and extend it to many other phenomena which are less stable and less localized in time and space than the Zeeman/Cotton situation. In the following pages, we use the concept for instances where research activities are orientated primarily toward technologies which facilitate both the production of scientific knowledge and the production of other goods. In particular, we use the term for instances where instruments and methods traverse numerous geographic and institutional boundaries; that is, fields distinctly different and distant from the instruments' and methods' initial focus.

We suggest that instruments such as the ultra-centrifuge, and the trajectories of the men who devise such artefacts, diverge in an interesting way from other forms of artefacts and careers in science, metrology and engineering with which students of science and technology are more familiar. The instrument systems developed by research-technologists strike us as especially general, open-ended, and flexible. When tailored effectively, research-technology instruments potentially fit into many niches and serve a host of unrelated applications. Their multi-functional character distinguishes them from many other devices which are designed to address specific, narrowly defined problems in a circumscribed arena in and outside of science. Research-technology activities link universities, industry, public and private research or metrology establishments, instrument-making firms, consulting companies, the military, and metrological agencies. Research-technology practitioners do not follow the career path of the traditional academic or engineering professional. They pursue "hybrid careers," shifting back and forth between different employers. Others, while remaining with a single employer, establish strong, albeit intermittent contacts with a variety of arenas which are not otherwise connected.

In conventional parlance, the analytic language used by sociologists and historians of science and technology often draws a distinction between technology and academic learning. The world of research-technology, we suggest, bridges the two. The bridging occurs with respect to knowledge, skills, artefacts, language and imagery, and their attendant interactions. In a research-technology frame, conventional oppositions such as theoretical and experimental, science or engineering, technology and industry are largely effaced. In this frame, the focus is neither on scientific practices, in the sense of theorizing about experimentally produced phenomena, nor on engineering practices, in the sense of constructing and producing definite end-user goods and services. Instead, the focus is on practices oriented toward the production and theorizing of open devices which potentially serve multiple spheres.

The research-technology perspective raises issues in three problem domains. Firstly, how can the research-technology phenomenon be situated with respect to the ongoing debate about the dynamic relationships of science and society? Secondly, how can it be situated with respect to a gradual scientization and increased occupational fluidity of engineering professions which characterizes the changing relationships between science and engineering? Thirdly, how can it be situated in the contemporary debates in philosophy and social studies of science over the relationships between theory and experiment? In this introductory chapter, we will briefly address each of these points before outlining the general analytic coordinates that structure the book.

Science and Society

The theme of "instrumentation between science, state and industry" does not square well with the venerable discourse which opposes "science" and "technology" in social studies of science. In this discourse, "technology" stands for the contrary of "science"; it represents the practical uses of science in society at large and is understood as separate from the somehow autonomous sphere of "science" (Layton 1971a). This vocabulary, widespread as it may be, is not very useful for our purposes, and, for that matter, for any inquiry into the role of instruments. Technology, in the sense of technical instruments and the knowledge systems that go with them, pervades all societal systems. There are technologies of science, of industry, of state, and so forth, and it would be ill-advised to assume that, in the end, they all flow out of "science." But even if the crude opposition of science and technology has little analytic value, the dual problem remains: how to effectively conceive the dynamic relationship between scientific spheres and other societal spheres, and how to conceive the role that technological matters play in this relationship.

Much of the debate surrounding these issues is framed in terms of "What drives what?" Does science drive technology (that is production technology, the field of utilitarian technology aimed at producing things for use outside science) or does technology drive science? Using "industry" and "state" as we do in this book as shorthand for extra-scientific social spheres, this translates into the question: Do science and its technologies drive those of industry and the state, or is it the other way around?

Schematically speaking, the relationship can take four forms: science drives industry/the state; industry/the state drives science; the relationship is independent; or it is dialectical. In terms of ideal types, these four positions have all had their protagonists. The current fashion seems to be a special version of the dialectical answer where science and industry/the state are inextricably interrelated (e.g. Latour 1992). In extreme formulations, the science/technology nexus has become a hybrid field of seamless webs where the distinction between them is no longer considered useful. According to this view, there is only technoscience, in which the boundaries between science and industry/the state are discursive artefacts that must be looked at in terms of their strategic utility. Moreover, these boundaries are in constant flux depending on the interests of dominant players.

The research-technology perspective does not accord with seamless analytical frames of this kind. We will argue that research-technology instrumentation is a phenomenon "in the middle" which does not coincide with either science or industrial production. We see it as a field of instrumentation outside both science and industry, yet important for both.

It is possible then to distinguish three spheres of instrumentation and instrument-makers: inside science, as in conventional studies of scientific instrumentation (Heidelberger and Steinle 1998; Heilbron, van Helden, and Hankins 1992; Löwy and Gaudillière 1998); inside industrial production, as in conventional studies of non-scientific technology, such as the assembly line (Noble 1984); and outside science and production, but for both. This third type belongs to research-technology. In other words, we wish to bypass one erstwhile notion whereby instrumentation in science and technology has two distinctly different sources, and another erstwhile notion whereby technology is an applied side of science.

The strong thesis that guides the analyses presented in this book is that research-technology generates broad fundamental impulses that drive scientific research, industrial production and technology-related state activities along their respective paths. Of course, the research-technology hypothesis does not deny that much instrumentation is conceived, developed and diffused within the strict confines of a narrow industrial (von Hippel 1988) or scientific (Edge and Mulkay 1976) context, nor does it imply that research-technology mechanisms account for all types of transfer from one sphere to another.

Science and Engineering

To better understand the emergence of research-technology, it is useful to see it against broad transformations in engineering practice and institutions. Historically, the knowledge base and professional practices of engineers in many fields have changed appreciably as technology has become ever more scientized. In the past, engineering was often associated with practical craft skills and with the application of technical recipes to concrete problems. Since at least the second World War, the intellectual and professional gap that separated science and engineering has gradually diminished. Emblematic of this rapprochement is the increasing use of the terms "engineering science" in the Anglo-American world, "Ingenieurwissenschaften" in German-speaking countries and "science physique pour l'ingénieur" in France.

The professional identity of engineering groups in civil engineering, mechanics, chemistry, electricity and electronics often entailed a demarcation from mathematized esoteric learning and disciplinary academic science, as well as a demarcation from the university departments that taught and researched such learning. While engineers trained in university schools of engineering, in many important respects they nevertheless stood outside of academia. Engineers' principal intellectual and professional identity instead lay with their industrial employers. Professional engineers generally centered their careers in non-academic organizations, where they usually remained (Layton 1971b). This traditional profile has changed appreciably, however. Today, engineering knowledge and practice increasingly bear the mark of high science as, in turn, academic disciplines depend increasingly on scientized engineering (Bucciarelli 1994).

The scientization of engineering is associated with growing cognitive specialization. New fields of academic learning have emerged, and many of them are directly relevant to engineering. Mastery of these fields by engineers often entails a grasp of advanced mathematics, as well as a firm grounding in academic science. Concurrently, many technical systems have become ever more complicated and large-scale, thereby requiring additional learning and skills. Beyond this, the scientization of engineering has involved significant professional changes. Engineers had long been envious of the luster of science and the high social status of scientists. The emerging links between engineering and academia have provided engineering professions with an opportunity to share the elevated status of academic learning. Also, scientized engineering involves enhanced career fluidity. Engendered by fast-moving technical frontiers, many practitioners move from project to project.

The last few years have seen the rise of two analytic schemata that focus on a convergence between scientists and engineers. In *The New Production of Knowledge*, Gibbons and his colleagues have suggested that the development of new knowledge-intensive economic spheres is accelerating the de-differentiation between scientists and engineers, and is producing a new category of cognitive and technical personnel whose point of reference is the solution of socially relevant problems (Gibbons et al. 1994). The Triple Helix perspective similarly hypothesizes a radical convergence between scientists and engineers – a convergence which putatively yields a historically new intellectual and technical breed expressed as a synthesis of the two professional groups (Etzkowitz and Leydesdorff 2000, 109–23; Leydesdorff and Etzkowitz 1998). This synthesis does not, however, take the form of a de-differentiation but instead a neo-differentiation (Shinn 1999). At first glance, research-technology might appear to belong to the New Production of Knowledge or Triple Helix schemata. However, it has to be established whether the kind of fluidity we associate with research-technology is of the same sort described in these two perspectives, particularly as regards the intellectual and social work connected with instrumentation.

Theory and Experiment

With few exceptions, students of science have long considered that experimentation was paramount in scientific research. Experimentation was seen as guiding theory, or even as governing it. This stance is reflected in many of the classical studies on Newton, Galileo, and Huygens, and it underpinned the work of philosophers in the logical positivist tradition (Suppe 1974; Westfall 1980). Pierre Duhem was among the first to question the dominance of experimental orthodoxy, and Kuhn successfully extended Duhem's thesis (Duhem 1915; Kuhn 1962). The relationship between theory and experimentation continues to be reassessed, and today many scholars believe that theory often guides, and even dictates experiments and their outcome (Bachelard 1951; Pickering 1984; Pinch 1986; Quine 1972, 1986).

Nevertheless, a handful of historians and sociologists question whether the relationship between theory and experimentation is as direct and unmediated as it is often made out to be. Peter Galison, for example, has argued that the old debate about the interplay of experiment and theory, and the attendant ideological debates about the epistemological correctness of idealist and empiricist positions, needs to be revised by introducing a third dimension; namely, instruments and the theories attached to them (Galison 1997). Galison does not suggest that instrumentation provides a panacea for establishing the validity of a knowledge claim; he instead indicates that instruments constitute a third reference against which statements can be tested, and are a semi-autonomous input into both experimentation and theory. Nevertheless, his approach also focuses predominately on the role of instrumentation inside science proper. It is a debate about science and technology in the procrustean framework of technology in and of science. Beyond Galison's influential contribution, one can observe a general renewal of interest in the technical, cognitive-epistemological and socio-cultural aspects of metrological devices throughout the field. How does the research-technology perspective fit into this debate?

In positing that research-technology is a specific kind of instrumentation, one which is explicitly characterized as poly-disciplinary and potentially extra-scientific in its purposes and effects, we confront the theory/experimentation problem from a different angle. It may safely be said that mainstream philosophical and sociological schools in the study of science have generally paid scant attention to boundary-crossing practices and representations of the sort common to research-technology where instrumentation transcends experimentation and the theory/experimentation matrix. This line of inquiry extends recent claims that independently of measuring and representing effects, experimental systems also perform controlling and productive functions for purposes beyond scientific knowledge and theory validation (Hagner and Rheinberger 1998: 355–73; Heidelberger 1998: 71–92).

A SPECIFIC KIND OF INSTRUMENTATION

Against the backdrop of ongoing debates around science/society relationships and theory/experimentation relationships, and changes in engineering practice and institutions, we can now turn our attention to the emergence and workings of research-technology. Referring back to the example of Jesse Beams and the ultra-centrifuge which opened this chapter, three major features of research-technology come to the fore. The first characteristic is its trans-community positioning, or as we say, its "interstitiality." Research-technologists wear many hats. Secondly, their devices exhibit a peculiar openness or "generic" quality. Research-technology devices branch outward to many spaces. Thirdly, research-technologies involve the development of standardized languages or "metrologies." Research-technologists create a lingua franca for theoretical and extra-theoretical uses.

The case histories presented in this book explore social interstitiality, generic instrumentality and metrological codification in a variety of trans-disciplinary, trans-science and extra-science settings. What accounts for this configuration and how do research-technologies acquire their distinct feature of travel between otherwise unconnected fields? How is it possible that local instrument achievements become global in the sense of a re-embedding in many other places, both inside and outside science?

Interstitial Communities

In what sense can one talk about research-technology communities? Jesse Beams, and to a greater or lesser degree the research-technologists who appear in this book, exhibit peculiarly "subterranean" modes of multi-lateral professional and institutional association which do not accord well with standard sociological notions of communities as ensembles of stable, institutionalized interactions. These research-technologists admittedly work within universities, industry, state or independent establishments, yet at the same time they maintain some distance from their organizations. In many instances, they pursue "hybrid careers," shifting back and forth between different employers or, while remaining with a single employer, lend their services to changing outside interests. We will also show that many research-technologists develop a personality make-up suited to sustain many-sided professional relationships and "multi-lingual" cognitive worlds.

Some sociologists will say that research-technology's social configurations should not, for these reasons, be called "communities," but rather non-communities, since research-technologists are not concentrated within one type of scientific, industrial or state organization which provides them with stable, recognized positions reserved for experts in generic precision instrumentation. Indeed, research-technologists' community identity cannot be mapped in terms of an organizational or professional referent. The referents of "academic scientist" or "industrial engineer" are not relevant to research-technology. Neither can the identity of research-technologists be based on the production of a definite category of fact (in science) or artefact (in production). Instead, we suggest that the shared project which conveys a semblance of community in the familiar sense of the term is their elaboration of diffuse, purposefully unfinished devices (not-yet facts and not-yet artefacts) to be distributed across the broadest possible landscape.

In cases where research-technology involves a shared project for groups of practitioners working within the same field of instrumentation, the term community, in the classical sociological sense, will be acceptable to most analysts. In other cases though, "shared project" merely means that research-technologists recognize each other's pursuits when they happen to meet. The term research-technology community refers here to something akin to the way tribesmen know they belong to the same tribe. In order to avoid confusion with other tribes, various insider/outsider affiliations are invoked. Rather than by tracing stable membership and hierarchical/promotional career structures, research-technologists can more easily be identified through specialized academic or trade journals and by their participation in national or international instrument fairs and expositions. Historically, instrument fairs have played a major role in the constitution of the research-technology movement (Shinn, chapters 3 and 5).

In connection with interstitiality we need to understand how research-technologists avoid standard forms of professionalization. What are the sources of their open and flexible group identities? Their interest as a class of experts seems to lie in expanding the sphere of unaffiliated, open-to-all, dispersed generation of devices that promise solutions to problems where precision detection and measurement, precision control of certain phenomena and even the controlled production of certain effects are crucial for success (Roqué, chapter 4). How do research-technologists manage to articulate and defend group interests in the absence of membership organizations with established boundaries? Separate as research-technology groups are from both conventional science and industrial engineering, yet parasitic on both, how do these quasi-communities assure community reproduction and growth? How do they sustain their autonomy in environments which have customarily rewarded monopolistic organizational linkages? (Nevers et al., chapter 6; see also Johnston, chapter 7.)

Generic Devices

We refer to the particular kind of technical artefacts research-technologists deal with as "generic devices." Research-technology communities first arose in the nineteenth century with precision mechanics and optics (Jackson, chapter 2) and today specialize in the invention, construction and diffusion of precision instrumentation for use both inside and outside academia. They develop packages or whole systems of generic detection, measurement, and control devices that focus on particular parameters which are potentially of interest to scientists, laboratory technicians, test personnel, production engineers, and planners (Gaudillière, chapter 9; Johnston, chapter 7, Rheinberger, chapter 8). Sometimes, as in the case of early lasers and masers, or in the case of laboratories producing new semi-conducting materials, research-technologists and their generic devices produce novel physical effects in order to explore their measurability and controllability.

In many instances, these devices are not designed to respond to any specific academic or industrial demand. Research-technologists may sometimes generate promising packets of instrumentation for yet undefined ends. They may offer technological answers to questions that have hardly been raised. Research-technologists' instruments are then generic in the sense that they are base-line apparatus which can subsequently be transformed by engineers into products tailored to specific economic ends or adapted by experimenters to further cognitive ends in academic research. Flexibility is part of the product. One could say that "interpretive flexibility" constitutes in itself a goal and an achievement. This is a precondition for research-technology's extended market that stretches from academia to industry and the state.

Roqué and Rheinberger, in chapters 4 and 8 respectively, show that research-technologists are typically involved in prototyping, in the sense that they avoid early closure of design processes that keeps devices generic. In connection with genericity we need to understand how research-technologists manage to maintain an instrument chain in which "core devices" are developed, that then spawn cascades of secondary apparatus, which are in turn used to solve a range of problems. How do generic devices make their way into both research and production?

Metrology

Metrologies can be seen as systems of notation, modeling and representation, including their epistemic justifications. Metrology is integral to the development of generic devices and the maintenance of interstitiality. Either the nomenclatures, units of measurement and standards of existing metrologies are refashioned in creating generic instrumentation, or else new ones are formulated. The lingua franca of metrology constitutes the vehicle that allows generic apparatus access to many audiences and arenas. At the same time, it preserves research-technologists from becoming caught up in the particular discourses of these audiences and arenas.

On one level, research-technologists may generate novel ways of representing, visually or otherwise, events and empirical phenomena. On a broader level, they may impose a novel view of the world by dint of establishing and legitimating new functional relations between recognized categories of elements that were previously perceived in a different light. In some cases, research-technologists' metrological work is instrumental in coalescing and crystallizing notations, analytic units and formulae into a corpus of rules or procedures which deserve to be called a methodology, and that eventually make their way into textbooks as state-of-the-art procedures. How is this achieved?

Ultimately, the issue of metrology includes questions concerning the particular epistemological stances, and even world views, associated with research-technology work. Do research-technologists sometimes even stylize and theorize their own practice and procedures in a manner that deserves to be called the advancing of a world view or episteme? (An example is the sweeping and comprehensive views of cyberneticists who see nature as a grandiose engineering feat, see Heims 1991.)

Dis-Embedding, Re-Embedding

One way of drawing together considerations of the institutional, instrument and metrological aspects of research-technology processes is to look at them in terms of an iteration of dis-embedding and re-embedding episodes in the far-flung trajectories of a particular device or prototype. Recent approaches in the philosophy and sociology of science and technology have consistently pointed to the situatedness, localness and embeddedness of all knowledge production. Arguments about instruments are at the core of these positions, whether they are framed in terms of tacit knowledge, craft, the bodies of experimenters, or science vernaculars (including Pidgin and Creole). At the same time, claims about universal standards of rationality in experimentation and engineering tend to be presented as mere representations or legitimations of scientific and technological practice.

In contrast, research-technology, as a distinctive mode of producing instrumentation for de-situated and trans-local uses both inside and outside science, appears as a distinct achievement of dis-embedding which lies outside the purview of such approaches. In this perspective, dis-embedding does not occur by default, as in diffusion theories, but is instead tied to specific skills and forms of representation. While admittedly all knowledge production, including instrument knowledge, is local, and all knowledge consumption is local too, the central question remains: how can knowledge be consumed far from its place of production, and how does it travel?

We suggest that generic instruments comprise a sort of dictionary that enables the translation of local practices and knowledge into diverging and multiple sites, and constitutes the transverse action of research-technology. Can something akin to universality arise through the sharing of common skills and representational systems located in something like a template, or "hub matrix?" Could one say that research-technologists design dis-embedded generic devices so that they can be readily re-embedded? Local re-embedding by engineers or scientists occurs within the limitations contained in the template of the generic instrument and also within the limitations of the local cultural and material context. Re-embeddings can thus differ considerably from one another, yet a certain fidelity to the hub template persists. To what extent does the use of a specific template by practitioners in different locales allow them to communicate effectively through the development of converging skills, terminologies and imagery? It may be this feature that makes research-technology the potent, universalizing motor that we take it to be.

THE BOOK

The instrument-related phenomena dealt with in this book may be seen as new in the sense that they have become more varied and broadly visible since World War II, yet it would be inappropriate to see research-technology as something radically new. Also, while research-technology may eventually increase in size and scope, this does not indicate that it is a new form of science. Instead, we consider research-technology as a new perspective, an alternative way of looking at instrumentation for social studies of science and technology. Since it is very much a phenomenon "in-between" and often relatively invisible to outside observers, it is not surprising that it has gone largely unnoticed by students of science and technology.

The episodes examined in this book span more than a century, beginning in the early 1800s and ending in the 1980s. Part I, on German optics and the *Zeitschrift für Instrumentenkunde*, traces early beginnings. In the remainder of the book contributions are organized according to the emphasis given to key analytic parameters set out in this introduction. To differing degrees and in different ways, the case histories explore how many interests, institutions, disciplines, and professions are traversed by generic instrumentation and its dis-embedders and re-embedders. Interstitiality provides the focus of Part II: While historians and sociologists generally concentrate on the genesis of stable relations and stabilizing structures, the very essence of research-technology is its fluidity and its operation between established institutions and interests. Part III explores instrument genericity: authors examine the trajectories of generic instrumentation systems and their specific applications. Part IV deals with metrological issues, in particular the roles played by generic instrumentation and interstitial communities in the work of standardization.

The ten studies presented in this volume explore the circumstances under which research-technology fields have emerged and evolved in light of changing demands inside and outside of science. Contributors deal with the places, times, and technological fields where research-technology occurs. They present the institutions, journals, meetings, forms of association, and the multi-professional and multi-personal identities that sustain research-technologies. In the concluding chapter we will situate research-technology in the landscape of social studies of science and technology and reflect on some of the broader societal corollaries of the research-technology movement.

[a]*Wissenschaftszentrum Berlin für Sozialforschung, Berlin, Germany*
[b]*GEMAS/CNRS, Paris, France*

REFERENCES

Bachelard, Gaston (1951). *L'activité rationaliste de la physique contemporaine*. Paris: PUF.

Brown, Frederick L. (1967). *A brief history of the Physics Department of the University of Virginia, 1922-1961*. Charlottesville, VA: University of Virginia Press.

Bucciarelli, Larry (1994). *Designing engineers*. Cambridge, MA: MIT Press.

Bud, Robert and Cozzens, Susan E. (eds.) (1992). *Invisible connections: Instruments, institutions and science*. Bellingham: SPIE Optical Engineering Press.

Duhem, Pierre (1915). *La théorie physique, son objet, sa structure* (1st ed.). Paris: J. Vrin.

Edge, David O. and Mulkay, Michael (1976). *Astronomy transformed: The emergence of radio astronomy in Britain*. New York: Wiley.

Elzen, Boelie (1986). Two ultracentrifuges: A comparative study of the social construction of artefacts. *Social Studies of Science* 16(4), 621–62.

Etzkowitz, Henry and Leydesdorff, Loet (2000). The dynamics of innovation: From national systems and "mode 2" to a triple helix of university-industry-government Relations. *Research Policy* 29(2, February): 109–23.

Galison, Peter (1997). *Image and logic: Material culture of microphysics*. Chicago: Chicago University Press.

Gibbons, Michael, Limoges, Camille, Nowotny, Helga, Schwartzman, Simon, Scott, Peter and Trow, Martin (1994). *The new production of knowledge. The dynamics of science and research in contemporary societies*. London: Sage

Gordy, W. (1983). *Jesse Wakefield Beams. Biographical Memoirs*, Vol. 54 (pp. 3–49). National Academy of Sciences of the United States of America.

Hagner, Michael and Rheinberger, Hans-Jörg (1998). Experimental systems, objects of investigation, and spaces of representation, in M. Heidelberger and F. Steinle (eds.), *Experimental Essays – Versuche zum Experiment* (pp. 355–73). Baden-Baden: Nomos

Heidelberger, Michael and Steinle, Friedrich (eds.) (1998). *Experimental essays – Versuche zum Experiment*. Baden-Baden: Nomos.

Heidelberger, Michael (1998). Die Erweiterung der Wirklichkeit, in M. Heidelberger and F. Steinle (eds.), *Experimental essays – Versuche zum Experiment* (pp. 71-92). Baden-Baden: Nomos.

Heilbron, John, van Helden, A., and Hankins, T.L. (eds.) (1992). Instruments (special issue). *Osiris* 6.

Heims, Steve J. (1991). *The cybernetics group*. Cambridge, MA: MIT Press.

Etzkowitz, Henry and Leydesdorff, Loet (2000). The dynamics of innovation: From national systems and "mode 2" to a triple helix of university-industry-government Relations. *Research Policy* 29(2), 109–23.

Hippel, Eric von (1988). *The sources of innovation*. Oxford: Oxford University Press.

Kuhn, Thomas S. (1962). *The structure of scientific revolution*. Chicago: University of Chicago Press.

Latour, Bruno (1992). *Aramis ou l'amour des techniques*. Paris: La Découverte.

Layton, Edwin T. (1971a). Mirror-image twins: The communities of science and technology in 19th century America. *Technology and Culture* 12, 562–80.

Layton, Edwin T. (1971b). *The revolt of the engineers: Social responsibility and the American engineering profession*. Cleveland, OH: Press of Case Western Reserve University.

Leydesdorff, Loet and Etzkowitz, Henry (eds.) (1998). *A triple helix of university-industry-government relations: The future location of research?* New York: Science Policy Institute.

Löwy, Ilana and Gaudillière, Jean-Paul (1998). *The invisible industrialist: manufactures and the construction of scientific knowledge*. London: Macmillan.

Noble, David (1984). *Forces of Production: a Social History of Industrial Automation*. New York: Knopf.

Pickering, Anthony (1984). *Constructing quarks: A sociological history of particle physics*. Chicago: University of Chicago Press.

Pinch, Trevor J. (1986). *Confronting nature: The sociology of solar-neutrino detection*. Dordrecht, The Netherlands: D. Reidel Publishers.

Quine, Willard V.O. (1969). *Ontological Relativity and other Essays*. New York: Columbia University Press.

Quine, Willard V.O. (1972). *Methods of logic*. New York: Holt, Rinehart and Winston.

Quine, Willard V.O. (1986). *Philosophy of logic*, 2nd ed. Cambridge, MA: Harvard University Press.

Shinn, Terry (1993). The Bellevue grand electroaimant, 1900-1940: Birth of a research-technology community. *Historical Studies in the Physical Sciences* 24, 157–87.

Shinn, Terry (1999). Change or mutation? Reflections on the foundations of contemporary science. *Social Science Information* 38(1, March), 149–76.

Suppe, Frederick (1974). *The structure of scientific theories*. Chicago: University of Illinois Press.

Westfall, Richard S. (1980). *Never at rest: A biography of Isaac Newton*. Cambridge, UK: Cambridge University Press.

PART I

ORIGINS OF THE RESEARCH-TECHNOLOGY COMMUNITY

CHAPTER 2

MYLES W. JACKSON

FROM THEODOLITE TO SPECTRAL APPARATUS: JOSEPH VON FRAUNHOFER AND THE INVENTION OF A GERMAN OPTICAL RESEARCH-TECHNOLOGY

One of Joseph von Fraunhofer's (1787–1826) greatest contributions to technical optics had its roots in ordnance surveying. Fraunhofer, the German territories' leading optician and optical-glass manufacturer of the period, needed an apparatus that could measure angles very precisely, as he wished to determine the refractive and dispersive indices of optical glass specimens as accurately as possible. Theodolites were the answer, as Georg von Reichenbach, co-owner of the Optical Institute and co-founder of the Mathematico-Mathematical Institute of Munich, was the builder of Europe's leading theodolites for the joint Bavarian and French Bureau of Topography and the Bavarian Bureau of Land Registry. Reichenbach had devised a new method for the dividing machine (*Teilmaschine*) that successfully divided a circle into equal parts with an unprecedented precision (Engelsberger 1969; Reichenbach 1821). Once the amazingly precise dividing machine had been constructed, Reichenbach turned to his main goal: the construction of precision theodolites for the ordnance surveying projects (and later Land Registry projects) undertaken in Bavaria during the first decade of the nineteenth century (Reichenbach 1804). These theodolites, as mentioned above, were used by ordnance surveyors in their triangulation projects. A base line was established (usually several miles in length), and by measuring angles subtended by the two ends of the base line and a tall object (a pole on a nearby hill top, for example) with a theodolite, distances could be calculated by using simple geometry. The maps were required for Napoleon's occupational forces as well as for King Maximilian I's new property tax program.

Up until 1813, theodolites were designed and used specifically and exclusively for ordnance surveying. In 1813, during his now famous Six Lamps Experiment, Fraunhofer adapted the theodolite for optical research. Fraunhofer redesigned Reichenbach's theodolite by placing a horizontal plate in front of the telescopic objective, which was pushed back slightly. This plate held a glass prism. The glass prism would refract white light into its constituent colors and the dark lines of the solar spectrum. By aligning the cross-hatchings of the theodolite with the dark lines (just as ordnance surveyors aligned the cross-hatchings with distant objects), Fraun-

17

B. Joerges and T. Shinn (eds.), Instrumentation: Between Science, State and Industry, 17–28.

hofer could obtain a very precise angle measurement for each dark line. The spectral apparatus had now been invented. Experimental natural philosophers throughout the German territories, France, and Britain used modified theodilites in their spectral research.

The first part of this essay explains how Fraunhofer, by constructing his spectral apparatus, successfully launched a research-technology in optics. The modified theodolite became the central link in a totally different network of investigators and experimental practices. It was not only used in testing achromatic lenses, but major alterations on the instrument made by Gustav Kirchhoff and Robert Bunsen later in the century situated the spectral apparatus in new research areas of physics, such as spectral analysis. One can indeed trace the contours of nineteenth-century physical disciplines by studying how the theodolite had been adapted by scientists for various uses. The second half of this essay discusses how German physicists of the late nineteenth century traced the emergence of their optical research-technology back to Fraunhofer.

FRAUNHOFER'S METROLOGY OF OPTICAL GLASS MANUFACTURING

Fraunhofer's essay *Bestimmung des Brechungs- und Farbenzerstreuungs-Vermögens verschiedener Glasarten in bezug auf die Vervollkommnung achromatischer Fern-röhren* ("Determination of the Refractive and Dispersive Indices for Differing Types of Glass in Relation to the Perfection of Achromatic Telescopes")[1] written from 1814 to 1815 was the culmination of his experiments on perfecting the manufacture of achromatic lenses (Fraunhofer 1888).[2] Fraunhofer's essay was certainly not an attempt to explain theoretically the nature of the solar dark lines (later to be called absorption or Fraunhofer lines) or the lamp lines (later to be called emission lines). The sole purpose of this essay was to publish his method for improving the construction of achromatic lenses for telescopes. Indeed, the essay commences by claiming that:

> The calculation of an achromatic object glass, and generally that of every achromatic telescope, necessitates a precise knowledge of the ratio of the sines of incidence and re-fraction [Snel's Law], and of the ratio of various types of glass which are used in the construction of telescopes. . . . Experiments repeated during many years have led me to discover new methods of obtaining these ratios, and I have therefore obeyed the wishes of several scholars [astronomers and experimental natural philosophers] in publishing these experiments, in the order I made them, with the necessary modifications that the experiments themselves forced me to introduce (Fraunhofer 1888: 3).

He concludes his piece by admitting:

---

[1] For a more detailed account of the physics of that essay, see Jackson (1996).

[2] Fraunhofer sent this paper to his good friend, the mathematician, ordnance surveyor and member of the Royal Academy, Johann von Soldner. Inclusion into this periodical meant instant recognition as a *Naturwissenschaftlter*, something that the working-class artisan desperately sought. All references to this work will be cited Fraunhofer, followed by the page number corresponding to Lommel's edited collection.

In making the experiments of which I have spoken in this memoir, I have considered principally their relations to practical optics. My leisure did not permit me to make any others, or to extend them any farther. The path that I have taken in this memoir has furnished interesting results in physical optics, and it is therefore greatly hoped that skilful investigators of nature would condescend to give them some attention (Fraunhofer 1888: 27).

Fraunhofer's essay provided opticians and experimental natural philosophers with a vastly more precise method for determining the refractive indices of glass samples than had ever before been attained. Previously, Fraunhofer himself had determined the relative dispersive and refractive indices of two kinds of glass by cementing them together, forming a single prism. If the two spectra produced by this compound prism appeared at the same place, without any reciprocal displacement, he concluded that their dispersive and refractive powers were the same and equal to the arithmetic average of the two extreme rays: red and violet. After the discovery of the lines, however, he quickly realized that two pieces of glass, which appeared to have the same refrangibility when employing the early method of testing, could actually have slightly different powers, as revealed by the existence in the overlap region of two lines where these should be only one. Fraunhofer had now accomplished what so many experimental natural philosophers had attempted earlier, namely to destroy the secondary spectrum resulting from the different partial dispersions of crown and flint glass. Indeed, so sensitive was this new method that a difference in refracting power was found not only in different types of glass or between the same glass sample taken from different levels of the same melting pot or crucible, but even between pieces taken from the opposite ends (top and bottom) of the same piece of a glass blank. A high concentration of lead oxide in flint glass resulted in glass blanks being denser at the bottom of the crucible, since lead oxide is much denser than the other ingredients. Such a density gradient would destroy any attempt to produce achromatic lenses. Fraunhofer had perfected a method of stirring so efficient that, in a pot containing over 400 pounds of flint glass, two specimens taken from the top and bottom had exactly the same refractive index.

The dark lines of the solar spectrum, then, provided Fraunhofer with a tool for gauging the efficacy of achromatic lens production. If the refractive indices determined by aligning the dark lines with Reichenbach's modified theodolite (now called the spectral apparatus) indicated that the glass was not suitable for constructing an achromatic lens (i.e. if another lens could not correct the first lens' chromatic aberration efficiently enough), Fraunhofer and his assistants would alter the recipe by adding more or less lead oxide, or by altering time of stirring or cooling, or by altering the ways in which the glass blanks were cut and ground into lenses. This portion of the process was trial-and-error, or an example of *pröbeln*, a Bavarian term similar to the German word *probieren*, meaning to test, or try out. *Pröbeln* was a process that, inherently, could not be precisely specified. It is a type of knowledge that, by its nature, does not lend itself to replication. It is part and parcel of artisanal knowledge since it is only the skilled artisan who knows when the trial has indeed been a successful one. It was through this trial-and-error technique, tested time and again by his solar-lines measurement technique, that Fraunhofer produced the world's most coveted achromatic lenses. In essence, he created a metrology for optical lens production, and became the new standard bearer of optical-glass manufacture.

THE PRACTICE OF SECRECY: PUBLIC AND PRIVATE KNOWLEDGE AT CLOISTER BENEDIKTBEUERN

Because Fraunhofer was an integral member of a business, the Optical Institute, strict measures were taken to differentiate between public and private knowledge. On the one hand, Fraunhofer wanted to increase the visibility of the Institute in order to increase sales. Hence, he needed to publicize aspects of his enterprise. This was accomplished by publishing lists of the Optical Institute's products as well as by entertaining visiting experimental natural philosophers and demonstrating to them his method for the calibration of achromatic lenses. Fraunhofer also wished to obtain credit for his work from the scientific community. This he was able to do by publishing his essay on the experiments that enabled such a calibration in the prestigious journal of the Royal Bavarian Academy of Sciences. On the other hand, certain forms of his artisanal knowledge needed to be kept secret. Hence, Fraunhofer policed both the written word and the social space. He published neither the procedures nor recipes for producing achromatic lenses, nor did he ever permit the experimental natural philosophers to witness the skills involved in lens production.

The reputation of Fraunhofer's legendary workmanship quickly spread throughout the German territories, as ordnance surveying projects were being conducted in Prussia, Saxony, Lower Saxony and Schleswig-Holstein during this period. Fraunhofer's workmanship was only matched by the entrepreneurial genius of Joseph von Utzschneider, director of the Optical Institute and Fraunhofer's employer. Utzschneider published the Optical Institute's price lists for achromatic telescopes, theodolites, borda circles, transit instruments, and optical lenses in journals affiliated with ordnance surveying and astronomy. These lists appeared, for example, in Gilbert's *Annalen der Physik und Chemie*, Franz Xavier von Zach's *Monatliche Correspondenz zur Beförderung der Erd- und Himmelskunde*, August von Lindenau and J.G.F. Bohnenberger's *Zeitschrift für Astronomie und verwandte Wissenschaften* (see, for example, Vol. II, 1816: 165–72) and even in optical texts such as J.J. Prechtl's *Praktische Dioptrik* (Prechtl 1828: 178).[3] Astronomers, ordnance surveyors, and mapmakers across Europe became familiar with the Optical Institute's products. Indeed, the Optical Institute served as a nodal point of a vast and complex network that provided the necessary instruments for charting both the globe and the heavens.

Experimental natural philosophers from Britain, France and the German territories would visit the Optical Institute to inquire whether Fraunhofer could manufacture lenses to their specifications. For example, Carl Friedrich Gauss travelled to Benediktbeuern in 1816 to place an order for astronomical instruments for his new observatory at the University of Göttingen. Other astronomers followed suit. Wilhelm Olbers from Bremen, Hans Schumacher from Hamburg/Altona and Copenhagen, and Friedrich Wilhelm Bessel from Königsberg, all ordered their astronomical

---

[3] Note that this list was published after Fraunhofer's death. But price lists were undoubtedly published in other optical texts from about 1818 onwards.

instruments from the Optical Institute. His reputation was also enhanced by the scores of articles published by Bessel, Olbers, and Friedrich Georg Wilhelm Struve announcing the discovery of stars hitherto unseen, but now made visible by Fraunhofer's refractors. And finally, of course, Fraunhofer's reputation soared as he continued to publish his research on light and physical optics in prestigious journals.

Visitors to Benediktbeuern were not restricted to astronomers and experimental natural philosophers. King Maximilian I; his son, King Ludwig I; Maximilian von Montgelas, the leader of Bavaria's reforms in the first two decades of the nineteenth century; and the King of Prussia all visited the secularized cloister to see the Optical Institute. The Czar Alexander I of Russia had originally planned to pay a visit on his return from the Congress of Vienna, but cancelled his trip at the last moment (Staatsbibliothek Preußischer Kulturbesitz, SPKB/2).[4] He did, however, place an order for achromatic telescopes for his Russian universities, including the renowned refractor at Dorpat.

Although there are numerous examples of distinction between public and private knowledge in Benediktbeuern, this section will briefly discuss four such examples. First, on 14 April 1810 the co-owner of the Optical Institute, Utzschneider, wrote to Fraunhofer: "Should the French, or some other foreigners now come to Benediktbeuern, do not show them the flint-glass oven. Also, be sure that they do not come in contact with Mr. Guinand. A certain Marcell de Seoren (?) has come for this reason. . . . One should show him nothing" (SPKB/1).

A second instance also involved Guinand, the glassmaker at Benediktbeuern who had taught Fraunhofer his trade. As a part of Guinand's contract of 20 February 1807, he received from Utzschneider an apartment for him and his wife and five cords of wood "on the condition that he never communicates the secret of the fabrication of flint and crown glass to anyone" (as quoted in Seitz 1926: 23). After numerous quarrels with Fraunhofer, Guinand left Benediktbeuern in December of 1813. Utzschneider agreed to pay him an annual pension as long as he and his wife promised not to divulge the secrets of flint and crown glass manufacturing, or to work for another optical institute which could rival Benediktbeuern (as quoted in Seitz 1926: 49–50). Such deals were commonplace. Guinand agreed, but broke his promise shortly thereafter. The third example of secrecy involved Johann Salomo Christoph Schweigger, Professor of Physics and Chemistry at the University of Erlangen (and later of Halle), who wished to visit Fraunhofer at the Optical Institute. Fraunhofer wrote Schweigger expressing his delight to entertain such a prestigious guest, but informed the experimental natural philosopher that although he would be glad to teach him how to use the dark lines of the solar spectrum to calibrate lenses, he would not be able to show the professor the glass hut or divulge any information on the manufacture of achromatic glass.[5]

Finally, the most informative example of secrecy involved Fraunhofer and John Herschel, Britain's leading experimental natural philosopher. On 19 September 1824, Herschel visited Benediktbeuern on behalf of the Joint Committee of the Royal Soci-

---

[4] This contradicts I. MacKenthun's claim that the Czar did indeed visit the Institute, although the point is really mute. See MacKenthun (1958: 131).

[5] Fraunhofer's letter to Schweigger, 2 August 1817. Deutsches Museum Archiv 7414.

ety and Board of Longitude for the Improvement of Glass for Optical Purposes (see Jackson 1994, see also Jackson 2000). British lensmakers and experimental natural philosophers had felt threatened by Fraunhofer's enterprise. Herschel hoped to take away information regarding Fraunhofer's glass making techniques. Imagine Herschel's surprise as representative of reformed, regal and industrial Britain where glass houses were located only in major cities, when he was greeted in a former cloister situated in the foothills of the Alps. He wrote to Charles Babbage, "I saw Fraunhofer and all his works but the one most desirable to see: his glass-house, which he keeps enveloped in thick darkness" (Royal Society of London Archives 1824: HS.2.199).

Although visitors increase the visibility and fame of an enterprise, they can also be potentially dangerous since they can witness any methods for glass manufacture. Fraunhofer and Utzschneider therefore needed to switch from controlling the written word to policing the social space. They controlled this space by prohibiting access to certain portions of their Institute. Such restrictions could be executed rather efficiently since Fraunhofer and Utzschneider could draw upon the pre-existing Benedictine architecture, which turns out to be an instantiation of public and private monastic life (see Jackson 1999: chapt. 6, see also Jackson 2000).

THE INVENTION OF A RESEARCH-TECHNOLOGICAL TRADITION

Fraunhofer was one of the first Germans to leave an indelible mark on technical optics and optical-glass manufacture. Indeed, his craftsmanship temporarily usurped the optical market from Britain. In 1825, the British experimental natural philosopher David Brewster wrote that it was to Fraunhofer's essay on the Six Lamps Experiment that one must turn for the "most accurate knowledge" in achromatic-lens calibration (*The Edinburgh Journal of Science* 1825: here 348). Later, Brewster defined "an excellent prism" as one "capable of showing Fraunhofer's lines" (Brewster 1831a: here 144). This was a change from the eighteenth-century definition of a superior prism: one which produced the Newtonian spectrum.

In his 1831 work entitled *A Treatise on Optics*, Brewster was explaining Newton's analysis of the colors of the spectrum. He confessed that, "no lines are seen across the spectrum . . ., and it is extremely difficult for the sharpest eye to point out the boundary of the different colours" while replicating Newton's *experimentum crucis* (Brewster 1831b: 67). He then diagrammatically compared the lengths of the colors of the spectrum as experimentally determined by Newton with Fraunhofer's results obtained by using Fraunhofer's flint glass (Brewster 1831b: 67–8). Indeed, his work on the calibration of achromatic lenses became a standard story in mid and late nineteenth-century optical texts published in Britain, France and the German territories.

By the end of the nineteenth century, the German optical industry, powered by the Carl Zeiss Company, was battling French and British competition in order to regain optical hegemony for Germany. In 1887, one hundred years after Fraunhofer's birth, Hermann von Helmholtz and Ernst Abbe invented a Germanic tradition, whereby Fraunhofer stood in a place of honor at the lineage's origin.[6] This part of my essay analyzes how Helmholtz and Abbe forged a research-technology in German optics.

On the eve of 6 March 1887, in the hall of the Berlin *Rathaus*, a packed audience that included some of Germany's leading scientific and political personalities was treated to an elaborate celebration of the one-hundredth year anniversary of Fraunhofer's birth. The spectators included Leopold Loewenherz, editor of *Zeitschrift für Instrumentenkunde* and soon to become the first director of the Technical Section of the Imperial Institute of Physics and Technology (or *Physikalisch-Technische Reichsanstalt*, henceforth, PTR); Rudolf Fuess, technologist, entrepreneur, and leader of the German Society for Mechanics and Optics; Wilhelm Foerster, director of the Royal Observatory in Berlin, professor of astronomy at the University of Berlin, co-founder of the Astrophysical Observatory at Potsdam, director of the Imperial Institute for Weights and Measures, ex-officio member of the Prussian Central Committee of Ordnance Surveying, and member of the International Bureau of Weights and Measures (Cahan 1989: 24); and Carl Bamberg, commercial advisor, entrepreneur and technologist. They were joined by leading mechanics and opticians throughout Europe. Other dignitaries joined the festivities: Berlin City Advisor D.J.G. Hartnack, Prussian Minister Heinrich von Boetticher, the Royal Minister of State and Finance Adolf von Scholz; Minister of the Interior Gustav von Gossler; Secretaries of State Heinrich von Stephan and Karl Herzog; Undersecretary Hermann von Lucanus; Director of the Ministry Greiff; the Bavarian Count Hugo von Lerchenfeld-Köfering; and the Bavarian Major General Robert von Xylander. The back wall of the auditorium was bedecked with flowers surrounding the bust of the Bavarian optician. A hush fell over the crowd as the Royal Music Director E. Schultz approached the podium in order to direct the Caecilia Choir of Berlin in the singing of the *Fraunhofer-Hymnus*, an anthem whose words had been written in 1831 by Eduard von Schenk and set to music by Schultz for the occasion.

Teary-eyed Germans with extended chests erupted in thunderous applause. A deafening silence ensued as Germany's spokesperson for physics, Hermann von Helmholtz, approached the podium. After formally greeting everyone to the celebration, he continued by explaining what exactly was being celebrated that late winter's eve.

> What we are celebrating is actually a day of remembrance for Germany's middle class (*Bürgerthum*), to which we proudly have occasion to call everyone's attention. Of all the various orientations of *bürgerliche Arbeit*, the art of practical mechanics holds an eminent place. . . . Mechanics stands at the top, striving for the highest level of exactness, pureness and dependability of its work . . . . I myself am one who, through much experience, can provide evidence on how high these first and most important values [exactness, pureness and dependability] of *bürgerliche Arbeit* have risen with the leading mechanics: how one always finds a master craftsman who has earned the greatest respect–not only in many university-cities throughout Germany, but also in some middle-sized cities without a university (v. Helmholtz 1887: here 115).

---

[6] For the notion of the "invention of tradition," see Hobsbawm and Ranger (1983).

Helmholtz continued by emphasizing that Germany respected the labor of the mechanic in general, and the *Handwerker* in particular. He claimed that he had learned much from mechanics, particularly instrument makers. Indeed, he wondered where physics, astronomy, atmospheric research, the navigational telescope, and electric telegraphs would be without "the intelligent assistance of practical mechanics" (v. Helmholtz 1887: 117). Helmholtz then turned to the subject of this oration.

> This class of *Bürger*, which in its quiet ways guards and sets in motion the best virtues of German *Bürgerthum*, commemorates today the memory of one of the first and most famous men from its own ranks. He rose from the poorest conditions using his strength and industry, worked against difficult obstacles, and became the owner of the most famous Optical Institute on earth at the time. His scientific discoveries have expanded our knowledge of the cosmos in ways previously unimagined, and has provided astronomers, physicists, and chemists with the task of completing the studies which these discoveries have sparked.
>   Fraunhofer has grown up and risen from the soil of the artisan (*Boden des Handwerks*) . . . (v. Helmholtz 1887: 117).

Wilhelm Foerster, who strongly supported Schott and Abbe's optical glass research, ascended to the podium. Once again the audience was treated to a short biography of Fraunhofer emphasizing the rescuing of the young child from the wreckage of a collapsed house as Prince Maximilian Joseph IV looked on. The imagery of light and darkness was employed to contrast Fraunhofer's optical research on light with the darkness of his financial position as an orphaned child. Foerster proceeded to link Fraunhofer's "technology of precision" (*Präcisionstechnik*) to a more encompassing talent of the German people. He claimed that when it came to the "rigorous arts" (*strenge Kunst*, i.e. precision technology as opposed to the aesthetic arts), the labor of German men and women stood side by side with important German intellectual merits (v. Helmholtz 1887: 122-3). Foerster traced this "talent of the German race" back to Fraunhofer. He offered a brief history of optics depicting England as the undisputed leader up until the beginning of the nineteenth century, or until Fraunhofer appeared on the international scene. Foerster underscored not only the fact that Fraunhofer's telescopes were being employed in observatories all around the world, but also that Fraunhofer's work gave a great boost to Germany's mechanical and optical research. Fraunhofer became the emblem of German precision measurement and mechanical and artisanal knowledge.

The Berlin celebration was one of several throughout Germany to commemorate the hundredth anniversary of Fraunhofer's birth. The Munich chapter of the German Society for Mechanics and Optics co-sponsored the Munich festival with city officials, while the Phyiscs Association in Frankfurt-am-Main offered its own day of remembrance (Fuess 1887).

On 5 March 1887, a day before the Berlin pageant, Abbe delivered a speech in the auditorium of the Physics Institute of the University of Jena, also in celebration of the one-hundredth anniversary of the birth of Joseph Fraunhofer (Abbe 1989). Although the pageantry of Jena never could quite rival that of Berlin, Abbe's oration was nevertheless very enlightening. Abbe, as Helmholtz, claimed that "under his [Fraunhofer's] impulse, a highly developed technical art arose from a business which still serves today as an unparalleled model of the inner collaborations of a pure science and "practical skill" (*praktischer Geschicklichkeit*)" (Abbe 1989: 320). Fraunhofer's relentless endeavoring combined with his "skilled hand" (*geschickte Hand*) resulted in the perfection of a method which enabled "research into the secrets of nature" (Abbe 1989: 320).

Abbe offered his listeners a brief history of Fraunhofer's life.

> As so many of our important men, Fraunhofer came from the soil of our people, from which the archetypal and unweakening strength (*Kraft*) continually renews the driving forces (*Triebe*) responsible for the intellectual blossoming of the nation (Abbe 1989: 321).

The Bavarian trio, Utzschneider, Reichenbach and Fraunhofer, reformed the mechanical arts. This reform led to astronomical discoveries and advances in disciplines requiring precision measurement. The audience was told how the labor of these three men, but Fraunhofer in particular, enabled the exciting research of German astronomers such as Struve and Bessel (Abbe 1989: 329). Thanks to the work of these three men, the mechanical arts had returned to Germany after centuries of absence. No longer did Germans need to turn embarrassingly to the instrument makers of the capital cities of their rival nations, France and Britain (Abbe 1989: 329). Munich, according to Abbe, became the capital of instrument builders for the entire scientific world. And since that day, Germans enjoyed a world-recognized lead in all branches of scientific industry. Fraunhofer's work was the advent of the "exact arts" (Abbe 1989: 329).

> If justifiable pride given to the successful participants of one's own people, which is conveyed to more encompassing cultural interests, is an undoubtedly good and worthwhile form of national pride and national ambition, then one may find gratification in such uncontrollable expression at a commemoration of Fraunhofer. Through him new paths have been pointed to and paved for our people upon which we can bring to bear with honor our natural gifts and the advantages of our raised standard of education to the peaceful competition of nations (Abbe 1989: 329).

Abbe concluded his oration by discussing his and Schott's research in the production of apochromatic lenses in the 1880s at the Carl Zeiss Works. By appealing to Germany's history of recognizing the importance of the artisanal knowledge of the mechanical arts, he was hoping the Prussian government would resume its subsidies for their optical-glass research at the Carl Zeiss Works, which they had supported for a two-year period, ending in 1886.

Abbe's historical account of Fraunhofer underscored Fraunhofer's humble beginnings and co-operation with Utzschneider in establishing a very profitable Optical Institute that became the envy of the scientific world. The entrepreneurial tradition, Abbe was arguing, was crucial to the formation of a late optical precision technology – just as crucial as the craft tradition epitomized by Fraunhofer. Interestingly, Abbe was embroiled in debates of the role of the Technical Section of the PTR during precisely this period. He argued, contra Werner von Siemens, that instrument makers should not be used by industrialists simply to undertake routine tests for quality control, but rather instrument makers should be permitted to have a greater input on the direction of research in the PTR. Hence, Abbe underscored the mutual relationship between Utzschneider, Reichenbach, and Fraunhofer: the managerial and craft traditions were both necessary for precision optical research. He clearly had vested interests in his construction of Fraunhofer. First and foremost, his history sent a clear message to Prussian officials present at his oration that governmental subsidy of science and technology had an historical precedence on German soil. Abbe praised the foresight of King Maximilian I's patronage of the self-educated, skilled Bavarian artisan. Germany could recapture world hegemony in optical technology if, and only if, Prussian officials would resume their support of the optical-glass manufacturing at the Carl Zeiss Works. Second, Abbe saw himself unifying technical knowledge with scientific theory, which was precisely what he claimed Fraunhofer had been doing earlier that century. Third, the importance of German skilled artisans to the German scientific enterprise was acknowledged some fifty years earlier. And as Germany was chasing Britain and France for world supremacy in scientific and technological industries, once again it was obvious that opticians, precision mechanics, and other skilled workers associated with those industries were prized possessions for the new Reich. Hence, his commemoration of Fraunhofer needs to be seen within the context of his support for Wilhelm Foerster's planned Institute for Precision Mechanics (*Institut für Feinmechanik*). Abbe, Foerster, and others feared that precision mechanics lagged woefully behind both France and Britain by the 1870s. The lead that the German territories had enjoyed throughout Fraunhofer's lifetime was being usurped by France and Britain, and many intellectuals feared that Germany would, once again, fall back into eighteenth-century obscurity. The recognition of the labor of skilled instrument makers, Abbe proclaimed, would propel German industry past their rivals just as the labor of skilled artisans, under the direction of entrepreneurs, assisted German science ascent to the forefront.

CONCLUSION

Joseph von Fraunhofer, by adapting an instrument designed specifically for ordnance surveying – the theodolite – to measure angles between the solar dark lines, had now introduced a new optical instrument, the spectral apparatus. His calibration technique of using those dark lines of the spectrum for determining refractive indices became the standard for optical-lens production until the combined efforts of Ernst Abbe and Otto Schott of the Carl Zeiss Company in the 1880s. His standard for calibration became routine for future optical research-technologies. This is powerfully evident from the Fraunhofer histories provided by both Helmholtz and Abbe.

Fraunhofer stood at the forefront of a German optical research tradition. His invention of the spectral apparatus gave rise to a precision metrology that set the world's standard for testing the efficacy of optical lenses. Fraunhofer's work on achromatic glass did not belong to a well-defined research tradition of the early nineteenth-century physical and geometrical optics, traditions that clearly existed in France and Britain at the time. Precisely because Fraunhofer was an artisan tying together different enterprises such as ordnance surveying, achromatic-lens making, precision optics, and astronomy, he could tap into several markets. His interstitiality was crucial to his success. Research-technologies shift the emphasis of historical explanation away from static accounts of theoretical commitments toward the way in which objects can travel across scientific and cultural boundaries, similar to the notion of boundary objects (Star and Griesemer 1989) and immutable mobiles (Latour and Woolgar 1986). But the concept of research-technology goes beyond the objects (such as instruments) themselves and explains how differing communities use them to determine standards and more precise metrologies. Research-technologies will illuminate the vastly complex, yet historically contingent and very informative, contours between scientists, engineers, private industry and governmental agencies: the very essence of Big Science.

Willamette University, Salem, OR, USA

REFERENCES

Abbe, Ernst (1989). *Gedächtnis, Gesammelte Abhandlungen Vol. II: Wissenschaftliche Abhandlungen aus verschiedenen Gebieten: Patentschriften, Gedächtnisreden.* Hildesheim, Zürich, New York: Georg Olms.

Brewster, David (1831a). Remarks on Dr. Goring's observation on the use of monochromatic light with the microscope. *The Edinburgh Journal of Science* 5 (New Series): 143–8.

Brewster, David (1831b). *A treatise on optics.* London: Longman, Rees, Orme, Brown, Green and John Taylor.

Cahan, David (1989). *An institute for an empire: The Physikalisch-Technische Reichsanstalt, 1871-1918.* Cambridge: Cambridge University Press.

Deutsches Museum, Archiv 7417. Fraunhofer's letter to Schweigger, 2 august 1817.

Engelsberger, Max (1969). *Beitrag zur Geschichte des Theodolits.* Dr-Ing. Dissertation der Technischen Hochschule München. München: Verlag der Bayerischen Akademie der Wissenschaften.

Fuess, Rudolf (1887). Vereinsnachrichten Deutsche Gesellschaft für Mechanik und Optik. *Zeitschrift für Instrumentenkunde* 7: 149.

Helmholtz, Hermann von (1887). Festbericht über die Gedenkfeier zur hundertjährigen Wiederkehr des Geburtstages Josef [sic] Fraunhofers. *Zeitschrift für Instrumentenkunde* 7: 113–28.

Hobsbawm, Eric and Ranger, Terence (1983). *The invention of tradition.* Cambridge: Cambridge University Press.

Jackson, Myles W. (1994). Artisanal knowledge and experimental natural philosophers: The British response to Joseph Fraunhofer and the Bavarian usurpation of their optical empire. *Studies in History and Philosophy of Science* 25: 549–75.

Jackson, Myles W. (1996). Buying the dark lines of the solar spectrum, in J. Buchwald (ed.), *Archimedes: New studies in the history and philosophy of science and technology I* (pp. 1–21). Dordrecht, The Netherlands: Kluwer Academic Publishers.

Jackson, Myles W. (1999). Illuminating the opacity of achromatic lens production: Joseph von Fraunhofer and his monastic laboratory, in P. Galison and E. Thompson (eds.), *The architecture of science* (chapter 6). Cambridge, MA: MIT Press

Jackson, Myles W. (2000). *Spectrum of belief: Joseph von Fraunhofer and the craft of precision optics.* Cambridge, MA: MIT Press.

Latour, Bruno and Woolgar, Steve (1986). *Laboratory life: The social construction of scientific facts.* Princeton, NJ: Princeton University Press.

Lommel, E. (ed.) (1888). Joseph von Fraunhofer's Gesammelte Schriften. Im Auftrage der Mathematisch-Physikalischen Classe der königlichen bayerischen Akademie der Wissenschaften. München: Verlag der königlichen Akademie in Commission bei G. Franz. It originally appeared in *Denkschriften der Königlichen Akademie der Wissenschaften zu München für die Jahre 1814 u. 1815*, Vol. V (not published until 1817).

MacKenthun, I. (1958). *Joseph von Utzschneider, sein Leben, sein Wirken, seine Zeit.* München: Universitäts-Druck-München.

Prechtl, Johann J. (1828). *Praktische Dioptrik als vollständige und gemeinfaßliche Anleitung zur Verfertigung achromatischer Fernröhren. Nach den neuesten Verbesserungen und Hülfsmitteln und eigenen Erfahrungen.* Wien: J.G. Heubner.

Reichenbach, Georg (1804). Nachricht von den Fortschritten der mathematischen Werkstatt in München. *Zachs Monatliche Correspondenz* 9: 377ff.

Reichenbach, Georg (1821). Berichtigung der von Herrrn Mechanikus Jos. Liebherr in München abgebenen Erklärung über die Erfindung meiner Kreiseinteilungsmethode. *Gilberts Annalen der Physik und Chemie* 68: 33-59.

Royal Society of London Archives (1824, 3 October). Herschel to Babbage, *Herschel Letters, HS.2.199*.

Seitz, Adolf (1926). *Joseph Fraunhofer und sein Optisches Institut*. Berlin: Julius Springer.

Staatsbibliothek Preußischer Kulturbesitz (1810, SPBK/1). Utzschneider Nachlaß Kasten 1, Briefe an Fraunhofer, 14 October.

Staatsbibliothek Preußischer Kulturbesitz (1814, SPBK/2). Utzschneider Nachlaß Kasten 1, Briefe an Fraunhofer, 9, 15 and 16 July.

Star, Susan Leigh and Griesemer, James (1989). Institutional ecology, 'translations' and boundary objects: Amateurs and professionals in Berkeley's Museum of Vertebrate Zoology, 1907–1939. *Social Studies of Science* 19: 387–420.

The Edinburgh Journal of Science (1825). 2: 344–8.

Zeitschrift für Astronomie und verwandte Wissenschaften (1816). Vol. II: 165–72.

CHAPTER 3

TERRY SHINN

THE RESEARCH-TECHNOLOGY MATRIX: GERMAN ORIGINS, 1860–1900

During the closing decades of the nineteenth century, the topography and substance of science and technology in Germany were significantly transformed through a re-alignment of the intellectual and material productions of academia, industry and the state. A new form of cognitive, instrumental and functional entity arose – the "research-technology" current – whose principal objective was the design, construction and diffusion of high performance devices for purposes of detection, measurement and control. In the 1870s and 1880s much of the initiative for community develop-ment came from Berlin's instrument craftsmen who prodded the new imperial gov-ernment into recognizing the nation's need for a strong precision instrument capacity and who militated in behalf of a formal instrument association. After a phase of short-lived reluctance, government responded eagerly to demands for the reinforce-ment of the country's precision technology potential.

Pressure for enhanced capacity in instrumentation came from numerous quarters. The Prussian army constantly demanded more and improved devices associated with railway transport, telegraphy, optical range-finding and long distance observation and munitions. Instrumentation was regarded as basic for the Schleswig-Holstein cam-paign and the Austro-Prussian and Franco-Prussian wars, but also for future military adventures. Industry similarly demanded ever more instrumentation.

Research-technologists working in small, traditional instrument-making shops, in industry, in state institutes or in universities sponsored an instrument society, the *Deutsche Gesellschaft für Mechanik und Optik*, as well as new periodicals for the purpose of developing a sense of group identity, sharing technical information and establishing extended markets for their material and intellectual output. Among the primary journals were the *Repertorium für Physikalische Technik für die Mathema-tische und Astronomische Instrumentenkunde*, the *Zeitschrift für Instrumentenkunde* and the *Zeitschrift für Vermessungswesen*. Germany's research-technology practitio-ners deployed two additional strategies to promote community stability and prosper-ity. They frequented innumerable industrial trade-fairs in Germany and abroad, where they demonstrated the effectiveness and applicability of technical devices. Parallel to this, they increasingly came to sponsor instrument exhibitions. On a dif-ferent level, one of research-technology's greatest successes occurred in 1889 when

B. Joerges and T. Shinn (eds.), Instrumentation: Between Science, State and Industry, 29–48.
© 2001 *Kluwer Academic Publishers.*

after years of effort a special instrumentation section was established within the huge and immensely influential national multi-disciplinary science association, the *Versammlung Deutscher Naturforscher und Ärzte*.

The genesis, role and identity of the nascent research-technology movement entailed paradoxical forces: on the one hand, reinforcement of intellectual and material distinctiveness, or the emergence of a kind of instrumentation field; but on the other, adherence to a pre-modern form of social organization that included the repudiation of forms of identity like "professionalism" which thrived through imposing boundaries and through exclusion. This dilemma of community identity was confronted in the early 1880s during the planning of the *Physikalisch-Technische Reichsanstalt*. The latter was to contain two sections, one for research in fundamental science and another for technical work. Three groups vied for control over the second section – industry, engineering and research-technology. The imposing force of industry and engineering resided in their professional status and influence. The strength of research-technology consisted precisely in its multi-localization with respect both to the employment organizations of practitioners and to markets. The research-technology movement prevailed.

In its dealings with the *Reichsanstalt*, the salient features of the research-technology movement emerged with utmost clarity. The movement exhibited three aspects: 1) Its vocation was the development of "generic" devices; that is, reliable, flexible core apparatus which could be put to a multitude of uses by a spread of clients. 2) It intended to generate "metrologies" – nomenclatures, measurement systems and units, precise values, and original representations of physical parameters. 3) Its organization was to be interstitial; that is, trans-institutional and trans-professional. It was to be a distinctive movement, not a separate, demarcated or distinct one.

This paper will sketch some of the major events that surrounded the emergence of Germany's metamorphosing instrumentation movement in the decades immediately before World War I. The empirical data is drawn from the *Deutsche Gesellschaft für Mechanik und Optik* and the *Zeitschrift für Instrumentenkunde*, the research-technology current's flagship periodical. This journal presents detailed information on the activities of the meetings of the precision technology movement. I have compiled a collective biography for 103 individuals associated with the *Zeitschrift für Instrumentenkunde* made up of the editorial staff and authors with the highest publication count.[1] An examination of the texts published in the journal provides insights into the work orientation and potential audiences of this group of research-technologists. The journal also contains often highly detailed minutes of the quasi-monthly meetings of the *Deutsche Gesellschaft für Mechanik und Optik*.

---

[1] Data for this prosopography is drawn from Poggendorff's *Biographisch-Literarisches Handwoerterbuch zur Geschichte der exacten Wissenschaften* (Poggendorff 1898–1940). This source contains biographical information for thousands of nineteenth and twentieth century mainly German but also many non-German scientists, engineers, instrument men, inventors, etc., as well as listings of their articles and books, data on society membership, awards and prizes, and sometimes patents.

THE BACKDROP

In the mid- and late 1860s, a handful of Prussian intellectuals grew concerned over the inadequate state of the country's instrument-making capacity. With the grudging approval of the Berlin Academy, and semi-covert backing of several key individuals in the Prussian Ministry of Education, a highly respected and influential Berlin mathematics teacher at the Royal Military College, K. Schellbach (1805–1892), assembled a commission to study the size, domains of competence, and achievements of Prussian precision technology. The Schellbach Commission completed its work in 1871 (Cahan 1989: 25–5, 30; Pfetsch 1970; Poggendorff 1898–1940). The Schellbach Memorandum argued that the Prussian instrument community was small when compared to those of France and even Great Britain. While Prussia was highly accomplished in some fields, there existed important technical lacunae. In its present condition, the Prussian instrument-making group was unable to satisfy the demands of the country's laboratories or expanding industries. The Schellbach Commission consequently recommended to the Berlin Science Academy that the latter intervene to set up a large and properly financed instrument institute. This institute, suggested the authors of the Memorandum, would benefit both academia and industry through the metrology devices that it would generate.

In 1872 the Berlin Academy responded publicly to the Schellbach program. The reply was negative. The Academy declared that instrument-making was too far removed from science to warrant the body's intervention. It was a matter for industry or for the instrument making occupation itself. This attitude, however, was met with consternation among powerful groups inside the Imperial German Education Ministry and by instrument men alike. W. Foerster (1832–1921), an eminent scientist and increasingly influential politician, believed it urgent that everything possible be done to accelerate Germany's precision technology potential (Cahan 1989: 24–6, 29–31, 54, 57, 70, 75, 85–6, 110, 195; Pfetsch 1970, 563–6; Poggendorff 1898–1940). Foerster held the chair of astronomy at the Berlin University. He was director of the Berlin Astronomy Institute and the Imperial Commission for Weights and Measures. Many viewed him as the most eloquent champion of a new form of science which entailed closer ties between academia and industry. In the early 1870s he held the powerful position of director of Berlin education. This meant that he enjoyed excellent relations with the court and government.

Through his personal connections Foerster arranged the establishment of a military commission to examine the preparedness of German instrument-making capacity. The Morozowicz Commission explored the ability of German instrument men to satisfy the nation's military needs, and its potential to derive fresh innovations. O. von Morozowicz was a highly respected Prussian general with considerable experience in military technology, and particularly telegraphic communications (Cahan 1989, 25–7). The conclusions of the Morozowicz Commission echoed the Schellbach Memorandum. The country was badly lacking in instrumentation. Government action alone could reverse the dangerous situation. On the basis of this evaluation, Foerster and his allies moved to establish a Berlin institute specialized in precision technology in the domains of mechanics and optics. The goal was to erect an agency similar to the *Conservatoire des arts et métiers* in Paris, where instrument research could be conducted and where future generations of instrument makers would be properly trained. The intended mechanics and optics institute would be to instrument-making what the Berlin Academy was to science. In 1874–75 this plan encountered continual stiff opposition from the Berlin Academy and from some circles within the ministry in charge of education. The hope for a publicly sponsored instrument institute gradually waned. Opponents divided learning and skills into two separate categories, science and engineering. Instrumentation, they argued, belonged to the latter. Foerster failed to convince them that instrument matters belonged to both domains, and because of this, he insisted, it would constitute a grave error to enslave it to either. While these commissions did not immediately re-orient public policy, they nevertheless reflected the advent of public concern over instrument questions. By the mid-1870s this concern had grown acute, particularly among some scientists. This preoccupation is expressed in a massive volume published in 1880 by the physicist Leopold Loewenherz (1847–1892) (Cahan 1989: 48, 114; Poggendorff 1898–1940). Loewenherz had a doctorate in physics and mathematics from the Berlin Astronomy Institute. His principal interest, however, lay in metrology. He was not a particularly gifted experimenter, nor was he strong in mathematics. His reputation was for coming-up with original instrument ideas and for their clever application to a broad range of domains within industry and academia. Loewenherz held a position on the Imperial Commission for Weights and Measures. His most notable contribution was the universal screw thread. During his seventeen years at the Commission, he also participated in metrology projects associated with thermometry. He acted as a consultant to the Siemens Company for over a decade. Again, his talent was in producing good ideas, and not in precise calculations. Loewenherz was an acknowledged advocate for a large and strong instrument community. From the late 1870s onward, he worked unendingly to establish a specialized instrument institute. He became the co-founder of the *Deutsche Gesellschaft für Mechanik und Optik* in 1879, and it was he who established the society's journal, the *Zeitschrift für Instrumentenkunde*, and who served as its editor and one of its main contributors until his premature death in 1892 ('Nachruf' in *Zeitschrift für Instrumentenkunde* 1892).

In his 1880 book, *Bericht über die Wissenschaftlichen Instrumente auf der Berliner Gewerbeausstellung im Jahre 1879*, Loewenherz described the many categories of instruments on exhibit at the 1878 London Universal Exhibition. He compared the situation of Germany with that of Britain and particularly France (Loewenherz 1880, particularly I–VII and 1–4). In his influential volume Loewenherz listed the major categories of instrument domains: mechanics, chronographs, optics (astronomy, navigation, spectroscopy and medicine), electricity (units of measurement, measurement devices, technical areas), geodesy, meteorology, physiology, organic, inorganic and physical chemistry. He compared Germany's achievements in each field with those of other nations, concluding that while in some areas, such as optics, German instrument-makers excelled, in many other domains Germany lagged behind. He called for the creation of societies for the promotion of instrument activities. Whether such bodies should be local or organized on a national level needed to be worked out. In his book, Loewenherz also reflected on the advisability of establishing a separate training program for precision instrumentation. The train of thought and questions raised by Loewenherz were to guide the emergent German research-technology movement for many decades ('Nachruf' in *ZIK* 1892).

Throughout the 1870s several prominent Berlin instrument craftsmen keenly monitored these discussions on their occupation. They were eager to expand their business opportunities but also wished to structure and strengthen precision technology. Their strategy envisaged the establishment of a specialized Berlin instrument society and the creation of specialized schooling.

THE *DEUTSCHE GESELLSCHAFT FÜR MECHANIK UND OPTIK*

In 1879 the *Deutsche Gesellschaft für Mechanik und Optik* was founded to pursue these objectives (Cahan 1989: 24, 26, 111, 114, 161). As indicated by the body's name, the intended purview lay in the traditional fields of mechanics and optics but soon widened to incorporate areas like electricity, magnetism and electro-magnetism. The Society's work included the organization of periodical meetings for the demonstration of new devices and discussions over technical innovations, sponsorship of instrument exhibitions, and the development and accreditation of instrument education. From the outset, the *Deutsche Gesellschaft für Mechanik und Optik* published a journal. It featured technical articles on recent instruments designed and constructed by society members, résumés of recent publications related to instrumentation, reports on the Society's training program and summaries of business meetings and debates. It also analyzed the successes and frustrations in spreading the influence of instrument interests inside Germany and in potential consumer markets.

It was instrument craftsmen who founded the society and constituted its backbone.[2] The major figures were R. Fuess, C. Bamberg, and H. Haensch. Fuess, born 1838 in Hanover, moved to Berlin as a young man where he became a highly skilled craftsman in precision mechanics at the instrument firm of Greiner and Geisler, which he purchased in 1876 (Cahan 1989: 43, 47–8, 56, 75, 79–80, 85–6; Poggendorff 1898–1940). Fuess specialized in meteorological devices and glass-related apparatus, and his name is associated with many devices for measurement of wind speed and direction, rainfall and humidity. He also innovated in barometry. The thrust of his endeavors lay in enhanced precision, and even more significantly, in automatic recording. The latter permitted additional measurements as well as the multiplication of devices. Fuess also specialized in apparatus for improvement of mineral samples for microscopic investigation. He was a strong ally of Loewenherz in his campaign for a mechanics and optics institute.

He also communicated with Foerster, and was a consultant for Siemens. He stressed the need for an "autonomous" instrument movement, one capable of resisting what he regarded as the often narrow, short-sighted demands of industry. For him, the *Deutsche Gesellschaft für Mechanik und Optik* comprised the foundational

---

[2] The origins and evolution of the *Deutsche Gesellschaft für Mechanik und Optik* is documented in considerable detail in the pages of *Zeitschrift für Instrumentenkunde*, which published the minutes of the semi-periodic meetings of the association. Henceforth, these minutes will be referred to by DGMO, followed by *ZIK* and year of publication. For an account of the origins of the association see: DGMO, *ZIK* 1881a, 1881b).

measure in the creation of the instrument community. From it would flow the necessary policies and programs (DGMO, *ZIK* 1882, 1887a, 1889a).

Bamberg and Haensch directed the routine aspects of the *Gesellschaft*, leaving policy to Fuess and his friend Loewenherz. Bamberg (1847–1892) was also an artisan instrument-maker and owner of his small company which specialized in astronomical instruments and apparatus used in geodetic work (Cahan 1989: 43, 47–8, 56, 75, 80; DGMO, *ZIK* 1892c; 'Nachruf' in *ZIK* 1892: 53). Bamberg's reputation was in applied optics. Like Fuess, Lowenherz and Foerster, he felt that the future of precision technology was linked to the growth of a specialized teaching program in instrument-making. His work at the society reflected this. Unlike Fuess and Lowenherz, however, Bamberg saw science rather than industry as a threat to the prosperity and independence of the German instrument movement. He feared that science would increasingly monopolize instrument work both for university laboratories and for industry. The influence exercised by Loewenherz in the society tipped the balance, he declared, too much in favor of university interests. Bamberg's death, occurring at the same time as Loewenherz,' went some way in dispelling apprehensions over eventual domination by the logic and demands of academia (DGMO, *ZIK* 1885e, 1897). Haensch (1831–1896) was referred to in his obituary as the "heart" of the *Deutsche Gesellschaft für Mechanik und Optik*, and the journal it produced beginning in 1881 ('Nachruf' in *ZIK* 1892: 192). Haensch was a highly skilled instrument mechanic. He did not own a company but worked alone and had no connections with political or industrial circles. He belonged to an earlier age, being motivated by the ideal of a tight-knit instrument brotherhood. Haensch willingly took on the work of organizing the society's meetings. He tended to the circulation of its monthly bulletin, and with the advent of the *Zeitschrift für Instrumentenkunde*, Haensch managed the periodical. His concern was not the position of instrument-making vis à vis academia and industry, its autonomy or the spread of instrument markets. He wanted a movement that emphasized a sort of communion between men of like mind. Haensch defined himself in terms of the devices he created, and for him contact with comrades of the same ilk was essential.

Artisans from Berlin formed the society in its first few years of existence, yet they were soon joined by instrument craftsmen from other cities: Hamburg, Hanover, Jena and Dresden in particular. A few industrialists and university scientists gradually penetrated its ranks. Their adhesion to the research-technology movement was organized around the *Zeitschrift für Instrumentenkunde*. During the 1880s, *Gesellschaft* meetings were dominated by the presentation, demonstration and evaluation of new or modified instruments and by discussion of their uses in academia and industry. Toward this end, Berlin academics and German industrialists frequently attended meetings either as guest speakers or guests of honor. In a few instances, these people subsequently became regular members of the society. Little by little, society participants began to realize that instrument innovation demanded the sharing of ideas and information. By 1886, the notion arose that instrument-making possessed sufficient internal consistency for practitioners to constitute a sort of self-referencing body. The body had core demands and interests. As well as being concerned with its standing in the external world, the Society concentrated on forging its own self-identity. This represented a modified version of the Haensch philosophy, and it constituted a victory for him.

In the early 1880s, the *Deutsche Gesellschaft für Mechanik und Optik* devoted a large portion of most meetings to educational matters. A private instrument school was set up in conjunction with an existing professional school. To furnish theoretical and practical training in mechanics and optics-related instrument building was the goal (DGMO, *ZIK* 1881c, 1881d). However, the venture soon collapsed. There were innumerable problems. The new school had to compete for students with already well established bodies. Since the instrument-making occupation was less clear-cut than many others – such as mechanics, electricity and metal work – potential pupils were not attracted in large numbers. The school demanded a small tuition fee, and although financial assistance was accorded to the most needy, this too constituted an impediment. Endless debates raged inside the society over course content. Many members, Fuess among them, called for an ambitious curriculum that included an important measure of mathematics and elementary physical theory. This proved unrealistic though, due to the low educational level of most of the pupils. Lowenherz had originally enthusiastically supported the idea of a training program. By the end of the decade, however, he as well as other society members began to doubt its wisdom, and even its feasibility. They wondered whether the capacity to design and construct innovative instruments was linked to an identifiable course content. Was it not advisable rather to encourage instrument-related subjects within existing teaching programs associated with mechanics, optics, electricity and chemistry? This strategy was at last adopted (DGMO, *ZIK* 1883, 1884, 1885b, 1885c, 1889b: 113, 1889c, 1890c). Thus while instruction remained an important issue for the emergent research-technology group, it strove to promote instrument building and instrument careers through influencing extant institutions.

An equal amount of energy went into consolidating and extending markets for instrumentation. Between 1881 and 1890 the society was involved in a sum of seventeen trade fairs or instrument exhibitions. The *Deutsche Gesellschaft* acted as the chief organizer of the 1884 Berlin instrument exhibition which displayed over 800 devices, most of them from the Berlin region or from other areas of Germany. It also helped organize instrument trade fairs in Brussels, Frankfurt, Munich, Hamburg and Jena (DGMO, *ZIK* 1883, 1885c and 1885d, 1886, 1888b). These areas were emerging as the principal poles of instrument development. In many instances, such as in Hamburg, Frankfurt or Brussels, the pallet of industrial endeavor was complete: mechanics, optics, electricity, chemistry and medical-related firms were represented, and the quickly *expanding Deutsche Gesellschaft für Mechanik und Optik* strove to present a device or devices for every category.

The society organized German exhibits in the 1893 Chicago Universal Exhibition. This represented a considerable victory for the society. It thereby became the semi-official spokesman of the nation's instrument movement. The Chicago venture sparked a lively debate in instrument circles over the structure and strategy of the growing community. Initially, the question was how best to present instrument exhibits. Were they to comprise a component within a domain of industrial activity, or instead to constitute an autonomous category of presentation? For example, should instruments appear on show as a small part of say electricity, optometry or land surveying, and thereby be scattered and fragmented? Perhaps better to concentrate them into a single big powerful and coherent cluster – an instrument theme. Throughout the 1880s, both approaches had been used, depending on the preferences of local trade/instrument organizers. It was the self-evaluation of Germany's performance at Chicago that brought this issue to a head. In a series of detailed society reports on Germany's competitive position in the 1893 Universal Exhibition, society observers identified a list of important shortcomings. American instrument capacity was already remarkable and continued to develop rapidly. It was particularly impressive in areas of mechanics and certain domains of electricity – such as telegraphy-related instrumentation and instruments associated with electrical production and diagnosis. French capacity was apparently declining, and Germany had little to fear from Britain in most spheres (DGMO, *ZIK* 1893, 1894a, 1894b, 1894c, 1894d and 1894e). Despite this, German instrumentation had not always received the recognition that it warranted in Chicago due to the manner in which it was exhibited. Too frequently, instrument demonstrations had been sacrificed to the narrowly industrial or technological aspects of firms. In many of the product displays, instruments were not given their fair share of attention by being made to appear as a subordinate component. However, if instruments were instead displayed all together, they would stand out more forcefully. Society observers insisted that the issue here was not merely one of representation. Something far more fundamental was involved. In their reports, they argued that instrumentation possesses an internal unity and coherence. By arranging devices next to one another, this unity was preserved and amplified. Hereby, the principles, capacities and potential of different devices for use in the broadest range of activities could emerge more definitely. By structuring exhibitions in terms of product-lines rather than instrument-lines, the public frequently failed to perceive that a given device could be profitably employed in an extended sphere of areas. Moreover, collective exhibition of instruments was of utmost use to instrument-makers themselves. By focusing on instruments rather than the niches in which they were to operate, the research-technologist could better appreciate the scope of application of a technical breakthrough and could identify where more endeavor was required (DGMO, *ZIK* 1893, 1894a, 1894b, 1894c, 1894d, 1894e). Indeed, one sees here the forerunner of an instrument logic.

PROSOPOGRAPHY

The *Deutsche Gesellschaft für Mechanik und Optik* has offered a glimpse into the origins, inner workings and early agenda of the German research-technology movement, yet the society's relatively narrow artisanal base and strong regional links have obscured other highly important features. A fuller appreciation of the sweep of practitioners can be acquired by an examination of the society's main journal, the *Zeitschrift für Instrumentenkunde*, established in 1881. The study presented here focuses on the men who published the most or who sat on the periodical's editorial board during its first twenty years. Such a strategy leaves unexamined almost 75% of authors and 65% of articles. Considerable work remains. The sample of 103 contributors has been divided into five categories: a) employees of state research centers, b) academics, c) men working in industry, d) artisans, e) engineers. In the first two decades of the journal's operation, the relative weights of several of these groups shifted, thereby suggesting changes in the research-technology movement's internal dynamics.

Staatliche Forschung

In their book, *Staatliche Forschung in Deutschland, 1870–1980*, P. Lundgreen et al. have argued that in the nineteenth century a new form of research, *Staatliche Forschung* or state research arose, evolved and came to maturity. According to these authors, the strategies and successes of this brand of endeavor has placed it on an equal footing with academic and industrial activities. State research was long centered in the numerous *Eichungskommissionen* of the German *Länder* (Lundgreen et al. 1986: 1–27), and near the end of the nineteenth century additional state research centers developed, like the Meteorology Institute, Sea Observatory and Astronomy Institute. The *Physikalisch-Technische Reichsanstalt* is frequently pointed to as the exemplar of state research (Cahan 1989: 39–42, 72–86, 121–5; Pfetsch 1970). All of these agencies were involved to differing degrees with establishing and imposing technical norms, certifying products and processes for public use, working out safety standards and assuring component inter-changeability. From our perspective, some if not all of these ventures had a bearing on research-technology.

One item emerges with particular clarity. Despite opinion to the contrary, to reduce the activities of research-technology to the *Physikalisch-Technische Reichsanstalt* constitutes a grave simplification of the instrument community, and it disregards its scope and complexities. Nevertheless, the events surrounding the *Reichsanstalt* and research-technology were foundational and revealing with respect to the latter's structure and strategies, and this will be dealt with in a later section of this chapter.

The career of A. Sprung (1847–1909) unfolded entirely in the framework of state research institutes (Poggendorff 1898–1940). Sprung's father was an elementary school teacher, and the youth's domestic environment encouraged the pursuit of intellectual matters. Sprung initially studied pharmacy, but in 1872 he began work at the University of Leipzig in mathematics and physics, completing his doctorate in 1876. The dissertation dealt with frictional effects on salt solutions. In 1876 Sprung received a junior posting at the German Sea Observatory in Hamburg, where he remained for a decade. He embarked on two studies. Sprung sought to apply mathematical tools to the description of wind currents. This took into account factors such as planetary rotation and friction generated by the earth's physical topography. In effect, Sprung was moving toward a mathematical model of atmospheric dynamics. At the same time, Sprung took up work on instruments, in particular devices related to precipitation – the amount of rainfall, its velocity and quantity of energy (mechanical and thermal) its relations with barometric pressure, humidity, wind speed and direction.

In 1886 he became director of the instrument division of the Prussian Meteorological Institute, where he collaborated closely with Fuess. Over the next five years, the two men designed and constructed sixteen new or highly modified devices, most of them associated with the metrology of rain and associated physical conditions. This joint venture spawned sixteen articles (most of them in the *Zeitschrift für In-*

strumentenkunde) and four patents (three of them in Fuess' firm). These successes heightened Sprung's growing instrument reputation, and in 1886 he was named head of the International Cloud Year. In 1892 he became director of the German Meteorological and Magnetism Observatory at Potsdam. His commitment to instrument endeavor persisted. In these years he concentrated on devices for the measurement of cloud phenomena. Through a highly original assembly of mirrors, wide-angle lenses and the use of delayed-timing apparatus, Sprung managed to collect quantities of precise information on dynamic properties of clouds. He measured altitude, direction, speed, cloud deformations and the impact of wind or barometric changes. Sprung attached great importance to his latest instrument innovations. In his articles in the *Zeitschrift für Instrumentenkunde* he stressed successes in capturing the dynamic qualities of physical events. Equally significant, Sprung considered that his instrument work warranted notice because he had managed to develop systems of remote sensing through the wise use of electrical impulses. In most of the texts that Sprung published between 1890 and 1905, he systematically emphasized the importance of remote reading and sensing. He considered that the further development of this work and its extension to many instrument domains would transform industrial processes and the experiments carried out in academic laboratories. In the case of the latter, it would permit scientists to investigate as never before physical interactions in the course of their occurrence and the dynamics of physical phenomena (Sprung 1881, 1882, 1886, 1899).

Staatliche Forschung often struck a balance between instrument innovation and the measurement of physical entities, and this profile is well elucidated by the career and outputs of H. Vogel (1841–1907) (Cahan 1989: 235; Moeller 1908; Poggendorff 1898–1940). Vogel's father was head of the Leipzig *Gymnasium*, and Helmholtz and Bunsen were friends of the family. Vogel entered the University of Leipzig in 1863 and in 1870 obtained his doctorate from Jena University. His dissertation dealt with the nebula assigned to the Leipzig Observatory by the International Star Mapping Project. In 1870 he was named director of a private observatory at Kiel. Vogel's dissertation had focused on new astronomical devices and techniques – photometry, photospectroscopy and photography. By all accounts, his exceptional manual dexterity in working with and calibrating instruments, and his expertise in contemporary instrumentation constituted his principal *atout* for winning the Kiel position. While there, he worked on the spectral emissions of Venus, Jupiter and Uranus, but above all he wanted to study the sun's rotation by spectral means.

In 1874 Vogel joined the new Potsdam Astronomical Observatory, where he was responsible for developing needed instrumentation. In 1876 he completed work on a new spectral device capable of generating precise data with very little light. Such an apparatus was required to study the most distant nebula. Thanks to his novel apparatus, Vogel succeeded in providing detailed information on several faint nebula during the years 1878–81. In 1882 he became general director of the Potsdam Observatory. He immediately turned his attention to the dynamics of fixed stars but unfortunately a lack of adequate instruments soon forced Vogel to abandon this pursuit. He began research on the photo-Doppler effect. He completed a high-performance apparatus in 1885, and immediately initiated observations on stellar velocity. These successes culminated in his appointment as director of the Paris International Congress of Astronomical Photography in 1887. Indeed, it was through the use of Vogel's latter device that spectral double-stars were identified and characterized. Vogel was renowned in German and international astronomy circles for achievement in the design and construction of refracting telescopes. But perhaps his greatest claim is for the reversible spectroscope of which he produced a small number for Berlin colleagues as well as for some foreign astronomers.

An inspection of Vogel's journal output reveals that he published in more than twenty reviews. About one third of his articles dealt directly with instrument innovation. The content of another third dealt with a mix of measurements of physical phenomena and instrument-related information.[3] While research-technology does not monopolize Vogel's endeavor, it nevertheless constituted an important, and perhaps even the predominant, or co-predominant, element. This appears to be the case for almost half of those individuals in our sample of *Staatliche Forschung*.

Academia

While the number of instrument men in state research institutes increased moderately between 1883 and 1900, academics, by contrast, flocked to the emergent research-technology movement. The percentage of academics climbed from 20% in the years 1880–90, to almost 33% during the following decade.[4] This may in part be attributed to the existence of new instrument-related journals which welcomed and even solicited academic contributors. While the *Zeitschrift für Instrumentenkunde* was the foremost mouthpiece for research-technology academic publications, there were certainly others, such as the *Zeitschrift für Vermessungswesens*, and numerous scientific journals that specialized in publication of experimental results and also featured articles on instrument topics. In her book on measurement physics in nineteenth century Germany, K. Olesko provides an another possible explanation for enhanced academic interest in the instrument movement (Olesko 1991: 432). German physics changed significantly in the decades preceding the turn of the century. In universities, physical and mathematical theory expanded, often at the expense of experimentation. Moreover, the focus of experimentation was changing. While in the past measurement methodology and instrumentation had comprised a central current in physical research, it was supplanted by a systematic search for new categories and expressions of phenomena – discovery as opposed to precision. Physicists raised in the mold of mid-century science saw in research-technology some of the practices, routines and goals that they themselves favored. By publishing in the *Zeitschrift für Instrumentenkunde* and similar periodicals, they sustained a certain academic legitimacy by dint of recourse to a scientific journal and the circulation of data in the public realm.

The career trajectory and practices of R. Helmert (1843–1917) highlight some of the salient traits of university-based research-technologists in the movement's early decades (Poggendorff 1898–1940). Helmert studied mathematics at Dresden from

---

[3] Sean Johnston shows that Vogel was a prominent figure in the colorimetry research-technology movement near the turn of the century. See Sean Johnston (1994: 44).

[4] Contrary to expectations, this growth took place mainly in German universities rather than institutions of higher technical training.

1859 to 1863. During part of this period he worked at the Saxony Office of Longitude. In 1868 he received his doctorate from Leipzig University, and in 1869 he became an assistant at the Hamburg Observatory. Helmert was appointed professor of geodesy and applied geometry at the *Aachen Technische Hochschule* in 1870, and in 1889 he became professor of geodesy at the University of Berlin. He earned another doctorate in 1902, this time in engineering. Helmert received a Berlin Academy prize in 1912 for his exceptional achievements.

This research-technologist designed optical instruments for the measurement of distances and the characterization of surfaces. In conjunction with this he improved the mathematical tools for correlating data supplied by his instrument set-ups and for interpreting it. Of particular importance, Helmert developed methodological instruments for error detection and its evaluation and management. The technologist marketed his numerous material and cerebral productions in a variety of spheres, as testified to by publications in over fifty different journals. This scope of outlets comprises one of the key features of research-technology. While accomplished and renowned experimenters sometimes succeed in publishing in a score of different periodicals, the demands, norms and profile of the experimental community, unlike research-technology, does not contain a logic of infinitely broad marketing. Helmert veritably proselytized his instrument innovations. He published texts that applied his devices to the measurement of distances on the earth's surface and dealt with items such as height, concavity and surface distortion. This is the essence of cartography. Beyond this, he extended his innovations to architecture and techniques employed in road and railway planning and construction. It was likewise extended to material and methodological problems in astronomy. The research-technologist sought to convince physicists and mechanics that his devices for measuring and characterizing surfaces were applicable in their domains as well. He contacted industrialists specialized in sugar, hoping to introduce an approach for superior quality control based on the precise description of crystals. Clearly, Helmert was not an academic in the ordinary sense of the term. Although a force within the university, his domains of impact and source of authority and legitimacy transcended academia. The logic of Helmert's operations was predicated on the permeability of the boundaries that regulate interactions between institutions and groups. Only through innumberable adapations and transactions could Helmert achieve the goal of re-embedding his generic instrument innovations in a maximum of sites – academia, architecture, building construction, civil engineering and manufacturing.

Industry

In the 1890s almost one third of our sample of research-technologists were industry employees. This was an increase of 10% over the previous decade, and it reflects the growing size of German industry, as well as expansion of the science-intensive sectors of the economy and the role played in general by science and technology (Faulhaber 1904). Moreover, by all accounts, links between industry, state research institutes and academia had become less restrictive and less prescriptive, but had instead become more numerous. Industry-based instrument experts published abundantly in the *Zeitschrift für Instrumentenkunde*, and there are few signs that this posed acute problems of industrial secrecy. This is possibly attributable to the fact that research-technologists went public but within a framework of core devices. However, why go public at all? Intercourse with other instrument specialists afforded satisfaction in terms of recognition for work well done, a measure of legitimacy, opportunities to see the "implantation" of a generic device, and not least of all, the opportunity to acquire ideas and information. It was perhaps this last aspect that fostered a spirit of tolerance toward extra-company communication on the part of industrial management. Through multiplied communication and channels of input, the firm might profit! For these and other complex reasons, individuals devoted to general, multi-purpose kinds of apparatus began to flourish within some areas of industry near the turn of the century.

Take the example of S. Czapski (1851–1907). He obtained a doctorate from the University of Berlin in physics in 1881. Beforehand, he had studied at Göttingen and Breslau. After receiving the doctorate, Czapski continued to do optics research in the capital until 1884 when his research came to the attention of E. Abbe – himself a noted research-technologist specializing in optics who headed the Zeiss Company. He at once went to work for the latter as one of Abbe's personal assistants. In 1891 Czapski became a science delegate on the directing board of the Zeiss shop (Poggen-dorff 1898–1940).

Czapski published five books and over fifty articles. Three of the books dealt with general instrument applications of Abbe's theories in optics (Czapski 1893, 1904, 1904 (with Otto Eppenstein)). In his articles Czapski focused on four principle topics: a) problems of chromatics in the presence of multiple lenses, b) the development of principles and techniques for polarizing microscopes and telescopes, c) techniques for reducing the thickness of lenses in refractometers, d) theories of stereoscopic magnification. Over his twenty-year career, Czapski struggled to disseminate his ideas, techniques and products across a broad spectrum of publics which included academic researchers (he published a book to this end in 1904), the Zeiss optical company, firms involved in mineralogical sampling and the development of appropriate instrument innovations, and the army. Near the end of his life, he wrote three treatises for the military which were intended to lay the groundwork for a more scientific approach to stereoscopic range-finders and telescopic sights. Linkage between research-technologists and the army was extremely common in Germany during this period, to such an extent that the military seems to have constituted a foundational component of research-technology in terms of an audience.

Artisans and Consultancy

As indicated above, artisans figured centrally in the birth of *the Deutsche Gesell-schaft für Mechanik und Optik* and the *Zeitschrift für Instrumentenkunde*, and therefore in the German research-technology movement. In the 1890s their involvement declined in relation to that of other groups. Based on the journal sample, artisans fell from about 25% to 15%. However, the total number of small firms specializing in instrument-making did not diminish; rather, the presence of alternative components swelled. This group contained two strands: on the one hand, independent artisans or very small companies (twenty to fifty craftsmen) which generated core devices and methods with broad applications, and on the other hand consultancy groups which sold instrument advice to specific clients – generally big companies. The importance of the latter remained low. Most of the individuals belonging to the artisan category exhibited traits very similar to those previously set-forth for Fuess, Bamberg and Haensch.

In the 1890s, H. Krüss (1853–1925) became director of the Hamburg instrument company of A. Krüss, which belonged to his father, E. Krüss, and before that to his grandfather (Poggendorff 1898–1940). H. Krüss received craft-training in mechanics and optics in Hamburg and then Munich. He attended courses at the Munich *Technische Hochschule*, and then transferred to the University of Munich where he earned a doctorate in physics in 1876. He immediately returned to Hamburg and served as an assistant to his father until the latter's retirement. Krüss specialized in four categories of instruments: photometry, colorimetry, spectroscopy, and automatization.

This research-technologist published nine books between 1890 and 1920. One important volume dealt with questions of switching; that is, how to automatically switch lenses of varying characteristics in a spectroscope, microscope or polarizer. Such an instrument system, wrote Krüss, had an infinity of applications, extending from the chemical industry (characterization of crystals of different kinds) to biology, mineralogy, optics and astronomy. Another book focused on the basic problems of light measurement, and how measurement was associated with both wave length and qualities of a substance's surface (H. Krüss 1913; H. and G. Krüss 1891; H. and G. Krüss 1892). Krüss designed and constructed devices that integrated his general instrument theories, but which were adapted to specific audiences. He engineered a particular type of projector for the military based on his methodological and technical innovations. The same principles underpinned the development of projectors for the film industry. Not satisfied with this, Krüss extended the instrument innovations to the field of ophthalmology and optometry. The latter ultimately resulted in additional instrument research for the navy. In addition to his many books, Krüss published over 100 articles in thirty-seven different journals. Some of the periodicals were industrial trade journals for the sugar and food industry, for the electrical apparatus and mechanical industries, and for pharmaceutical and medical industries. Krüss also published in Germany's most renowned physics and chemistry academic periodicals where he explained the characteristics and uses of his new methods and devices. During the two decades that preceded his death, Krüss was the most influential instrument-maker of the "old school" (artisans) in the *Deutsche Gesellschaft für Mechanik und Optik*. He believed that the best instrument innovations were to be generated in independent instrument workshops rather than in big companies. The latter, he said, constricted the instrument imagination and required that too much attention be given to matters of production rather than generalist devices having a multi-market potential.

Engineering

In my sample fewer than 5% of the individuals were engineers, that is only one person in the 1880s, and by stretching the definition of "engineer," perhaps three people in the 1890s. Why this was so remains obscure, and at this juncture one can do little more than advance hypotheses. At the end of the nineteenth century, the German engineering community was in full metamorphosis (König 1993). The expansion of industry, and particularly science-intensive industry, generated an acute and protracted demand for engineers in a spread of areas – mechanics, optics, electricity and chemistry. Engineers occupied a specific niche, and there was little incentive to extend their purview. The issue was to organize the occupation and to transform it into a full-fledged, legitimate profession. In conjunction with this, tensions arose between certain strands in academia and the accelerating engineering movement. German engineers desired high academic training and certification, of the sort traditionally reserved for academic scientists, attorneys and doctors. The battle for this raged during much of the last half of the century. There seems to have been little energy or interest in matters associated with things instrumental.

Another fundamental issue also has to be raised. Engineering education at that time, and in some instances still today, was characterized by precise skills in problem-solving of a particular kind. Engineer training is sometimes likened to learning of recipes and prescriptions. While this description is surely an exaggeration, it nevertheless contains a grain of truth (Bucciarelli 1994). An antagonism possibly existed between the reasoning and technical skills adapted to the design and construction of generic instruments and that suited to effective engineering. While there is a "closed" element to traditional engineering, the nature of research-technology is fluidity and "openness." If this hypothesis is correct, a mismatch thus occurred between the dynamics inherent in engineering practices and those of research-technology instrument making and diffusion.

INSTRUMENT POLITICS

Two events occurring in the mid- and late 1880s gave added impetus to the German research-technology initiative. The first revolved around the *Physikalisch-Technische Reichsanstalt*. From the early 1880s onward, the program of Siemens to establish a big science and technology center gained momentum. At the outset, however, the plan was indefinite and relatively unstructured. Siemens, with the advice of his ally and friend Helmholtz, nevertheless clearly expressed the wish to erect an outstanding facility for fundamental research in the physical sciences. Helmholtz was to direct this operation. There was also to be a second section of the *Reichsanstalt*. Its organization and purposes, however, were heatedly debated, and it was precisely here that the instrument movement asserted itself.

Three groups vied for control of the *Reichsanstalt*'s technical section. Industrialists called for a facility that would undertake routine tests associated with quality control, standardization, safety and certification. In this project, the technical section would serve as a responsive agent to urgent industrial needs. The second group competing for control consisted of engineers. The *Verein Deutscher Ingenieure*, which represented many of the country's professional engineering groups, was an increasingly powerful force. It strove to enhance engineers' public image and status, and to reinforce their institutional clout. Domination of the technical section of the already prestigious *Reichsanstalt* would further this cause. F. Grashof, head of the *Verein Deutscher Ingenieure*, waged a campaign within the *Reichsanstalt*'s organizational board to convince it that the advancement of engineering comprised a necessary resource for the future progress of both science and industry (Cahan 1989: 54–6, 75). He argued that the technical section's proper function was to develop engineering-linked knowledge, to solve problems introduced by industry and science, and to formulate improved educational programs.

The *Deutsche Gesellschaft für Mechanik und Optik* and its allies did not agree! The *Physikalisch-Technische Reichsanstalt*'s board included several prominent research-technologists, among them, Fuess, Bamberg and Loewenherz. They vigorously combated the offensives of the industrial and engineering lobbies. Industry, they insisted, should not be admitted to the *Reichsanstalt*. Industrial tests were necessary and legitimate, but it was not the correct place for them. Testing belonged to industry, and the latter should organize itself accordingly. The *Reichsanstalt* was supposed to be an institute for "general" knowledge rather than partisan, self-interested knowledge. The same held for engineering. It certainly needed to protect and advance its professional and partisan goals, but the *Reichsanstalt* should not be the theater for this. Moreover, *Gesellschaft* spokesmen on and off the *Reichsanstalt* board pointed out that engineering activities were often only very loosely connected with research, as engineers simply "executed" plans and projects. This had never been the intended role of the new institute.

By contrast, the design and construction of precision instruments fitted the *Physikalisch-Technische Reichsanstalt* agenda on numerous grounds (DGMO, *ZIK* 1887b, 1889b: 112, and 1889d: 258). Generic instrumentation was inherently a domain of research and not a matter for recipes or execution. Furthermore, the instruments arising out of the technical section's endeavors would be of utmost utility both to science and industry, and could also benefit engineers. The precision technology movement was not a vested interest like industry and engineering. Its purview and scope transcended narrow, local concerns and in this respect, so it was claimed, it resembled science. As a consequence, the direction of the technical section should be given to an individual close to instrument circles. After long and stormy deliberation, the section's top position went to Loewenherz. Although his premature death in 1892 interrupted his efforts to install research-technology as the sole or dominant activity, he nonetheless went far in this direction. It was not coincidence that the *Deutsche Gesellschaft für Mechanik und Optik*'s instrumentation journal, the *Zeitschrift für Instrumentenkunde*, was saved from closure by the active intervention and financial subsidies of the *Physikalisch-Technische Reichsanstalt*'s technical section.

A second turning-point in the history of Germany's research-technology movement was reached in 1889. Since 1883 many scientists and artisans committed to the precision instrument movement had fought to gain recognition and special status for their initiatives inside the *Versammlung Deutscher Naturforscher und Ärzte*. The latter body was Germany's biggest science association. Founded at mid-century, it boasted almost 5,000 members in a host of fields ranging from the physical and life sciences to medicine. Loewenherz spearheaded the drive for the establishment of an instrument section inside the *Versammlung*. He was a member in good standing and influential in physics and technology-related commissions.

Loewenherz' task entailed not merely the establishment of an additional committee, but indirectly, the re-shaping of the logic that governed the Association. In the past, proceedings had been structured around academic disciplines and fields, or around professions, like pharmacy and medicine. In this configuration, instrumentation was present, yet remained "instrument-bound." Apparatus currently existed as a small part of each discipline and activity. But this arrangement was not satisfactory, for while instrumentation was acknowledged as fundamental, it nevertheless enjoyed little visibility or status in its own right. Loewenherz, the *Deutsche Gesellschaft*, and the *Zeitschrift* wanted this to change. The arguments often echoed those associated with the Chicago exhibition (DGMO, *ZIK* 1886a, 1886b, and 1886c, 1889d: 224).

Loewenherz suggested that generic instrumentation possessed an internal logic, and that this logic was jeopardized by merging apparatus into a myriad extant fields. Of course, he continued, the aim of instrumentation was to be of service to various fields, yet this was best achieved through the acquisition of a circumscribed form of instrument autonomy. Such autonomy would facilitate the coming-together of apparatus experts and would heighten communication among them. Research-technologists would thus be in a better position to discern instrument needs of different potential consumers and to satisfy those needs. To maintain a configuration inside the *Versammlung Deutscher Naturforscher und Ärzte* where precision technology was subsumed under orthodox disciplines meant that instrument-makers' understanding of what was possible and required of them necessarily remained narrow and truncated. This was unacceptable (DGMO, *ZIK* 1886a, 1886b, and 1886c, 1889d: 224; 1890b). In 1887 and 1888 Loewenherz' petitions in favor of a new instrument section were examined by the *Versammlung*'s central committee. Although initially viewed with skepticism, little by little the Association's position shifted, and in 1892 an instrument division was opened. This event was heralded in the *Deutsche Gesellschaft* and the *Zeitschrift für Instrumentenkunde* as a key victory for precision instrumentation (DGMO, *ZIK* 1890a: 224; 1892a, and 1892b). While continuing to service the whole of science and industry, instrumentation would henceforth comprise a vocation that was embodied in an identifiable community.

CONCLUSIONS: TRAJECTORY AND STRUCTURE

To what extent and in what ways do the trajectories of our sample reflect the structures of generic devices, metrologies, and interstitial community which underpin the research-technology perspective? Virtually all of the biographies presented above suggest that these instrument men dealt in generic devices, although of course some more fully than others. The University of Berlin professor, Helmert, developed fundamental techniques and apparatus which he believed would subsequently be endowed with additional refinements, according to the needs of the particular domain of their implantation. His devices were not designed with a specific, local end-user in mind. Rather, Helmert elaborated broad instrument capacity. Thus, his apparatus could be projected into a maximum of consumer sectors, extending from navigation to the characterization of crystals in the food industry, and from cartography to the analysis of error in the physics or medical laboratory. The remarkable breadth of application of these generic devices was apparently linked more to accomplishments in methodology than material instruments.

All of the other men examined here also generated generic instruments rather than devices primarily intended for a specific, local client. Nevertheless, the picture is a complicated one. Krüss, for example, was a research-technologist who frequently followed through with a generic apparatus, in the sense that he himself adapted it for a specific end-user, such as in the case of his projectors. Indeed, Krüss adapted the basic optical system for a series of markets like the military and the film industry. Despite this, he systematically resumed the vocation of generic instrument development. In this instance then, we discern someone who constantly circulated between generic activities and the re-embedding function.

The issue of interstitial community was similarly prominent in the German research-technology instrument movement. In the battle for control of the *Physikalisch-Technische Reichsanstalt*'s technical section, Loewenherz, Fuess and Bamberg stressed the broad-based, extra-institutional character of their kind of instrumentation, as opposed to the much narrower, partisan, professional dimension of engineering, and for that matter also industry. Here, the research-technology movement was represented as something relatively singular and unique in terms of organization and function. Moreover, the interstitial structure is indirectly testified to by virtue of the innumerable audiences and clients between which instrument men navigated. They belonged to none of their clients, yet served all. Haensch, for example, insisted on the need to sustain a generic instrument autonomy. He saw merit in an instrument community where practitioners would recognize one another through the devices they produced rather than the clients they served. Generic instrumentation would itself, he argued, provide the foundations for the community – a community whose chief characteristic lay in its ability to link up with other groups while remaining apart.

This was, however, balanced by another consideration. Both at the Chicago Exhibition and the *Versammlung Deutscher Naturforscher und Ärzte*, research-technology observers and spokesmen clamored for the establishment of a separate generic instrument section. Their argument was twofold. Firstly, that instrumentation possessed an inherent, internal logic which required that devices be clustered together; and secondly, that by dint of a measure of separateness, there would emerge a greater appreciation of instruments on the part of clients and a clearer picture of the total market for particular devices. Is this stance inconsistent with interstitial community? Perhaps not altogether so. Interstitial structure requires that a community attains and sustains an equilibrium between fluidity on the one hand, and on the other hand, attachment to a conventional, embedded social/institutional entity. An interstitial body is probably by nature unstable, and hence must oscillate between entropy and involvement with existing organizations and institutions. To this end, complex strategies may intervene.

Finally, to what extent is the metrology principle active in the German research-technology movement? Our response to this question is nuanced. In our conceptual schema, Bernward Joerges and I have described "metrologies" with respect to three elements: standards and units of measurement, representations, and paradigms (chapters 1 and 12 of this volume).

Many of the individuals encountered in my research-technology sample were highly active in technical areas related to units of measurement and standards. For example, Krüss introduced several novel luminosity units and set forth general techniques for generalizing them, and Czapski suggested a new system for the treatment of colored light in a range of lenses. The sample also includes at least one instance of a research-technologist who transformed a representational frame of reference in physics and beyond. Helmert strove to organize analysis and understanding of many physical manifestations on the basis of surface features – a sort of physiognomy of the physical world. For him, through the characterization and classification of surface structures, it was possible to identify universal similarities ranging from crystals to geodesy and from cartography to architecture and infrastructure. Here, Helmert recommended a sort of new unified vision which subsumed extensive areas of knowledge. He suggested that light absorption was best understood as a function of surface properties, meaning that light could then be studied from this vantage point.

Some of the work of Sprung and Fuess can likewise be regarded as representation transformation. Their devices and techniques that dealt with automatic registering and switching by means of electrical impulses profoundly affected instrumentation in the broadest sense of the term. Due to this, scientific research turned increasingly to the investigation of dynamic properties of phenomena. Automatic registering and switching fueled this by making it easier to measure and describe events in ever shorter and more complete time intervals. The relevant scale of time, and with it a component of the concept of time, was altered. In effect, the intellectual and material instrument productions of Fuess and Sprung participated importantly in recasting the relations between time, physical phenomena and perceptions of the dynamic properties of the latter. Moreover, switching and automatic registration very rapidly moved beyond meteorology and the academic laboratory. Soon, industry assimilated the changing perception and technique into production processes, particularly those associated with fluids (especially gases), foodstuffs (brewing and other fermentation-based products), and, somewhat later, petroleum products and plastics. A strong case can be made that such changes in production processes had a subsequent impact on concepts and practices of industrial organization, management and worker consciousness and strategy ("the new working class"). Whether these changes constitute a shift of representation, or perhaps something larger, is not fully clear. Whatever the response, research-technology instruments certainly played a central role in the dawning of a novel repertoire of portrayal and control. Ultimately, the new devices of Fuess, Sprung and colleagues contributed to a fresh paradigm where dynamic properties, instability and uncertainty provide the framework for seeing the world and for acting upon it.

GEMAS/CNRS, Paris, France

REFERENCES

Bucciarelli, Larry (1994). *Designing engineers.* Cambridge, MA: The MIT Press.
Cahan, David (1989). *An institute for an empire: The Physikalisch-Technische Reichsanstalt, 1871–1918.* Cambridge, MA: Cambridge University Press.
Czapski, Siegfried (1893). *Theorie der optischen Instrum. nach Abbe*, gr. 8°. Breslau: Trewendt.
Czapski, Siegfried (1904). *Grundz. der Theorie der opt. Instrum. nach Abbe*; 2. Aufl. Leipzig: Barth.

Czapski, Siegfried (with Otto Eppenstein) (1904). *Mitarb. an Bilderzeugg. in opt. Instrum. v. Standpkt. d. geom. Optik v. M. v. Rohr.* Berlin.

Deutsche Gesellschaft für Mechanik und Optik (= DGMO). Minutes in *Zeitschrift für Instrumentenkunde* (= *ZIK*): (1881a). *ZIK* 1 (Febr.): 66; (1881b). *ZIK* 1 (April): 302–03; (1881c). *ZIK* 1 (May); (1881d). *ZIK* 1 (Sept.); (1882). *ZIK* 2 (Dec.): 46; (1883) *ZIK* 3 (Febr.): 116; (1884). *ZIK* 4 (June): 280; (1885a). *ZIK* 17 (Jan.): 5; (1885b). *ZIK* 5 (March): 103; (1885c). *ZIK* 5 (Sept.): 313; (1885d). *ZIK* 5 (Oct.): 370; (1885e). *ZIK* 5 (Nov.): 400; ; (1886a). *ZIK* 6 (April): 137; (1886b). *ZIK* 6 (Sept.): 311; (1886c). *ZIK* 6 (Oct.r): 351; (1887a). *ZIK* 7 (Jan.): 39; (1887b). *ZIK* 7 (Dec.): 39; (1888a). *ZIK* 6 (Oct.); 351; (1888b). *ZIK* 8 (Nov.): 428; (1889a). *ZIK* 9 (Jan.): 35; (1889b). *ZIK* 9 (Febr.): 112, 113; (1889c). *ZIK* 9 (April): 152; (1889d). *ZIK* 9 (May): 224, 258; (1889e). *ZIK* 10 (Aug.): 293; (1890a). *ZIK* 10 (May): 224; (1890b). *ZIK* 10 (Aug.): 293; (1890c). *ZIK* 10 (Oct.): 33; (1892a). *ZIK* 12 (May): 167; (1892b). *ZIK* 12 (July): 247–8; (1892c). *ZIK* 12 (Aug.): 290; (1893). *ZIK* 13 (Nov.): 33; (1894a). *ZIK* 14 (June): 211; (1894b). *ZIK* 14 (Aug.): 330-7; (1894c). *ZIK* 14 (Oct.): 367; (1894d). *ZIK* 14 (Nov.): 405; (1894e). *ZIK* 14 (Dec.): 421–6; (1897). *ZIK* 17 (Jan.): 5.

Faulhaber, C. (1904). Die optische Industrie, in *Handbuch der Wirtschaftskunde Deutschlands, Vol. 3* (455–72). Leipzig: P.G. Teubner.

Johnston, Sean (1994). *A notion or a measure: The quantification of light to 1939.* PhD Dissertation. Leeds, UK: University.

König, Wolfgang (1993). Technical education and industrial performance in Germany: A triumph of heterogeneity, in R. Fox (ed.), *Education, technology and industrial performance, 1850–1939* (pp. 65–87). Cambridge, MA: Cambridge University Press.

Krüss, Hugo (1913). *Neue Wege und Ziele naturw. Arb.* Hamburg: Friedrichsen.

Krüss, Hugo and Krüss, Gerhard (1891). *Kolorimetrie und quantitative Spektralanalyse in ihrer Anwendung in der Chemie,* 2. Aufl. Hamburg, Leipzig: L. Voss.

Krüss, Hugo and Krüss, Günter (1892). Quant. Spektralanalysen und Kolorimetrie. *Zeitschrift für Anorganische Chemie,* No. 1.

Loewenherz, Leopold (ed.) (1880). *Bericht über die Wissenschaftlichen Instrumente auf der Berliner Gewerbeausstellung im Jahre 1879.* Berlin: Springer.

Lundgreen, Peter, Horn, Bernd, Krohn, Wolfgang, Kueppers, Günter, Paslack, Rainer (1986). *Staatliche Forschung in Deutschland, 1870–1980.* Frankfurt, New York: Campus.

Moeller, G. (1908). Nachruf. *Schriften der Astronomischen Gesellschaft* 42: 7.

Nachruf (auf Leopold Loewenherz). (1892). *Zeitschrift für Instrumentenkunde* 12 (December, gebundene Jahresausgabe): 401.

Nachruf (auf Leopold Loewenherz). (1892). *Zeitschrift für Instrumentenkunde* 12(12): 53, 192.

Olesko, Kathy M. (1991). *Physics as a calling. Discipline and practice in the Königsberg seminar for physics.* Ithaca, NY: Cornell University Press.

Pfetsch, Frank (1970). Scientific organisation and science policy in imperial Germany, 1871–1914: The Foundation of the Imperial Institute of Physics and Technology. *Minerva* 8(4), 557–80.

Poggendorff, Johann Christian, ed. (1863–1919; 1925–1940). *Biographisch-literarisches Handwörterbuch zur Geschichte der exacten Naturwissenschaften.* 6 Bände: Bde. 1–2: Leipzig 1863. Bd. 3: 1858-1883. Teil 1–2. Hrsg. v. W. Feddersen u. Arthur von Oettingen. Leipzig 1897–98. Bd. 4: 1883–1904. Teil 1–2. Leipzig 1904. Bd. 5: 1904–1922. Teil 1–2. Biographisch-literarisches Handwörterbuch für Mathematik, Astronomie, Physik, Chemie u. verwandte Wissenschaftsgebiete. Red. v. P. Weinmeister. Leipzig. Berlin 1925–26. Bd. 6: 1923-1931, Teil 1–4. Red. v. Hans Stobbe. Berlin 1936–1940.

Sprung, Adolf (1881). Neue Registrierapparate für Temperat. und Feuchtigkeit der Luft. *Zeitung für Instrumentenkunde (ZIK),* Vol. 1.

Sprung, Adolf (1882). Continuirl. Registrirung der Robinson'schen Schaalenkreuzes. *Zeitung für Instrumentenkunde (ZIK),* Vol. 2.

Sprung, Adolf (1886). Discussion der Aufzeichn. der Sprung-Fuess'schen Thermobarographen in Spandau. *Zeitung für Instrumentkunde (ZIK),* Vol. 6.

Sprung, Adolf (1899). Photogrammetr. Wolkenautomat und seine Justirung. *Zeitung für Instrumentenkunde (ZIK),* Vol. 19.

PART II

INTERSTITIAL WORLDS

CHAPTER 4

XAVIER ROQUÉ

DISPLACING RADIOACTIVITY

No one could be further from a research-technologist than Marie Curie, the quintessentially pure scientist, it seems. And yet, Curie's views on radioactive science accorded a central role to metrology and industry, centrally important to our subject. We need not be too puzzled by this, as long as we do not take Curie's persona at face value. Curie may have been a pure scientist – but one with a view of radioactivity that encompassed industry, metrology, and therapy, a view that ultimately derived from a self-imposed mandate to identify and accumulate radioactive elements. Curie's pure, powerful sources allowed her and her co-workers to operate in most radioactivity-related domains – quite as research-technologies do. The uses of accumulation are analyzed in the first section of this paper.

Radioactive substances, however, were not to be had easily: They were buried in tons of thorium and uranium ores, and Curie realized early in her career that industrial resources were essential for acquiring radioactive material. This led her to assist in the birth of the French radium industry in the 1900s and feed it with expert advice and technicians through to its demise in the late 1920s, while cultivating contacts with radium producers worldwide. Curie's sustained relationship with the radioelements industry is discussed in the second part of the paper. Rather than being of peripheral significance, Curie's industrial activities were an integral part of her research effort.[1] Such pursuits seem to undermine prevailing representations of her as a pure scientist, yet these are not without value, particularly as they were cultivated and propagated by Curie herself. She extolled the commitment to research and "the atmosphere of peace and meditation which is the true atmosphere of a laboratory." According to one of her collaborators, the laboratory "was for her a unique place of work and meditation, isolated from the world" (M. Curie 1923: 67; Guillot 1967). Building on a long-standing rhetoric of disinterestedness and detachment, these statements actually underlined Curie's campaign to provide her institute with "industrial means of action" and contributed to the ongoing debate on state support of science in France.

---

[1] Curie's biographers have paid scant attention, if any, to her industrial pursuits (see E. Curie 1938; Pflaum 1989; Quinn 1995; see, however, Reid 1974).

B. Joerges and T. Shinn (eds.), Instrumentation: Between Science, State and Industry, 51–68.
© 2001 *Kluwer Academic Publishers.*

This study, however, is not just about Curie and her collaborators but also about her laboratories, above all that at the *Institut du Radium*, the institution that came close to realizing Curie's all-inclusive vision of radioactivity. Besides research laboratories, the institute comprised a metrological unit which acted as a national laboratory for radioactivity; biological and medical sections; and eventually industrial facilities in the outskirts of Paris. The industrial dimension is perhaps the most characteristic and intriguing feature of Curie's scheme, as I have argued elsewhere, yet this was a multipurpose institution through and through. The *Institut du Radium* was such a propitious ground for radioactive research-technologists, and such an effective platform for displacing radioactivity, that I suggest it can usefully be regarded as a "generic institution" (see Roqué 1997[2]).

THE USES OF ACCUMULATION

The accumulation of radioactive substances was the backbone of Marie Curie's work on radioactivity. This she conceived as "the physico-chemical study of radioelements," an autonomous new science whose explicit aim was the identification, isolation, and concentration of these elements, as opposed to, for instance, the investigation of the innermost constitution of matter. Adrienne Weill listed the acquisition of "samples that are pure or of maximum concentration, even at the cost of handling enormous quantities of raw material" as one of the policies of Curie's research, "to which she always remained faithful" (Weill. 1981: 500). The chemical and physical characterization of radioactive elements Curie regarded above all as a means of devising concentration procedures, in order to produce them efficiently in substantial amounts. These alone would allow their properties to be linked. This circular reasoning lent accumulation its scientific and industrial meaning: "as we must study the radioelements, we need to manufacture them" (CP, n.a.f. 18394: 281).[3]

We can see the logic of accumulation at work from the discovery of radium. Pierre and Marie Curie's attempts to isolate and identify the new element were repeatedly frustrated by "great misfortunes" such as spillage: "In the course of a delicate operation with radium we lost an important quantity of our stock, and we still cannot understand the cause of the disaster. On this account, I find myself forced to put off the work on the atomic weight of radium." Important as Curie's early misfortunes are in understanding her decision to grow radium, we need not give them undue weight. Radioactive sources were essential to the conduct of the new science, and their crucial import has previously been discussed, most fully by Jeff Hughes (1993). No wonder, then, that the Curies's colleagues marvelled at the strength of their radioactive samples: "It goes without saying that your researches are more effective, and that you can observe phenomena that are not even perceptible here" (CP, n.a.f. 18434). The radium collectors held a clear advantage in discovering liminal phenomena.

---

[2] on which this paper is largely based. I also draw heavily on Soraya Boudia's paper in the same issue (1997b), and her PhD dissertation (Boudia 1997a).

[3] The following abbreviations have been used: AC: Archives M. Curie at the Institut Curie, Paris; CP, n.a.f. (=nouvelles acquisitions françaises): Pierre and Marie Curie Papers, Bibliothèque Nationale, Paris; RP: E. Rutherford Papers, Cambridge University Library, Add MS 7653.

In 1903, the couple negotiated with the Austrian government the sale of several tons of residues from a rich uranium mineral (pitchblende), which would allow them to "prepare a greater quantity of this luckless substance. For this," Curie wrote her brother Joseph in December that year, shortly after being awarded her first Nobel Prize, "we need ore and money. We have the money now, but up to the moment it has been impossible for us to get the ore. We are given some hope at the moment, and we shall probably be able to buy the necessary stock which was refused to us before. The manufacture will therefore develop."[4] Marie Curie was in direct contact with the Vienna Academy of Sciences, which controlled the uranium-rich St. Joachimstahl mines in Bohemia; she was the only scientist outside Vienna who managed to exploit the residues from uranium extraction after the Austrian government placed an embargo on their export in 1904. Between 1904 and 1906, Curie was able to get some 10 tons of residues from St. Joachimstahl, which bore 400 mg radium chloride and a fresh determination of the atomic weight of radium. In 1910, Curie's quest for pure radium metal was successfully completed, earning her a second Nobel prize. The isolation of radium, stated Curie in her Nobel lecture, was "the cornerstone of the entire edifice of radioactivity," a view echoed by Jean Perrin on the 25th anniversary of the discovery of radium (M. Curie 1907, 1966: 202–12; M. Curie and Debierne 1910b; Perrin 1924).

Curie gave the same treatment to other radioactive elements, like polonium and actinium. Between 1902 and 1906, the identity of polonium was contested by German chemist Willy Marckwald, who claimed his "Radiotellur" was not Curie's "polonium" – as it turned out to be. In her 1903 dissertation, Curie had stressed the need "to prepare a small amount of [polonium] of the highest possible concentration," and this remained a concern for years to come. A researcher of the laboratory, Paul Razet, studied the methods of extraction, trying various acids and timing the process (Boudia 1997a; M. Curie 1903; Razet 1907). In 1911, Curie and André Debierne reported to the *Caisse des Recherches Scientifiques* about the use of a grant for studying polonium, with a view "to prepare [it] in a highly concentrated state, and to study its properties." Polonium having a half-life of about five months, this meant vast amounts of raw material had to be manipulated in a rather short time span. Typically, the work "involved several workers in a factory during several months." Also typical is the conflation of entrepreneurial and scientific aims: Curie was interested in the element polonium and its manufacture, but she was well aware that for fellow Nobel radioactivist Ernest Rutherford "a weighable quantity of polonium in pure state" was needed to settle "a matter of very great interest and importance," namely, whether polonium changed into lead, as claimed by disintegration theory (M. Curie and Debierne, CP, n.a.f. 18436: 8–15, 1910: 38–40; Maurice Curie 1925: 222–3; Rutherford 1910).[5]

Actinium, discovered by Debierne in 1899, tells a similar story. Between 1902 and 1905, no one knew for sure whether Friedrich Giesel's "Emanium" was actually Debierne's element, nor whether it belonged to the disintegration series of uranium

---

[4] Marie to Józef Skłodowski, n.d., and 23 December 1903, respectively, quoted in E. Curie (1938: 226). See Roqué 1994.

[5] Maurice Curie, chemist and engineer, was Marie Curie's nephew.

(Kirby 1971, finds in Giesel's favor). Both issues pushed forward its isolation and concentration in a pure state, not just at Curie's laboratory. In 1907, Bertrand Boltwood, a prominent American radioactivist, sought Rutherford's assistance to get an active actinium preparate. Rutherford managed to get some 500 kg of pitchblende from S. Meyer in Vienna, yet the ore was eventually processed in Paris, most likely following a procedure devised by Debierne (Armet de Lisle 1906; Boudia 1997a: 126–8). The concentration of actinium was Curie's last major scientific undertaking. Catherine Chamié, Curie's assistant between 1920 and 1934, recalls her efforts to organize the concentration of actinium, to which we return below (Chamié 1935; M. Curie 1930). The enabling capacities of Curie's actinium came to match those of Aimé Cotton's *grand électroaimant*: In 1929, Salomon Rosenblum placed a strong actinium source prepared by Curie herself in the poles of the strong Bellevue magnet, revealing the existence of alpha spectra (M. Curie and Rosenblum 1931; Quinn 1995: 406–8; Shinn 1993).

While Curie's colleagues were fully aware of the uses of accumulation, they have not fared so well with historians. Curie's efforts to isolate and purify radioactive elements have been dismissed as "obsessive," suggesting they did not relate to a legitimate scientific concern. In *Grand Obsession*, Rosalynd Pflaum echoes the view put forward by Robert Reid in his 1974 biography: "Marie Curie was obsessed . . . with the idea that the scientific world in general was sceptical that what she claimed to be completely new elements, previously unknown to mankind, were in fact elements at all . . . Curie set about [producing radium metal] with the same resolution and obsession with which she had attacked her first sacks of pechblende" (Reid 1974: 106, 162; also I. Curie 1954). Statements like these not only understate the difficulties attending the identification of radioactive substances but also the actual resistance met by the emerging discipline.[6] Besides, they are relevant to our subject because they reveal a certain inability to come to terms with a research-technology agenda.

Accumulation, at any rate, led Curie into production, as the concentration of radium in uranium minerals is exceedingly low. Ten tons of rich ore yielded a mere gram of radium after a costly treatment involving some fifty tons of chemicals and hundreds of tons of water; in the case of low grade ore, like autanite, hundreds of tons had to be processed (M. Curie 1912; Maurice Curie 1925: 173).[7] These figures clearly demanded the facilities of a factory, and some researchers were indeed led into production by the scarcity of radioactive products. Frederick Soddy and Otto Hahn, for instance, were involved in the manufacture of mesothorium, a cheaper substitute for radium, while B. Boltwood worked as a technical advisor for the Standard Chemical Company (Badash 1979: 140; Freedman 1979; Hahn 1948; 1962). Most radioactivists, however, relied on colleagues and commercial suppliers to get their radium, never attempting its separation on a large scale. No other radioactivist was so consistently concerned with the production of radioactive elements as was Curie (on

---

[6] The point is made in Hughes (unpublished).

[7] The figures for autanite are: 800 tons of mineral, 300 tons of chemicals, 200 tons of coal, 15,000 tons of liquids.

the radium industry see Badash 1979: chapt. 10; Landa, 1981; Vanderlinden 1990. Contemporary sources include Henrich 1918, and Maurice Curie 1925: chapt. 4).

THE CURIE LABORATORIES AND THE RADIUM INDUSTRY

Pierre and Marie Curie sought assistance from a chemical company shortly after announcing the discovery of radium. In July 1899, the Société Centrale de Produits Chimiques, the company that marketed some of the instruments patented by P. Curie, undertook the treatment of several tons of pitchblende residues from St. Joachimstahl. The company provided reagents and workers in exchange for the commercial disposal of part of the radium produced. According to a recently uncovered agreement, the Curies would get 3,500 francs *à titre de droits d'auteur* in case the company sold the products of the treatment of one ton of pitchblende (Boudia 1997a: 95–7[8]; Hurwic 1995: 132; see also Suess 1899, CP, n.a.f. 18434). The Curies' righthand man, Debierne, organized the treatment (P. and M. Curie 1900). These activities were partly financed through prizes and grants from the Académie des Sciences, including a substantial 20,000 franc grant in 1902; next year, the Curies' Nobel prize (70,000 francs) helped to defray mounting expenses (see Crawford 1980; Hurwic 1995: 124, 189, 198, 220). At the time, only two other companies were selling radium salts, E. de Häen of List, near Hanover, and Giesel's employer, the quinine factory of Buchler in Brunswick. Radium was first produced in Austria in 1905 at the Auer company in Atzgersdorf, under the supervision of radioactivists S. Meyer and E. von Schweidler (Badash 1979: 136; Boudia 1997a: 97; Elster 1899, RP; Hessenbruch 1994).

The Curies' readiness to approach industry was less unusual than may seem at first sight. They were based at the *École Supérieure de Physique et de Chimie Industrielles* (ESPCI), which according to T. Shinn was an exception to the lack of empathy with industry in the French education system (Shinn 1980, 1985: esp. 164; on science and industry in france see also Nye 1986; Paul 1985: chapt. 4; Pestre 1984). Moreover, Marie Curie's first piece of research (a study of the magnetic properties of tempered steels supported by the Society for the Advancement of National Industry) had put her in touch with a number of engineers and directors of manufacture in the metallurgical industry. She preferred this task "mi-scientifique, mi-industrielle" to teaching[9] (M. Curie 1898).

From 1903 on, medical applications turned radium into a valued commodity. With the growth of a sizable market, the industrial extraction of radium could be envisaged as a profitable business. In France the opportunity was seized by the son of a quinine industrialist, Émile Armet de Lisle, who early in 1904 extended the family business with a factory at Nogent-sur-Marne, close to Paris, devoted to the manufacture of radium salts. Armet de Lisle's main problem remained the obtaining of uranium minerals. Austrian pitchblende was not to be had after 1903, so he began an extended search for alternate ores. This, among other commercial reasons, led Armet to launch a journal devoted to the marvel substance, *Le Radium*. The first issues in-

---

[8] The deal is recalled in Frédéric Haudepin to Marie Curie, 3 February 1922 (AC, 3146).
[9] Marie Curie to Józef Sklodowski, 23 November 1895, quoted in E. Curie (1938: 206).

cluded directions for amateur prospectors for radioactive minerals, and discussed the techniques of manipulation and instrumentation.[10]

From the outset, close ties connected Armet de Lisle's *Sels de Radium* and the Curies' laboratory. Each uranium ore required specific treatment. Besides providing technical advice on the treatment of the available ores, the laboratory assistants formed the company's sole pool òf technicians. These included Frédéric Haudepin, who had gained some experience at the *Société Centrale des Produits Chimiques*; Jacques Danne, who divided his time among the laboratory, the factory, and the edition of *Le Radium*; and later on, Georges Gabriel Robillard and Paul Razet. The company also benefited from the popularity and prestige of the Nobel couple, which extended to her co-workers. The laboratory got crucial logistical support in return, eventually setting up a small office right inside the company's works. This *annexe* let many of Curie's assistants gain a foothold in industry: "Il est à remarquer que la petite usine occupe des jeunes gens qui trouvent ensuite des emplois dans l'industrie du radium à laquelle ils rendent de grands services" (Armet de Lisle 1908, CP, n.a.f. 18443; Boudia 1997a: 132, 1997b: 250–3).

People from Curie's laboratory also staffed the two other radium-producing companies established in France before the war. Albert Laborde, a graduate of the ESPCI who had published several papers together with Pierre Curie, supervised the "service of measurement and purification" at the *Société Anonyme des Traitements Chimiques*, set up by Henri de Rothschild in 1910; he later worked as an engineer for the Compagnie Générale de Radiologie. J. Danne parted with Curie in 1909 to set up a school-technical laboratory complex of his own at Gif-sur-Yvette. Curie's imprint was apparent in the comprehensive layout of Danne's *Laboratoire d'Essais des Substances Radioactives*, which covered every aspect of radioactivity, from research to dosage through mineral prospecting, instrument making, and documentation (as his editor, Danne kept the important library of *Le Radium*). In 1912, together with his brother Gaston, likewise a ESPCI graduate and former laboratory researcher, Danne established the *Société Industrielle du Radium*. As for other radioelements, the only company producing mesothorium in France, *the Société Minière Industrielle Franco-Brésilienne*, was assisted by Curie and Debierne themselves (M. Curie 1909, CP, n.a.f. 18435; Boudia 1997a: 136–43).[11]

The incorporation of H. de Rothschild's and Danne's companies was facilitated by the discovery of uranium deposits in Portugal and England. The owners of the South Terras mines, in Cornwall, where pitchblende had been found, turned to Danne to organize the extraction and production of radium, while commissioning a report on the ore's worth from the Curie laboratory. Curie wrote the report on the basis of field work carried out by her assistant, Jean Danysz. Armet de Lisle faced competition by securing the rights to exploit Portuguese deposits and drafting an ambitious project of expansion. Curie, Debierne, and Erich Ebler, professor of chemistry in Heidelberg, were to form the "scientific council" of the new "General Radium Production Company (Armet de Lisle's Radium Works)." According to a draft contract,

---

[10] Émile Armet de Lisle (1853-1928) graduated in chemistry in Paris and inherited the Société du Traitement des Quinquinas in 1878. See Marie Curie (1929, CP, n.a.f. 18435); Pelé 1990.

[11] Monazite, a rich thorium ore, was mined in Brazil.

Curie and Debierne would assist in the production of radium and the development of new applications, in exchange for the extension of their own premises within the factory and discretionary use of its facilities. In the end, the French scientists withdrew their support because they doubted the efficiency of Ebler's patented treatments, which were to supplement Curie's extraction method. Besides, Curie did not want to taint her image by blessing commercial ventures, not even those of a generous industrialist like Armet. A few months earlier, she had seen her name and authority abused by South Terras Mines Ltd., which freely quoted her report in press releases and ads[12] (see Boudia 1997a: 140–3; CP, n.a.f. 18435; Fawns 1913). During the negotiations with Armet de Lisle, Curie and Debierne instructed him not to use their names in any kind of advertisement.[13]

Increasingly uneasy about her dependence on commercial firms, Curie strove after World War I to provide the newly completed *Institut du Radium* with industrial facilities of its own, to be discussed below. Her laboratory's relationship with industry changed in other ways, too. Up to Armet de Lisle's death in 1928, the relationship with his company remained a key asset.[14] French radium producers, however, processed foreign ores, and after the war they faced severe shortages and competition from American companies exploiting the rich carnotite deposits of Colorado. Curie adapted to sweeping changes in the radium industry by extending and diversifying her contacts with producers, in France and elsewhere. Taking advantage of her wartime contacts in office, she channelled her demands through the *Comité militaire de Corps radio-actifs* which, on Curie's behalf, asked French producers to assist her institute materially and logistically. In 1924, Rothschild's *Société Anonyme de traitements chimiques* offered to Curie "le local de notre labo de purification et de fractionement."[15] *The Institut du Radium* also dealt with the *Société du Radium* (1926–1934), *Minerais & Métaux* (1925), The Radium Company of Colorado (1919–1923), the *Société Française d'Energie et de Radio-Chimie* (1920–1934), the *Deutsche Gasglühlicht-Auer Gesellschaft* (1930-1934), the *Société Minière Industrielle Franco-Brésilienne* (1911–1934), and Siemens France (1930–1934).[16] While the relationship of the laboratory with these companies was often limited to the supply, either at cost or without charge, of various materials, Curie's dealings with the *Union Minière du Haut-Katanga*, the company exploiting the rich mineral deposits of the Belgian Congo, were far more substantial. Having tested the first samples produced by Union Minière, in 1924, in between sessions of the 4th Solvay Conference, she negotiated with the company the study of a number of radioactive elements, notably those in the actinium series. Ever faithful to her amassing policy, Curie stressed that it was "extremely important, *from a scientific point of view*, to extract these elements in fair

---

[12] CP, n.a.f. 18435: 176–92.

[13] See corrected draft of contract, and Marie Curie and André Debierne to Émile Armet de Lisle, 2 May 1914 (CP, n.a.f. 18436: 206-13); Émile Armet de Lisle (CP, n.a.f. 18436: 197-205).

[14] Correspondence with the company *Sels de Radium*, 1920–1929 (AC).

[15] Note by Curie on visit to the factory, 29 April 1925 (AC, 3142); report of treatment, n.d. (AC, 3136). This laboratory appears to have been closed down without explanation (Marie Curie to Henri de Rothschild, 9 January [1926], CP, n.a.f. 18461).

[16] Correspondance avec les sociétés industrielles de radium (AC). Years refer to the period covered by the extant correspondence.

amounts." The company would provide the mineral and the means of extraction; Curie would study the products and assist in the improvement of extraction techniques. This arrangement persisted until the early 1930s.[17]

THE METROLOGY OF RADIOACTIVITY

Far from being utilitarian and one-sided, Curie's relationship with the radioelements industry had profound effects on the layout and activities of her laboratories – at the Sorbonne from 1906 and the *Institut du Radium* from 1918. The development of extraction techniques posed problems of its own, defying the skills of first-rate radioactivists like Hahn: "My work on Mesothorium is awfully annoying as the problem of concentration is a very hard one."[18] Curie's concern with purification found its way into papers, textbooks, and even Nobel lectures, where detailed descriptions of industrial procedures could be found. Students and collaborators were easily sold on the ethos of purity and accumulation. Before the discovery of isotopes, for instance, the chemical identity of some radioactive elements chiefly worried members of the laboratory because it made it "very difficult to find a simple reaction which might serve as a method of concentration" (Szilard 1908).[19] Also reflecting the existence of scientists' and industrialists' common concerns, members of the laboratory produced and regularly updated tables with the characteristics and distribution of uranium and thorium minerals around the world (Kolowrat 1910, 1913; Szilard 1909a). Prospecting remained a concern at Curie's laboratory well after her death. At the Commisariat de l'Energie Atomique after the Second World War, Irène Joliot-Curie directed the division of raw materials and prospecting, which trained prospectors to "send them hunting for uranium throughout France and the French overseas territories" (Orcel 1996; Weart 1979: 224).

Instrumentation received constant attention at the laboratory, especially in connection with dosage, testing, and prospecting. Electrometers were widely used in radioactive research, yet the electrometer was "not at all a portable instrument," and simpler electroscopes sufficed when ease of use and not precision was required (Honoré 1926: 92; on the electrometer, see M. Curie 1906; Lattes 1909; on instrument design at the Curie laboratories, see Boudia 1997a: 144–62). Pierre Curie's portable electroscope, improved upon by ESPCI graduates Charles Chéneveau and A. Laborde and marketed by the *Société Centrale des Produits Chimiques*, proved successful in industry, prospecting, therapy and agriculture (Chevenau and Laborde 1907, 1909; Colomer 1926). Its designers had set out to build an all-terrain instrument "dont les avantages pratiques puissent permettre à n'importe quel physicien ou chimiste d'obtenir aisément, au laboratoire ou sur le terrain, la valeur de la radioacti-

---

[17] Correspondence with the 'Société Générale Métallurgique de Hoboken (Union Minière du Haut Katanga, 1923-1934 (AC, 3552-3794); Vanderlinden 1990: 100, 105. Union Minière sought the assistance of other prominent radium institutes, particularly of the one in Vienna (Stephan Meyer to Ernest Rutherford, 31 August 1923, RP).

[18] Otto Hahn to Ernest Rutherford, 12 July 1908 (RP, Add. 7653). Cf. Badash (1979: 139): "Each mineral posed its own special difficulties and many competent chemists would fail to establish successful commercial processes for concentration, conversion into chlorides or bromides, and final purification."

[19] On research carried out at Curie's laboratory up to World War I, see Davis 1995.

vité d'un corps, quelle qu'on soit la nature." Another researcher of the laboratory, Bela Szilard, devised a portable electrometer in which the delicate gold leaves were replaced by a robust needle, a system he was to perfect at the *Laboratoire des Produits Radioactifs*, the industrial laboratory that employed him (Szilard 1909b). Even Armet de Lisle joined in, producing a portable electroscope for the field analysis of radioactive ores (Armet de Lisle 1905, 1906). Such electroscopes allowed the activity (and thus the value) of radioactive minerals to be established "in a few minutes, with no chemical manipulation" – a "precious first indication" when it came to sort out the richest mineral on site (Maurie Curie 1925: 126–8). Irène Joliot-Curie designed an electroscope to control the activity of radioactive fertilizers; commercialized by Brewer, it was adopted by the *Laboratoire Central de la Répression des Fraudes*, which referred to the *Institut du Radium* for calibration (I. Curie 1922; Maurice Curie 1925: chapt. 7).[20]

Standardization and metrology were central to this scheme and were tightly controlled by Marie Curie, to the extent that her laboratory became the institution of reference when it came to certifying radioactive sources and instruments in France. This should come as no surprise, for metrology has often mediated between science and industry – as in the case of electrotechnology in the nineteenth century. Indeed the establishment of robust units is of utmost concern to growing industries. Work on radioactive standards had the multidisciplinary, fluid nature of research technologies: it affected the practice of radioactivity in the laboratory; it vindicated and certified the therapeutical value of the new science; and it served the needs of the burgeoning radium industry. At issue was the economic value of radium, which imposed demands for fastidious accounting on scientists, physicians, and manufacturers alike. Transactions, whether scientific or commercial, required the precise comparison of minute amounts of radium – in short, a radium standard (Schaffer 1991; Wise 1995).

Curie was much concerned with this problem. Precision measurement dominated her research work from the outset, and she came to share these interests with her husband, a skilled instrument designer and maker. They had both been engaged for months in the precise dosage of radium emanation (radon) by the time of Pierre Curie's death (E. Curie 1938: 251). When the various demands for the establishment of a radium standard were finally met by the radioactivists in 1910, many took it for granted that Marie Curie would play a leading role. This much is well known and need not be recalled here (on the establishment of the International Radium Standard, see Boudia 1997a: 182-202). More significantly, standardization brought a certain division of labor. In Britain and Germany, sections for radioactive standards were set up within existing national metrological institutions, liberating laboratories from the burden of calibration and testing. However, Curie could not imagine a measurements service not subjected to the discipline of a research laboratory. After preparing the standard, she decided to create a permanent measurements service, which would issue certificates of the radium content of the products submitted to it by institutions or individuals. This section of the laboratory was partly born out of the need to control the radium content of the radioactive salts manufactured in France (M. Curie, CP,

---

[20] Radioactive fertilizers were prepared with residues from radium manufacture. Their activity was between 3% and 10% that of a commom uranium oxide, U_2O_5.

n.a.f. 18435: 48–50). Soon the institute's *Service de Mesures* (officially established in 1914) was effectively acting as the French national metrological laboratory for radioactivity (see Boudia 1997b: 260; M. Curie 1912: 286, 1935: 510). Indeed, in her posthumous *Radioactivité*, Curie explicitly ranked the Service de Mesures alongside the British National Physical Laboratory, the German *Physikalisch-Technische Reichsanstalt*, and the American National Bureau of Standards (M. Curie 1935: chapt. 24; Maurice Curie 1925: 106).[21] She wrote to justify the creation of the service that "I have been led to think that there is a public service to be organized (measurements) which I cannot ignore, and that it could not have been properly established without myself, and my laboratory's participation. Actually, for measurements concerning such novel questions, only a laboratory in which related research was constantly going on would be able to resolve any difficulties which might arise." More than anything else, Curie's attitude shows her determination to participate in all aspects of radioactivity in France, an ambition legitimized by the public good, for the *Service de Mesures*, like the International Radium Standard, had "un caractère de haute utilité générale" (Boudia 1979b: 260; M. Curie 1912: 826). The fact that radioactivity was a new, sparsely populated scientific field; that Curie was so closely associated with radium herself; and that a modern central metrological institution was missing in France, all played into Curie's hands.

Curie's metrological activities are obviously related to key features of research-technologies, such as the definition of standards and protocols. Neither such activities nor Curie's enduring involvement with industry concerns seem to match her heroic image as a pure scientist, and yet this can be traced back to Curie's long-standing campaign to provide the *Institut du Radium* with "industrial means of action," to which I turn.

THE *INSTITUT DU RADIUM* AS A "GENERIC INSTITUTION"

Literally as well as scientifically, radium meant capital. It has been estimated that by 1914 "there was probably no single piece of research apparatus in the larger physical institutes approaching the value of their radium stocks" (M. Curie and Regaud, CP, n.a.f. 18436: 99; Forman, Heilbron, and Weart 1975: 89).[22] By 1910, Curie had amassed her first gram of radium – or her first million francs, as she valued it in a report of the radium held in her laboratory. Of this amount, a mere tenth corresponded to actual manufacturing expenses, the rest representing "a *plusvalue* origi-

---

[21] See Shinn (1980: 626) on the embarrassment of French representatives, at a 1908 meeting in London where industrial standards for electricity were to be agreed on, at France having no national laboratory. As in the case of radioactivity, a private laboratory (Janet's Laboratoire Central d'Electricité) acted as a surrogate for such an institution in France.

[22] There is conflicting evidence regarding the price of radium. Most sources agree that prices mounted between 1904 and 1914 as demand vastly exceeded supply (see among others Landa 1981: 163; Maurice Curie 1925: 171–2). Armet de Lisle, however, appears to have sold radium bromide at 400 francs per mg throughout this period (see ads in Le Radium and Stefan Meyer to Ernest Rutherford, 18 November 1911, RP, Add. 7653). Between 1911 and 1914 the Austrian government did not change the price of radium, which according to Meyer "was always lower than the French or German one" (Meyer to Rutherford, 17 February 1914, RP, Add. 7653).

nating in the value of the discovery, plus our [Pierre and hers] personal work over a number of years." Curie capitalized on this value to lay out the policies of her long-dreamed-of *Institut du Radium*, now finally under construction, created by the University of Paris and the Pasteur Institute in 1909 (Boudia 1997a: 129–30). Stressing the need to centralize "as much radium as possible," in 1912 Curie donated half of her radium to the new center, on the condition that it should not just allow the institute to lead in radioactive research, but also "contribute to the development of the industry of radioactive substances in France," and "assist the progress of biological and medical researches" (M. Curie 1912, CP, n.a.f. 18436: 17–25; Forman, Heilbron, and Weart 1975: 89). [23]

As intimated in this document, Curie had high hopes of making the *Institut du Radium* into a truly multipurpose institution, catering to the needs of research, industry, therapy, and education. She wanted the institute to realize her all-inclusive view of radioactivity by becoming a national center for radioactivity. This ambitious scheme was precisely defined as soon as the institute resumed operations after the war. The *Institut du Radium* was in two sections, a physico-chemical one headed by Curie and a medical one headed by Claude Regaud. During the war, both of them had learned to deal with government officials and draw on public sources of funding. Following a meeting of the institute's council late in November 1918, Curie and Regaud wrote a report on the expansion of the institute's services. The 47-page document dealt with education; the establishment of both an experimental station for the biology laboratory and a therapeutic section; the supply of radioactive sources; and the creation of a "Laboratoire d'études et d'essais industriels" (M. Curie and Regaud 1918, CP, n.a.f. 18436: 43–88. Quotations in following paragraphs belong to this report). J.-P. Gaudillière and I. Löwy have pointed out that the multidisciplinary structure of the institute's medical section was "very unusual for a French medical establishment," yet this will not concern us here. I will briefly discuss the other features in turn (Gaudillière and Löwy 1998). [24]

The training in Curie's laboratory was not geared to a career in academic science, but preserved rather the technical character of the *grandes écoles de sciences appliquées*, such as the *Ecole Supérieure de Physique et Chimie Industrielles*. Indeed, many of Curie's students came from the ESPCI. As one of them recalls: "Le passage de l'École de Physique et de Chimie à l'*Institut du Radium* n'avait rien d'exceptionnel. Bien d'autres l'avaient pratiqué avant moi. Les Curie puis Mme Curie … s'étaient entourés d'anciens élèves de l'École" (Goldschmidt 1987: 14). The courses on radioactivity offered by the *Institut du Radium* were not just opened to students of the University of Paris, but also to "chimistes et physiciens qui dans un but industriel ou autre désirant suivre les exercises de manipulation sans passer d'examen." These courses bred researchers and technicians that moved freely between the laboratory and the factory, a passage that was doubtless facilitated by Curie's extensive contacts in industrial milieus and by the attachment of *the Service de Mesures* to her institute.

---

[23] In 1911 the Austrians valued the radioactive preparations of the Vienna Radium Institute at 1.7 million marks (some 1.4 million francs, ignoring fluctuations in the exchange rate of 1900).

[24] On the origins and institutionalization of radiotherapy in France, see Vincent (1997: 293–305), Bordry and Boudia 1998, and Pinell 1992.

As for radium supplies and the industrial laboratory, Curie and Regaud said in their report that the latter conformed "to current trends, according to which science and its applications must advance together. The very nature of our work, involving the treatment of large amounts of material, demands industrial means of action, otherwise the laboratory will be completely unable to carry out its task." As a first step towards the provision of sources, in December 1918 Curie donated most of her radium, one gram, to the institute. At the same time, she wrote an alarming "Report on the Suitable Extension of the Institute of Radium" that painted a grim picture of the state of radiotherapy and the radium industry in France. The science of radioactivity, "born in France, out of the work of French scientists," now thrived in other countries. This situation demanded a close cooperation between science and industry, "more necessary [in the domain of radioactivity] than anywhere else." The creation of an industrial laboratory was seen as "an essential element of progress in the study of radioelements and their applications" (CP, n.a.f. 18436: 24–5 and 101–27, respectively).

Curie had a specific center in mind: a suburban site consisting of a chemical works, hangars in which to keep the minerals, and a small laboratory for measurements. The center would perform the industrial treatments demanded by the research at the *Institut du Radium*, improve the techniques for the extraction of radium, in order to make efficient use of minerals and undercut prices; manufacture new radioactive elements, especially those in the thorium family ("d'autres corps radio-actifs, qui ne sont pas encore sortis du domaine des laboratoires spéciaux, trouveront, sans aucun doute, des applications importantes quand, par des méthodes de préparation rendues industrielles, on aura pu les fabriquer en suffisantes quantités"); study new industrial applications of radium, and, last but not least, provide the institute with 200 mg RaCl a year. The center would cost 600,000 francs to build, and 100,000 francs a year to maintain.

Throughout the 1920s, Curie relentlessly advocated the creation of such a place, as part of her campaign for a grand radioactive center. She used every public speech to put forward her vision of a multipurpose institute prepared to deal with several hundred kilos of radioactive material. In Madrid, in 1919, she told a gathering of physicians that "a physics laboratory equipped with industrial means of action is indispensable for the development of the technique of the radioelements." At the *Conservatoire National des Arts et Métiers*, in March 1920, she stressed that once completed with an industrial laboratory, the *Institut du Radium* "pourra rendre des services de plus en plus importants à la Science pure, à l'enseignement, à l'industrie des radioéléments, et à la lutte contre la souffrance humaine." Curie often made her point with the help of a historical argument, stressing the harsh conditions under which radium had had to be discovered and also that it could not have been prepared without industrial means. She used to show pictures of the "hangar of the discovery" to great effect. The practice was well in place by 1921, when Curie toured America for one gram of radium. In her speech on receiving the William Gibbs medal at the American Chemical Society, for instance, she told her audience: "I may show you now the pictures of my first laboratory which was very primitive and poor. The conditions of my work have already improved in my new laboratory, and I believe that they will be made more efficient by the interest which your country has so generously proved [sic] to me" (CP, n.a.f. 18394: 39, 112 and 189, respectively). Somewhat ironically, it was in this context that Curie's heroic image took shape. The latter draws heavily on her own biography of Pierre Curie, which features the vivid description of the heroic times at the laboratory of the ESPCI that has been adopted by most biographers and commentators. Curie's complaints about the lack of resources suited the promotion of a mid-size industrial laboratory (M. Curie 1923: 146; E. Curie 1938: 179).[25]

Curie's autarchic ideal became a pressing concern when Armet de Lisle's factory changed hands at his death in 1928. The new owners made their support of Curie's laboratory conditional on their using her name in advertising, which infuriated Curie. Precious residues made available by Union Minière were being treated *chez* Armet, a work Curie regarded as "a constituent part of the workings of my laboratory." With the support of Georges Urbain, head of the neighboring *Institut de Chimie Appliquée*, Curie renewed her case for an independent, non-commercial, middle-size industrial laboratory before the University of Paris and the *Direction de l'Enseignement Supérieur* (CP, n.a.f. 18443: 220–7). She finally managed to get a "Laboratoire de gros traitements" built at Arcueil, where semi-industrial treatments were undertaken after May 1933 (M, Curie 1934, CP, n.a.f. 18436; Vanderlinden 1990: 104; Weart 1979: 114, on the uses of the Arceuil laboratory for the Joliot-Curies).

CONCLUSION

The development of radioactivity in France, as seen from Curie's *Institut du Radium*, shows significant parallels with the subject of this book. As this may not be readily apparent, I offer in conclusion a brief iconic argument in two parts, the first relating to Curie, the second to her institute.

Curie's all-inclusive vision of radioactivity defies categorization. Her career belies the existence of rigid boundaries between pure and applied science, science and technology, or science and industry. However, Curie's interstitiality was predicated on the existence of such boundaries. Purity and disinterest guaranteed that her dealings with industrialists would be free from any taint of commercialism. What Heilbron and Seidel describe as the "inhibitions of the academic scientist against securing a personal interest in his discoveries or inventions" were particularly strong among French savants, and remained strong in the 1930s "despite the increasing integration of academic science and industrial development." Even the left-wing scientists who, from the 1930s on, promoted what T. Shinn has called the autonomous science system were caught by the contradiction between making extensive use of industry and finding it "problematic and a little disgraceful" (Heilbron and Seidel 1989: 111–2; Shinn 1994: 76). M. Curie was not so much disturbed by practical matters as by private interests, which "would tend to [corrupt her]" and "would always be on the verge of spoiling the general interest and the public good" (Pestre 1997; also Pinault 1997). Thus, her statements on pure science and her unequivocal identification with its purported values belonged in a broad discourse about the social value of science and the need for state support. A reviewer of Marie's biography of Pierre Curie was

---

[25] The Curie myth is discussed in Bensaude-Vincent (1994) and Gubin (1990).

clear about the book's message: "Neither public powers nor private generosity actually accord to science and to scientists the support and the subsidies indispensable for truly effective work" (Harrow 1924).

On the other hand, Curie's solidly pure persona matched the reassuring academic façade of the *Institut du Radium*. The institution behind it was active in several domains besides research. What granted the institution this generic quality? And what allowed Curie to displace radioactivity so effectively? The answer, I believe, lies in a tight control over radioactive sources, which played the role of generic instruments in Curie's hands. Curie's exceptionally pure and powerful radioactive samples generated impulses that drove scientific research and industrial production along their respective paths. Intensity and purity were the physical parameters that, once certified by the *Institut du Radium*, made them potentially attractive to research scientists seeking support for disintegration theory, production engineers willing to improve extraction procedures, medical doctors establishing protocols for treatment, and test personnel bound to calibrate sources and instruments. Radioactive sources meant all of these things for Curie, yet they came first and had a logic of their own. Like instruments, radioactive sources allowed Curie "to stabilize and organize phenomena without intruding on the configurations and theories which govern them," they were "a semi-autonomous input" (Joerges and Shinn in their introductory chapter). Some historians have taken this enabling capacity for the misleading effect of a positivistic philosophy of science, while others have simply asked whether Curie had occasion, after the discovery of radium, "to use her head or only her hands" (Badash 1975; Malley 1979). Curie's allegiance to radioactive sources, like Aimé Cotton's allegiance to instruments, appears "eccentric or recalcitrant" – perhaps an indication of a common concern with open-ended, research-enabling technologies (Shinn 1993: 180).

Universitat Autònoma de Barcelona, Spain

ACKNOWLEDGMENTS

This work has been partially supported by the Spanish DGES, research program No. PB96-1169. – For permission to quote from archival material, I thank Cambridge University Library.

REFERENCES

Archives M. Curie at the Institut Curie, Paris. Pierre and Marie Curie Papers, Bibliothèque Nationale, Paris:
  AC, 3146. Frédéric Haudepin to Marie Curie, 3 February 1922.
  AC, 433. Chamié, Catherine (1935). Unpublished 5-page reminiscence *A la mémoire de Madame P. Curie*.
  AC, Correspondence with the company *Sel de Radium*, 1920–1929.
  AC, 3142. Note by Curie on visit to the factory, 29 April 1925.
  AC, 3136. Report of treatment, n.d.

AC, 3552–3794. Correspondence with the "Société Générale Métallurgique de Hoboken (Union Minière du Haut Katanga), 1923–1934.

Armet de Lisle, Émile (1898). Propriétés magnétiques des aciers trempés. *Bulletin de la Société d'Encouragement à l'Industrie Nationale*, published in I. Joliot-Curie (ed.), *Oeuvres de Marie Sklodowska Curie* (pp. 3–42). Warsaw: Panstwowe Wydawnictwo Naukowe.

Armet de Lisle, Émile (1905). Nouvel électroscope pour la recherche des minéraux radioactifs. *Le Radium* 2: 217–18.

Armet de Lisle, Émile (1906). Électroscope pour la recherche des minéraux radioactifs. *Le Radium* 3: 62.

Badash, Lawrence (1975). Decay of a radioactive halo. *Isis* 66(4): 566–8.

Badash, Lawrence (1979). *Radioactivity in America: Growth and decay of a science*. Baltimore: John Hopkins University Press.

Bensaude-Vincent, Bernadette (1994). Une robe de cotton noir. *Cahiers de Science & Vie* 24: 76–85.

Bordry, Monique and Boudia, Soraya (eds.) (1998). *Les rayons de la vie. Une histoire des applications médicales des rayons X et de la radioactivité en France, 1895-1930*. Paris: Institut Curie.

Boudia, Soraya (1997a). *Marie Curie et son laboratoire: Science, industrie, instruments et métrologie de la radioactivité en France, 1896-1914*. PhD dissertation, Université de Paris VII. To be published by Éditions des Archives Contemporaines.

Boudia, Soraya (1997b). The Curie laboratory: Radioactivity and metrology, in S. Boudia and X. Roqué (eds.), *Science, medicine and industry: The Curie and Joliot-Curie laboratories. History & Technology* (special issue) 13(4): 249–65.

Chevenau, Charles and Laborde, Albert (1907). L'eléctromètre à quadrants. *Le Radium* 4: 145–54.

Chevenau, Charles and Laborde, Albert (1909). Appareils pour la mesure de la radioactivité, d'après la méthode électroscopique. *Journal de Physique* 8: 161–174.

Colomer, Félix (1926). *Manuel pratique du radium à l'usage des agriculteurs, chimistes, ingénieurs, médecins et prospecteurs*. Paris: Éditions Actualités.

CP, Pierre and Marie Curie Papers, Bibliothèque Nationale, Paris (n.a.f. = nouvelles acquisitions françaises):

CP, n.a.f. 18394. *Conférence sur l'Institut du Radium à l'Université des Annales*, 23 February 1925 (pp. 277–99).

CP, n.a.f. 18434. Eduard Suess to Pierre Curie, 13 June 1899.

CP, n.a.f. 18434. Friedrich Giesel to Pierre and Marie Curie, 12 January 1900.

CP, n.a.f. 18435: Marie Curie, draft obituary of Armet de Lisle, 1929.

CP, n.a.f. 18435. Marie Curie to Jacques Danne, 18 June 1909.

CP, n.a.f. 18436. Armet de Lisle, Émile. *Notice sur la* General Radium Production Company (Armet de Lisle's Radium Works Limited) (pp. 197–205).

CP, n.a.f. 18436. *Marie Curie, Exposé relatif au radium qui se trouve actuellement dans mon laboratoire* (pp. 17-25), 3 March 1912.

CP, n.a.f. 18436. Marie Curie and Claude Regaud. *Appel du Comité de Patronage*.

CP, n.a.f. 18436. *Dossier Rapport sur la richesse des mines South Terras (faisant usage autorité Marie Curie)*, (pp. 176–92).

CP, n.a.f. 18436. Marie Curie and André Debierne, *Rapport relatif à la subvention accordé par la Caisse des Recherches Scientifiques* (pp. 8–15).

CP, n.a.f. 18436. Corrected draft of contract. Marie Curie and André Debierne to Émile Armet de Lisle, 2 May 1914 (pp. 206–13).

CP, n.a.f. 18436. Marie Curie and Claude Regaud, *Rapport et propositions concernant l'extension des services de l'Institut du Radium* (pp. 43–88). (1918).

CP, n.a.f. 18436. Marie Curie, *Rapport sur l'activité du laboratoire de gros traitments d'Arcueil*, 25 June 1934.

CP, n.a.f. 18443. Émile Armet de Lisle to Marie Curie, 21 September 1908.

CP, n.a.f. 18461. Marie Curie to Henri de Rothschild, 9 January 1926.

Crawford, Elisabeth (1980). The prize system of the Academy of Sciences, in R. Fox and G. Weisz (eds.), *The organization of science and technology in France, 1808-1914* (pp. 283–307). Cambridge, UK: Cambridge University Press.

Curie, Eve (1938). *Madame Curie*. Paris: Gallimard.

Curie, Irène (1922). Électroscope pour la mesure de la radioactivité des engrais. *Annales de la Science Agronomique Française et Étrangère*.

Curie, Irène (1954). Marie Curie, ma mère. *Europe* 32: 89–121.

Curie, Marie (1898). Propriétés magnétiques des aciers trempés. Bulletin de la Société d'Encouragement à l'Industrie Nationale (1898). Published 1954 in Irène Joliot-Curie (ed.), *Oeuvres de Marie Sklodowska Curie* (pp. 3–42). Warsaw: Panstwowe Wydawnictwo Naukowe.

Curie, Marie (1903). Über den radioaktiven Stoff "Polonium". *Physikalische Zeitschrift* 4: 234.

Curie, Marie (1906). Eléctromètre à fil de quartz. *Le Radium* 3: 202–3.

Curie, Marie (1907). Sur le poids atomique du radium. *Comptes Rendus de l'Académie des Sciences* 145: 422.

Curie, Marie (1912). Les mesures en radioactivité et l'étalon du radium. *Journal de Physique* 2: 715–826.

Curie, Marie (1923). *Pierre Curie*. New York: Macmillan.

Curie, Marie (1930). Sur l'actinium. *Journal de Chimie et Physique* 27: 1.

Curie, Marie (1935). *Radioactivité*. Paris: Hermann.

Curie, Marie (1966). *Nobel lectures in chemistry, 1901-1921*. Amsterdam: Elsevier.

Curie, Marie and Debierne, André (1910a). Sur le polonium. *Le Radium* 7: 38–40.

Curie, Marie and Debierne, André (1910b). Sur le radium métallique. *Comptes Rendus de l'Académie des Sciences* 151: 523.

Curie, Marie and Rosenblum, Salomon (1931). Spectre magnétique des rayons du dépôt actif de l'actinium. *Journal de Physique* 10: 309.

Curie, Maurice (1925). *Le radium et les radio-éléments*. Paris: Baillière.

Curie, Pierre and Curie, Marie (1900). Les nouvelles substances radioactives et les rayons qu'elles émettent. *Rapports présentés au Congrès International de Physique* 3: 79.

Davis, J.L. (1995). The research school of Marie Curie in the Paris Faculty, 1907–1914. *Annals of Science* 52: 321–55.

Fawns, Sydney (1913). Le radium, sa production et ses usages. London: *The Mining Journal*: 46–50.

Forman, Paul, Heilbron, John, and Weart, L. Spencer (1975). Physics circa 1900. Personnel, funding, and productivity of the academic establishments. *Historical Studies in the Physical Sciences* 5: 1–185.

Freedman, Michael I. (1979). Frederick Soddy and the practical significance of radioactive matter. *British Journal for the History of Science* 12(42): 257–60.

Gaudillière, Jean-Paul and Löwy, Ilana (1998). Disciplining cancer: Mice and the practice of genetic purity, forthcoming in J.-P. Gaudillière, I. Löwy, and D. Pestre (eds.), *The invisible industrialist: Manufacturers and the construction of scientific knowledge* (pp. 209–49). London: Macmillan.

Goldschmidt, Bertrand (1987). *Pionniers de l'atome*. Paris: Stock.

Gubin, Eliane (1990). Marie Curie et le radium: L'information et la légende en Belgique, in Université Libre de Bruxelles (ed.), *Marie Curie et la Belgique* (pp. 111–29). Brussels: ULB.

Guillot, Marcel (1967). Marie Curie-Sklodowska (1867–1934). *Nuclear Physics* A 103 (October): 1–8.

Hahn, Otto (1948). Einige persönliche Erinnerungen aus der Geschichte der natürlichen Radioaktivität. *Naturwissenschaften* 35: 67–74.

Hahn, Otto (1962). *Vom Radiothor zur Uranspaltung. Eine wissenschaftliche Selbstbiographie*. Braunschweig: Vieweg.

Harrow, B. (1924). One of an immortal pair. Mme. Curie on the struggles of a scientist. *The New York Times*, 6 January 1924.

Heilbron, John L. and Seidel, Robert (1989). *Lawrence and his laboratory*. Berkeley, CA: University of California Press.

Henrich, Friedrich (1918). *Chemie und chemische Technologie radioaktiver Stoffe*. Berlin: Springer.

Herschfinkel, H. (1910). Sur le radio-plomb. *Le Radium* 7: 198–200.

Hessenbruch, Arne (1994). *The commodification of radiations: Radium and X-Ray standards*. PhD dissertation. Cambridge: University of Cambridge.

Honoré, F. (1926). *Le radium*. Paris: Gauthier-Villars.

Hughes, Jeff A. (1993). *The radioactivists: Community, controversy and the rise of nuclear physics*. PhD dissertation, University of Cambridge.

Hughes, Jeff A. (no year). The radium wars: Physics, chemistry, and the "political economy" of radioactivity, 1898-1914. Unpublished.

Hurwic, Anna (1995). *Pierre Curie*. Paris: Flammarion.

Kirby, H.W. (1971). The discovery of actinium. *Isis* 62(3): 290–308.

Kolowrat, Léon (1910). Tables des substances radioactives. Revues et completées. *Le Radium* 7: 1–3.

Kolowrat, Léon (1913). Tables des substances radioactives. *Le Radium* 10: 1–4.

Landa, Edward R. (1981). The first nuclear industry. *Scientific American* 247: 154–63.

Lattes, C. (1909). Méthode de mesure des faibles courants. *Le Radium* 6: 73–4.

Malley, Marjorie (1979). The discovery of atomic transformation: Scientific styles and philosophies in France and Britain. *Isis* 70(252): 213–23.

Nye, Mary J. (1986). *Science in the provinces. Scientific communities and provincial leadership in France, 1860-1930.* Berkeley, CA: University of California Press.

Orcel, Jean (1996). Quelques souvenirs des Curies, in H.-J. Schubnel (ed.), *Histoire naturelle de la radioactivité* (pp. 40–1). Paris: Muséum National d'Histoire Naturelle.

Paul, Harry W. (1985). *From knowledge to power: The rise of the science empire in France, 1860-1939.* Cambridge, UK: Cambridge University Press.

Pelé, J. (1990). L'usine Armet de Lisle à Nogent. *Bulletin de la Société Historique et Archéologique de Nogent-sur-Marne*: 66-75.

Perrin, Jean (1924). La radioactivité et son importance dans l'univers, in *Le Radium. Célébration du vingt-cinquième anniversaire de sa découverte (1898-1923)* (p. 16). Paris: Presses Universitaires de France.

Pestre, Dominique (1984). *Physique et physiciens en France, 1918-1940.* Paris: Éditions des Archives Contemporaines.

Pestre, Dominique (1997). The moral and political economy of French scientists in the first half of the 20th century, in S. Boudia and X. Roqué (eds.), *Science, medicine and industry: The Curie and Joliot-Curie laboratories. History & Technology* (special issue), Vol. 13, No. 4: 241–8.

Pflaum, Rosalynd (1989). *Grand obsession. Madame Curie and her world.* New York: Doubleday.

Pinault, Michel (1997). The Joliot-Curies: Science, politics, networks, in S. Boudia and X. Roqué (eds.), *Science, medicine and industry: The Curie and Joliot-Curie laboratories. History & Technology* (special issue), Vol. 13, No. 4: 307–24.

Pinell, Patrice (1992). *Naissance d'un fléau: Histoire de la lutte contre le cancer en France (1890-1940).* Paris: Métailié.

Quinn, Susan (1995). *Marie Curie. A life.* New York: Simon & Schuster.

Razet, Paul (1907). Sur la concentration du polonium. *Le Radium* 4: 135–9.

Reid, Robert (1974). *Marie Curie.* London: Collins.

Roqué, Xavier (1994). La stratégie de l'isolement. *Cahiers de Science & Vie* 24: 46–67.

Roqué, Xavier (1997). Marie Curie and the radium industry: A preliminary sketch, in S. Boudia and X. Roqué (eds.), *Science, medicine and industry: The Curie and Joliot-Curie laboratories. History & Technology* (special issue), 13(4): 267–81.

RP, E. Rutherford Papers, Cambridge University Library, Add MS 7653:

    RP, Add. 7653/E13. E. Suess to Pierre Curie, 13 June 1899.

    RP, Add. 7653/E13. Julius Elster to Ernest Rutherford, 27 June 1899.

    RP, Add. 7653/PA37. Émile Armet de Lisle to the president of the Royal Society, 14 November 1906 (6)

    RP, Add. 7653/PA376. Armet de Lisle, Émile (1906).

    RP, Add. 7653. Otto Hahn to Ernest Rutherford, 12 July 1908.

    RP, Add. 7653. Stephan Meyer to Ernest Rutherford, 18. November 1911.

    RP, Add. 7653. Stephan Meyer to Ernest Rutherford, 17. February 1914.

    RP, Add. 7653. Stephan Meyer to Ernest Rutherford, 31 August 1923.

Rutherford, Ernest (1910). Properties of polonium. *Nature* 82: 491–2.

Schaffer, Simon (1991). Late Victorian metrology and its instrumentation: A manufactory of ohms, in R. Budd and S.E. Cozzens (eds.), *Invisible connections: Instruments, institutions, and science* (pp. 23–56). London: SPIE Optical Engineering Press.

Shinn, Terry (1980). The genesis of French industrial research, 1880–1940. *Social Information* 19: 607–40.

Shinn, Terry (1985). How French universities became what they are. *Minerva* 23: 159–65.

Shinn, Terry (1993). The Bellevue *grand électroaimant*, 1900-1940: Birth of a research-technology community. *Historical Studies in the Physical and Biological Sciences* 24(1): 157–87.

Shinn, Terry (1994). Science, Tocqueville, and the state: The organization of knowledge in modern France, in M.C. Jacob (ed.), *The politics of western science, 1640-1990* (pp. 47–80). New Jersey: Humanities Press.

Szilard, Bela (1908). Un étude sur le radio-plomb. *Le Radium* 5: 299-301.

Szilard, Bela (1909a). Tables des principaux minerais d'uranium et de thorium. *Le Radium* 6: 233–40.

Szilard, Bela (1909b). Sur un appareil destiné à la mesure de la radioactivité. *Le Radium* 6: 363–6.

Szilard, Bela (1910). Tables des substances radioactives. Revues et completées. *Le Radium* 7: 1–3.

Vanderlinden, Jacques (1990). Marie Curie et le radium "belge", in Université Libre de Bruxelles (ed.), *Marie Curie et la Belgique* (pp. 91–109). Brussels: ULB.

Vincent, Bénédicte (1997). Genesis of the pavillon Pasteur of the Institut du Radium, in S. Boudia and X. Roqué (eds.), *Science, medicine and industry: The Curie and Joliot-Curie laboratories. History & Technology* (special issue), Vol. 13, No. 4: 293-305.

Weart, Spencer (1979). *Scientists in power*. Cambridge, MA: Harvard University Press.

Weill, Adrienne (1981). Curie, Marie (Maria Sklodowska), in Ch. Gillispie (ed.), *Dictionary of scientific biography*, Vol. 2 (pp. 497–503). New York: Scribner's.

Wise, Norton (ed.) (1995). *The values of precision*. Princeton: Princeton University Press.

CHAPTER 5

TERRY SHINN

STRANGE COOPERATIONS
THE U.S. RESEARCH-TECHNOLOGY
PERSPECTIVE, 1900–1955

The U.S. instrument community changed considerably during the first half of the twentieth century. The community was composed of two principal components: narrow-niche instrument designers, constructors and users, and instrument men whose interest lay in the theory and the development of generic multipurpose devices. Narrow-niche instrument practitioners built apparatus in response to specific problems. They worked within a professional setting in academia, industry or the state, and they rarely crossed professional or institutional boundaries. In contrast, instrument generalists, that is research-technologists, designed and built multipurpose generic devices. They focused on instrument theory and designed apparatus in such a way that basic principles could be readily dis-embedded from this equipment. The production of generic devices required them to work with many diverse audiences and thus occupy the interstices between established institutions throughout their careers. These generic devices were then tailored to specific user needs and thus re-embedded in local environments.

The U.S. research-technology movement developed fitfully and often under considerable duress between 1900 and 1955. Some of America's first practitioners worked in universities, but they often stood on the margins of the academic system. Their peripheral position stemmed in part from a preoccupation with the operation of instruments rather than the operation of nature. In the 1920s and 1930s, state metrology services, industry and the military grew increasingly dependent on advanced sophisticated devices, yet they remained by and large indifferent to the generic instrumentation proposed by research-technology practitioners, and hence insensitive to the fortunes of the struggling embryonic research-technology current. In the end it was a combination of new technical fields (nucleonics, electronics, and control engineering among others), the post 1945 scientization of industrial production, and rapid military expansion that constituted the immediate material context for the often problematic rise of research-technology in America. The creation of the Instrument Soci-

B. Joerges and T. Shinn (eds.), Instrumentation: Between Science, State and Industry, 69–95.

ety of America in 1946 and the *Instrument Society of America Journal*[1] in 1954 can be seen as emblematic of the research-technology community's hard won consolidation.

Before 1946, America's research-technologists had to struggle constantly against narrow-niche instrument designers, constructors and users, and the relations between these two groups were often uncomfortable and even acrimonious. The emerging notion of generic devices based on broad metrological principles and operating in an interstitial environment clashed with the practices and values of erstwhile instrument practitioners who were allied to single application devices and who often worked within a narrow profession for a single institution or employer. As consciousness about the social and institutional implications of research-technology and its special potential gradually developed, American research-technologists set up a lattice of strange cooperations. Working with like-minded instrument groups, in 1935 they established the Instrument Publishing Company, which acted as a sort of unofficial coordinating center for efforts to secure an open, heterogeneous instrument sphere, and battled against groups that wanted to squeeze out interstitial play.

The establishment of the Instrument Society of America and its journal stimulated research-technology in two ways. They fostered productive relations between generic instrument groups and other instrument bodies committed to narrow-niche devices for use in restricted environments. In pursuit of this end, the Society and the Journal refined a vocabulary that distinguished between various instrument movements along functional, cognitive, and institutional lines. By the 1950's, a climate of coexistence and cooperation had begun to develop.

This chapter explores the slow and often difficult emergence of America's research-technology movement. My claims are in part based on prosopographical data drawn from America's foremost instrument journal, the *Review of Scientific Instruments* (founded in 1930), and from a variety of biographical dictionaries and other biographical sources. For the years 1930 to 1960, I identified the 600 instrument men who published most frequently in the *Review of Scientific Instruments*. Of this sample, standardized information was gathered for 204 people. Through screening based on practitioners' involvement with generic instruments and metrologies, and based on their commitment to an interstitial career, this number was reduced to 137.[2] Qualitative information on the emergence of the American research-technology current, its structures and evolution, and its often turbulent relations with narrow-niche instrument practitioners, was principally gleaned from the *Instrument Society of America Journal*, *Instrument Magazine*, and the *Review of Scientific Instruments*.

---

[1] The *Proceedings* of the Society were first published in 1946, but the Society's formal organization was not fully elaborated until 1948.

[2] Three criteria were used to distinguish research-technology articles from other categories of texts: 1) the inclusion of theoretical considerations in instrument work; 2) a preference for dis-embedding and re-embedding projects; 3) a career trajectory characterized by movement between numerous institutions, interests and employers.

SUCCESS AND PARADOX, 1900–1930

American industry, education and research developed dramatically during the last decades of the nineteenth century and the first decades of the twentieth. Instrumentation grew at a corresponding rate, however more so in the field of narrow-niche instrumentation than generic instrumentation and research-technology. Nineteenth century American science and technology boasts an impressive number of inventors and researchers, some of whom exhibited something akin to the research-technology spirit, amongst them Alexander Graham Bell and Thomas Edison (Bruce 1973; Hughes 1976; Pasachoff 1996). Before 1900 the men who designed, built and diffused generic devices appear to have worked in isolation from one another and to have been barely aware, or entirely unaware of the sorts of concerns that we later associate with research-technology. This transition towards group activity amongst instrument men first occurred predominantly outside academia. Consequently in America, just as in Germany where the first research-technology movement emerged, links between universities and research-technology were extremely weak. The lack of interest amongst American universities in generic projects was prompted by very different circumstances than those which lay behind the German academic indifference to research-technology in the mid and late nineteenth century.

Germany had a long and impressive tradition in fundamental research, built in part on the initiatives and structures of the Berlin Academy of Science, founded in 1700. The fundamental science current gained further legitimation through the Humboldt movement and the creation of Berlin's Friedrich-Wilhelms-Universität in 1809, and through the vigorous research efforts introduced in Germany's innumerable regional universities during the nineteenth century (Ben-David 1991). While much competent engineering and applied research emerged throughout this era, the emphasis was consistently on pure science (König 1980). When in the 1860s and 1870s attempts were made to introduce instrumentation to the Berlin Academy of Science on a par with basic research, most German scientists balked. They invoked their strong and successful achievements in fundamental investigations. While for them instrumentation was important, it was only a means, and did not constitute an end in itself (Shinn this book, chapter 3).

In comparison with many European countries, the U.S. had few universities for much of the nineteenth century. The first highly regarded technical universities were West Point (1802) and Rensselaer Polytechnic Institute (1824) who focused mainly on applied subjects. Johns Hopkins University, America's first research university, was established only in 1876, and although several other universities such as Princeton, Yale, Harvard, and Chicago also soon turned to fundamental inquiry, for a long period the number of such institutions remained relatively small. The 1862 Land Grant Act drove American higher education away from pure research and towards a form of knowledge that served the immediate needs of agriculture, industry and commerce. Higher state education received government aid with the proviso that instruction and research benefit the local economy. In response to this policy, in the 1870s, 1880s, and 1890s, state universities sprung up across the country, many with sizable and prosperous departments in agriculture and engineering. By the 1890s, the

U.S. had over thirteen top-rate schools of mechanical, civil, and electrical engineering, with a total enrollment of over 8,000 students per annum.

The American economy leapt forward during these decades, with an approximate average growth rate of over 3%. Growth was spectacular in chemistry and electricity. Many new firms developed, and existing companies expanded massively. This was not only true for firms associated with transport and metallurgy, but also for firms associated with electricity production and electrical goods, chemical products and optical equipment (Noble 1977).

The fact that the meshing of American academia and industry in that period surpassed that of any other industrially advanced nation was the outcome of an additional factor: namely, that the American economy and educational system took shape and consolidated during the crucial period of the "capitalization of knowledge" (Braverman 1974). The capitalization of technical and scientific knowledge was marked by heightened consciousness among industrialists of the profit value of formal learning as opposed to craft learning. The capitalization of knowledge was set in motion through business initiatives to control knowledge developed outside business per se, and through fostering and regulating a specific type of knowledge within firms – the bureaucratization of industrial learning (Chandler 1977). In this spirit, many American companies built impressive research centers during the early decades of the twentieth century (Noble 1977). Of utmost significance for U.S. research-technology, the capitalization of knowledge in America often occurred with reference to specific short term technical demands. Companies wanted knowledge and problem solving capacity to be focused on urgent, clearly posed dilemmas. In view of this circumscribed agenda, there was little space for the development of undirected, far reaching instrumentations capable of addressing myriads of problems in manifold environments. While the capitalization of knowledge indirectly stimulated Germany's research-technology movement by opening an interstitial landscape, in the U.S. the capitalization of scientific and technical knowledge instead reinforced an already strong alliance between academia and enterprise. This transformed the linkage into a procrustean structure which tended to smother the interstitial arena that is so fundamental to designing and diffusing generic apparatus. Stated differently, the alliance between enterprise and academia overdetermined the latticework of educational, economic, cognitive and technical relations. Through overcrowding, this alliance temporarily impinged upon the space where research-technology practices could emerge.

A Narrow-Niche Ground-Swell

The occupational body entitled The Scientific Apparatus Makers of America was founded in 1902. Headquartered in Pittsburgh, Pennsylvania, and numbering over 200 charter members, the purpose of this organization was to bring together individuals and enterprises involved in the design, construction and sale of a broad range of measurement and control apparatus. The Scientific Apparatus Makers of America was the first body in the U.S. that represented instrument interests and sought to forge bonds between instrument men. By 1920, the organization could claim almost 2,000 members. However, the number of member enterprises in instrument design and construction remained small. These companies rarely had more than 30 employees, and many had no more than three or four. While membership was national, most members were concentrated along the eastern seaboard and the mid west. Pittsburgh, Philadelphia, Boston, Chicago and Detroit stood at the head of the list. This geographic distribution reflected the industrial vigor and size of these regions.

The Scientific Apparatus Makers of America provided potential instrument customers with technical information about products supplied by member companies. This society published catalogs that listed the address of instrument manufacturers and indicated instrument specifications and uses.[3] The intent was principally economic – the circulation of a form of publicity in order to stimulate instrument sales. But the association also strove to extend the field of application of devices by demonstrating novel uses for them. The society furthermore wanted to establish lines of communication between instrument-makers. To achieve this end, it held irregular meetings in its Pittsburgh offices, where practitioners could come together and display their recent findings and innovations. This facet of its operations remained strictly regional however, and soon disappeared from the agenda.

The importance of the Scientific Apparatus Makers of America lies in its intention to establish connections between people involved with scientific instruments and to kindle a measure of occupational identity. However, the association's purview and goals never embraced research-technology. The focus remained end-user equipment. Issues like the generic properties of devices did not arise, nor was there awareness of a potential for interstitiality, whereby a distinctive space lying between various established institutions, interests and clients could be carved out for the purpose of designing a special category of instrumentation. The concept of introducing new metrologies was similarly lacking. Practitioners wanted to apply existing metrologies in the best possible manner in order to satisfy the short term demands of impatient customers. While the Scientific Apparatus Makers of America constitutes a significant precursor to American research-technology, the movement's cognitive direction, goals and identity had a very different source and labored in a very different direction.

Discipline-related scientific and technical learning constituted a second source of American instrument endeavor. This current was at first predominantly centered in optics, and ultimately a substantial portion of this learning developed inside America's universities. Optics has long been one of the fields where instrumentation has flourished – the telescope, microscope, navigational devices, theodolite and spectroscope are good examples (in this book see Jackson, chapter 2; Shinn, chapter 3; Johnston, chapter 7, and Mallard, chapter 11).

In the U.S. the Optical Society of America, founded in 1915, was the unofficial focal point of much research. In 1917, the Society established a specialized periodical, *The Journal of the Optical Society of America*. The journal published articles penned by authors from numerous backgrounds – amateur scientists and technologists, inventors, craftsmen, industrialists, engineers and scientists. Soon however, articles written by university researchers predominated. A comparison of texts published in 1920 and 1930 shows that the content of the Journal had begun to change.

---

[3] I examined seven catalogs produced by the Scientific Apparatus Makers of America Society published during the first decade of the organization's existence.

Instrument-related articles, as opposed to articles on experimentation and theory, represented an increasing proportion of the texts of this broadly circulated journal. Indeed, by 1922 pressure on the journal to publish ever more instrument articles had grown so intense that a separate supplement was devoted exclusively to pieces on instrumentation. Authors dealing with instrumentation constituted a distinctive subset of the journal population in which non-academics were preponderant. Nevertheless, there were many academic contributors. They differed from their university colleagues as academics who wrote instrument articles were often from the lower ranks of the university hierarchy. They were also professionally mobile, frequently moving from college to college and between the university and industry (Shimshoni 1970).

Some academics in experimental and theoretical physics protested against the Journal's increasing instrumental orientation.[4] They argued that while instruments occupy an important place in optics, they are nevertheless a means and not an end in themselves. Care should thus be taken to guarantee that instrumentation articles not overwhelm the *Journal of the Optical Society of America*.

In the journal, many instrument articles initially focused on optics. Soon however, texts related to mechanics and heat, and to instrumentation in chemistry, electricity and electronics began to appear. By 1930, these articles made up more than half of the total, and those focusing on electricity and electronics had more than quadrupled. This dramatic move away from optical devices was another cause for concern among some U.S. academic optics specialists.

During this period, heightened interest in instrumentation was also occurring in American academic reviews specializing in mechanics, chemistry, and electricity. By the 1930s, instrument-related research and writing had become so abundant and widespread that the abstracting service of the American Chemical Society began to list and report on instrument articles. In 1933 it catalogued over 600 articles. In 1935 the number climbed to over 900, and it rose to over 1,000 the following year.[5] Instrument work had became pervasive. Yet only a very small amount of this research appears to have dealt with generic devices.

A NEW FOCUS

American instrument research spilled outward between 1930 and 1955, moving into innumerable spheres – nuclear and atomic physics, military-related areas, cosmology, space and rocketry, cybernetics, computers and molecular biology. A growing body of instrument practitioners turned their attention to the fundamental instrument laws which underpinned the operation of broad families of devices. The careers and instrument work of Jesse Beams, W.W. Hansen, Winfield Salisbury and Hewlett and Packard are illustrative of this. These men thought in terms of template devices, whose generic characteristics would enable end-users to adapt the master device to many environments. The ultracentrifuge, rumbatron, and oscilloscope are a few examples. Successful endeavors were made to bring like-minded instrument men into

---

[4] OJA resistance to instrumentation 1917.
[5] *Chemical Abstracts* of the American Chemical Society 1936.

closer contact, where the goal was to share ideas and information or to anticipate instrument needs.

This breed of instrument designer cautioned against turning generic instrumentation into another scientific or engineering specialty. The transverse character of generic instrument production precluded this specialization. Specialty status, with its high visibility and constant demand for privileges and prerogatives, could ultimately alienate generic instrument researchers from their clientele, who could come to see the importation of devices from outside as a challenge to their material, cognitive and professional autonomy, and who could also perceive this importation as a threat to their own internal instrument making capacity. Instrument people concerned with generic devices thus advocated an "in between" posture. Rather than building their identity around an institution, profession or employer, they would instead operate out of whatever convenient organization came to hand, or simultaneously, they would operate out of a variety of organizations. Their allegiance and identity lay in the instrument vocation per se. This line of reasoning, and its concomitant instrument practices, accompanied the founding of the *Review of Scientific Instruments* in 1930, the Instrument Publishing Company in 1935, and the creation of the Instrument Society of America in 1948 and its eponymous journal in 1954. Each of these bodies will be examined in turn in the following sections.

The Review of Scientific Instruments

Founded in 1899, the American Physical Society rapidly became the national coordinating body for numerous disciplines in the physical sciences. It handled matters related to specialist journals and national and regional science conferences, and acted as a lobby in dealings with government (Kevles 1971: 77–81). During the 1920s, the Society's governing board became worried about the growing discontent inside the *Journal of the Optical Society of America* over the increasing number of instrument articles, many of which were unrelated to optics. The board began to consider ways to resolve the instrumentation problem. In 1930 the American Physical Society established a separate, fully independent periodical devoted exclusively to scientific instrumentation – the *Review of Scientific Instruments*. The Society was chiefly motivated by two considerations. 1) The recent growth in the number of instrument articles surpassed by far the capacity of the *Journal of the Optical Society of America* to publish them, and as already indicated, an increasing share of texts dealt with instrumentation unrelated to optics. 2) More decisively, the American Physical Society saw the *Review of Scientific Instruments* as a trans-disciplinary and trans-specialty vehicle that might rekindle interest for shared perspectives and methodologies in an otherwise increasingly specialist-minded and fragmented national scientific community.

In an editorial appearing in the first issue of the new *Review* it was argued that this goal might realistically be attained, as certain kinds of instruments serve in a variety of scientific fields. By basing their practices on such devices, scientists might begin to think convergently, and thereby surmount the barriers of in-communication and incomprehension. Of course, not all of the devices to be presented in the new journal would have the potential for inducing technical and intellectual convergence,

Table 5.1. Employers of Instrument Authors, 1930-1955

	1930	1935	1940	1945	1950	1955
University	40%	45%	55%	45%	40%	40%
Business	30%	20%	20%	25%	25%	30%
Public Agencies	15%	20%	15%	15%	10%	5%
Independent	15%	20%	5%	5%	5%	5%
National Research Centers	0%	0%	0%	10%	20%	20%

but the editorial board believed that a meaningful number of them would do so, thereby warranting a policy of transverse instrument technology (F.K.R. 1930a,b). It is worth noting that the trans-community objectives expressed by the American Physical Society and presented in the *Review of Scientific Instruments* substantially echoed the aspirations formulated in the newly founded British *Journal of Scientific Instruments* and the French *Revue d'optique théorique et instrumental*.

The *Review of Scientific Instruments* appeared monthly. It presented three categories of information: technical articles, news about instrument congresses and workshops, and advertising for new or improved devices. Authors came principally from academia, industry, and public metrological services. Fewer texts were contributed by technical consultants and inventors. Between 1930 and 1955 the distribution of these groups remained reasonably stable (see Table 5.1). There was one exception to this, however. A new employment pole emerged after World War II; namely, the national research centers associated with the military and atomic investigations – Los Alamos, the Argonne Laboratory, Brookhaven and Oak Ridge. These sites reflected the transfiguration of post World War II American science – as money poured in, new agencies opened and thousands of new jobs were created (Kevles 1971: 324–48). The emergence of this employment sector, and its rapid increase in size and influence, resulted in a decrease in the percentage of *Review of Scientific Instrument* authors based in America's universities. In contrast, the proportion of industry-based authors and inventors remained roughly constant.[6] Instrument men thus came from all quarters of the country's employment and intellectual landscape.

During the three decades beginning in 1930, the types of instrumentation reported on in the *Review of Scientific Instruments* changed appreciably. This corresponded to shifts in the American disciplinary landscape. Articles on mechanical and optical devices partly gave way to texts which dealt with apparatus from the fields of electricity, magnetism, electronics, and nucleonics (see Table 5.2). This transformation bespoke two changing circumstances. First, magnetic and electronic detection, measurement, and control parameters interfaced with more physical phenomena and with a broader range of technology than mechanical and optical parameters. Moreover,

---

[6] This analysis is based on information concerning the employers of authors that is included in most *Review of Scientific Instruments* articles.

Table 5.2. Instrument Domains

	1930	1935	1940	1945	1950	1955
Mechanics	35%	30%	25%	20%	15%	15%
Optics	25%	25%	25%	20%	20%	20%
Electricity	20%	20%	15%	15%	10%	10%
Magnetism	5%	5%	5%	10%	10%	10%
Electronics	15%	15%	25%	30%	35%	30%
Nucleonics	0%	5%	5%	5%	10%	15%

magnetic and electronics-based instrumentation frequently proved more precise than competing metrological systems. Second, as the industrial sectors connected with radar and telecommunications (vacuum tubes, semi-conductors, oscilloscopes), and with remote sensing and aviation grew, niches for these kinds of instruments rose correspondingly. Additional novel technical domains, such as computers, emerged during this time, thus further extending the demand for magnetic and electronic apparatus. By the mid-1950s, up to 30% of the devices presented in the *Review* dealt with the apparatus of the "new physics" in conjunction with industry, medicine, metrology, university research, and the military.

Instrument Citation

Scientometric analysis of citation patterns of journal articles reveals distinctive referencing practices for instrument practitioners in general as opposed to academic science and industrial engineering groups on the one hand, and for research-technologists as opposed to narrow-niche instrument men on the other. Referencing practices in articles in the *Review of Scientific Instruments* by research-technology practitioners and narrow-niche instrument men were compared. Global referencing practices of the Review's articles were then compared with those in academic publications such as the *Physical Review* and engineering journals. Citations which appeared in the *Review of Scientific Instruments* fell into six categories: 1) scientific journals, such as *Physical Review* and *The Journal of Applied Physics*; 2) patents; 3) engineering journals, such as *Electronics, Production Engineer, Nucleonics* and *Communications and Electronics*; 4) instrument journals, such as the *Journal of Scientific Instruments* and *Instrument Practice*; 5) restricted circulation industry bulletins. These include in-house documents, such as the *Philips Technical Review*, the *Bell Technical Journal*, and the *General Electric Technical Bulletin*; 6) undisclosed private communications. These documents often take the form of reports prepared in the context of a consulting contract.

In the *Review of Scientific Instruments*, references to restricted circulation in-house industrial bulletins, to confidential communications, and to patents occur more frequently than they do in specialized academic journals. A comparison of the *Physical Review* and the *Review of Scientific Instruments* shows that references to academic journals account for over 90% of the citations of the former, while they make up no more than 55% in the latter. In contrast, when the three rubrics of patents, technical documentations, and confidential reports are taken together, they rival the number of academic references in the *Review of Scientific Instruments*.

Comparison between research-technologists and narrow-niche instrument practitioners who published in the *Review of Scientific Instruments* is equally striking. Research-technologists introduce citations into their articles four times more frequently than narrow-niche instrument specialists. Moreover, only 30% of the citations of narrow-niche instrument practitioners are to academic periodicals, while the figure is almost two times greater for research-technologists. Research-technologists refer to patents three times more frequently than do narrow-niche instrument men (7% as opposed to 2%). Narrow-niche instrument-makers refer mainly to in-house documents.

The demarcations between research-technologists and all other categories of science and technology groups, including narrow-niche instrument practitioners, emerge more fully when citation practices are divided into "public" and "private" channels. Public communications include journal publishing and patenting, and private communications comprise in-house industrial documentation and confidential reports. For research-technologists, a relative balance exists between the two channels – 65% public against 35% private. This configuration contrasts with scientists who operate overwhelmingly in the public sphere (over 95%) and it contrasts with narrow-niche instrument men who operate primarily in the private sphere (85% private and 15% public). This publication pattern reinforces our contention that the instrument landscape associated with research-technology takes place in a framework of many audiences, information outlets, and end-users. The topography is unrestrictive. While cognitive and institutional boundaries persist, they are nevertheless readily and frequently traversed. Transverse movement, or boundary crossing, is routine. Although this is the essence of research-technology, some transverse practices also apparently characterize, albeit to a much smaller degree, instrument-makers whose identity, vocation and function is linked to narrower instrument horizons.

Finally, this analysis of citation practices casts light on why past prosopographic studies of scientists have accorded such little attention to research-technology endeavors. Whether intentionally or not, instrument work is often systematically underrepresented among research-technologists who for various reasons have become equated with the science profession. Research-technologists who have a reputation in academic science frequently refrain from listing their own publications in instrument journals, including the *Review of Scientific Instruments*; and they almost never list publications in in-house bulletins or refer to their work in confidential reports. This bias occurs, for example, when academic scientists write the official biographies of research-technologists. The latter are often portrayed as "scientists," rather than as the purveyors of generic instrumentation. For instance, Jesse Beams, father of the generic ultracentrifuge, is presented in the *American Academy of Science Memoirs* as a gifted and productive physicist who published over 140 articles. Digging into Beams' technical accomplishments however, it quickly emerged that he produced over 50 additional articles and reports that deal with instrumentation and technology which are not listed in his official biography. Upon examination of his "scientific" articles it becomes clear that most deal with instrument matters rather than with experimentation or physical phenomena. Innumerable other instances of systematic silence toward instrumentation work could be cited.

RESEARCH-TECHNOLOGY VERSUS NARROW-NICHE INSTRUMENTS

While the *Review of Scientific Instruments* constituted an important forum for the rallying of research-technologists, this vehicle often strongly overlapped with academia. Some research-technologists perceived the need to involve additional groups in their struggle to affirm their movement. This need was of the utmost importance since narrow-niche instrument groups were extending their grasp over America's firms, universities and technical bureaucracies.

Narrow-Niche Initiatives

Between 1932 and 1940 a host of narrow-niche instrument journals were founded, the most prominent being *Instrument Magazine*, and *Instrumentation*. These and similar reviews concentrated on the narrow-niche instrumentation spectrum. The research-technology perspective was rarely represented. The reviews presented articles dealing with improvements to commonly used niche devices. Articles performed a variety of commerce-based functions, which included acting as a kind of showcase. When authored by personnel working in an instrument manufacturing company, the texts provided visibility for the firm's recent product lines. When penned by employees of technically minded instrument end-user firms, the articles attested to the company's engineering performance and capacity to capitalize on instrument applications. They showed what kind of work was done and the high standards to which it was accomplished.

This cluster of journals conveyed abundant information about the availability of new devices, and specified their technical strengths and deficiencies. While part of this data took the form of company advertisements, another portion took the form of independent consultancy evaluations of equipment. Linked to this, there were reports on where devices could effectively be employed – mainly in industry – and where they should be avoided. A smaller amount of space was given to announcements of local and regional instrument workshops and congresses.

These journals were thus principally concerned with equipment refinements and end-uses. The main theme was the further modification of already embedded devices in order to fit them into new contexts, or to upgrade performance in existing contexts. *Instrument Magazine* and *Instrumentation* were connected to industrial interests, and they had few links with academia, state research and metrology services, or the military. Nor was their focus research-technology. Of the 137 people constituting our American research-technology prosopography (presented below), only sixteen individuals published an article in this category of instrument review. This stemmed from their disinterest in the dis-embedding and re-embedding dynamic. In many key respects the perspective of these two journals was an energetic and upbeat version of the Scientific Apparatus Makers of America.

The Instrument Publishing Company

The development of a distinctive research-technology current inside the U.S. required the emergence of an alternative framework from bodies like *Instrument
Magazine* and *Instrumentation*, and furthermore a framework with a very different
perspective. In 1934 a heterogeneous group of academics, instrument-making companies, and industrialists met to establish America's first publishing house specialized in books and technical manuals dealing with instrumentation. The Instrument
Publishing Company was located in Pittsburgh, which was emerging as the country's
unofficial instrumentation capital. Many instrument-making firms were implanted on
the eastern seaboard, as were large numbers of firms that employed their services and
equipment. Moreover, instrument-oriented universities, like MIT and the Carnegie
Institute, were located there as well. Finally, Pittsburgh had been the home of the
Scientific Apparatus Makers of America, which still figured in the minds of some as
the patron of American instrument efforts. The Instrument Publishing Company
quickly became a bridgehead for research-technology militants who wanted to reinforce generic instrument programs, and if it were necessary, at the expense of the
erstwhile community attached to narrower instrument applications. During its first
decade, the Instrument Publishing Company annually published over a dozen books
and scores of technical manuals. Half of them were in the area of electricity and electronics, and around 20% were in mechanics, with another 20% in optics. The remaining 10% dealt with instruments in chemistry and medicine for example.

The ambitions of the Instrument Publishing Company went beyond mere publication. From its inception, the company strove to organize and shape the nation's embryonic instrument movement – to instill it with an identity and to protect its collective interests. This agenda included strengthening patent protection for novel devices
and giving support to preferential tariff barriers. The publishing house sought to coalesce the broadest possible spectrum of instrument-maker interests and currents –
academia, the military and public metrological services, manufacturing, instrument
design and production, instrument consultancy, and of course, instrument consumers
themselves.

For the Instrument Publishing Company, two groups were highly significant,
academics and control engineers. The involvement of academics was of utmost importance. Academics had in large part worked outside the pale of the Scientific Apparatus Makers of America, and few academics published in journals like the *Instrument Magazine* and *Instrumentation*. Yet university personnel were becoming
increasingly central to instrument invention and innovation. It was deemed crucial to
make them an integral component of the workings of the new publishing house's
activities. Control engineers had a strong presence in the Publishing Company. In the
late 1930s and 1940s, almost 40% of the texts printed by the Publishing House concerned aspects of control engineering practices, or the detection, measurement and
control devices involved in control processes.

Sponsorship of instrument workshops and congresses figured prominently in the
Publishing House's agenda. Between 1935 and 1940, it participated in 53 meetings
involving over 400 speakers. Meetings fell into three categories: 1) The biggest
number of encounters dealt with end-user applications of existing devices – most

often in chemistry, metallurgy or electricity and electronics. Consultants, instrument company engineers, or sometimes academics presented devices and methods to end-user clients in large firms. 2) This category proved far more restricted in scope, only involving encounters between instrument inventors and makers. While rubric one entailed down-stream re-embedding, the theme here was the elaboration of broader uses for equipment already in operation. Dis-embedding mingled subtly with re-embedding. 3) The final kind of meeting sponsored by the Instrument Publishing Company was for academics, or occasionally also for important instrument person-alities in enterprise. Academic meetings often took place at a university venue – MIT, Harvard, Princeton, Columbia or Pittsburgh University. Two topics prevailed: a) new designs in generic instrumentation, and b) how to glean information about existing narrow-niche equipment for the purpose of designing new generalist instru-ment systems. In this setting research-technology was at last fully acknowledged as a distinctive and fundamental activity, and it was vigorously promoted.

The Publishing House agenda advocated two kinds of transverse activity. The first was cognitive. Many instrument workshops and conferences encouraged multi-disciplinary and interdisciplinary approaches to instrumentation based on fields such as electronics, chemistry, optics and mechanics. Disciplinary boundaries were per-ceived as restrictive, and even as destructive to instrument progress. Effective inno-vations were thought to be grounded in multi-disciplinary and multi-skill integra-tions. Control engineering was frequently held up as a fine example of this. The second kind of transverse interaction was invoked when groups from the university, instrument manufacturing, consulting and multiple end-users were brought together. The goal was greater dis-embedding and re-embedding effectiveness; and with it, greater instrument-making efficiency. However, this objective often proved illusive.

The quasi-explicit promotion of research-technology by the Instrument Publish-ing Company, including its interfacing with end-user groups, was frequently stone-walled. The blockage had two sources, both of which were the outgrowth of struc-tural aspects of American education and industry. It was difficult to persuade oc-cupation-defined, employer based professionals to think about instruments in terms of their origins, development and application in a framework that transcended indi-vidual user interest and a device's narrow technical niche. This was in part because instrument work took place in a circumscribed employment environment – often a single company or economic sector, and in a single technical field. In America, tech-nical personnel and engineers often had a task-based mentality rather than a broader function-based perception.[7] Furthermore, in the U.S. more than any other industrial culture, a sense of professional identity was particularly acute among technical cadre. It was in America that professional organizations in innumerable fields of engineer-ing emerged quickly and became highly assertive. Professional organizations exer cised considerable sway over their members due to the success of their lucrative or-thodoxies and this meant that there was little incentive for technical cadre to operate elsewhere.

---

[7] Emblematic of this orientation is the birth of Taylorism inside the U.S. and its rapid and pervasive development and influence.

This already pre-programmed arena of instrumentation was further demarcated by the nature of relations between industry and the university. Industry dominated university departments of engineering, and in many instances science departments as well (Noble 1977; Servos 1980). Hence, even for instrument men in the university, academia seldom functioned as a haven where generic devices could safely be designed, and from which these devices could effectively be diffused. That the U.S. university system per se could at this time cradle the nation's research-technology movement was an institutional impossibility, although academia would eventually play a crucial role.

The Instrument Society of America

The Instrument Society of America was founded in October, 1948. The first step in its foundation took place in March, 1937, when the Instrument Publication Company and the Metallurgy Department of the Carnegie Technical Institute combined forces to organize a national congress in Pittsburgh devoted to scientific devices. The meeting was promoted and underwritten by the Brown division of the Honeywell Company, located in Minneapolis, Minnesota. The Brown division's specialty was control engineering, and in particular, continuous flow systems. For several years there had already been a connection between America's protean research-technology effort and control engineering groups. The meeting attracted participants from academia, government, the military, industry and commerce. In 1939 the Carnegie Technical Institute Chemistry Department and the Instrumentation Publication Company sponsored a second national congress, again held in Pittsburgh. It attracted an even larger public. As a direct spin-off of these meetings, the American Society for Measurement and Control was set up in 1941. It continued the efforts initiated in 1938 and 1939 to bring more people into the instrument sphere. The American Society for Measurement and Control was more tightly coupled to academia than its predecessor, the Scientific Apparatus Makers of America, had been. Academic science and engineering departments were enrolled in the instrument cause. This was in part made possible because of the growing scientization of engineering. University engineering departments were developing stronger ties with discipline-based departments. New domains of engineering were emerging, such as chemical engineering, electronics engineering and most recently nuclear engineering, whose technical processes incorporated increasingly higher levels of scientific knowledge and skills, and which relied on ever more intricate and interconnected instrument systems.

The Instrument Society of America was set up in the fall of 1948 in response to fears that narrow-niche instrument manufacturers and users might soon dominate America's instrument landscape. Among its founders were W. Chatles (director of instrumentation of the Kentucky Carbon and Carbide Company), A.H. Peterson (director of the Instrument Division of General Electric), and L. Porter (director of the Dow Chemical Company Instrumentation Division). Even though these men came from industry, they were not proponents of narrow-niche instrumentation. They expressly acknowledged the role and centrality of research-technologists (Rimbach 1954). Spokesmen from The Instrument Publishing Company, like the influential Richard Rimbach who was one of the company's founders, had long striven to extend the purview of the generalist instrument-maker in America. Rimbach recognized that research-technology could survive and prosper only if it were accepted by potential competing and hostile end-user instrument groups. He had long envisaged an instrument collective in which end-users figured prominently, and where research-technologists could also prosper. It was this vision that spurred Rimbach to participate in the creation of the Instrument Society of America, where he long served as one of the body's guiding directors. By bringing together the two groups of sometimes adversarial instrument practitioners, (research-technologists and narrow-niche instrument makers) Rimbach hoped that America could construct an effective instrument arena. But more than anything else, this would require greater acknowledgment and acceptance of the crucial role of research-technologists.

Rimbach, the Instrument Society of America, and the Instrument Publishing Company concurred that control engineering constituted an exceptional window of opportunity for the reinforcement of research-technology in America. Control engineering entails the use of a variety of elaborate scientific instruments, as well as their effective integration. Devices include detectors, complex coupling systems, signal processors, and automated responses. Control engineering fit nicely into the embryonic practices and philosophy of cybernetics (Mindell 1996). Cybernetics entails both locally tailored devices and generic equipment. They considered that the crusade on behalf of research-technology could benefit from control engineering, because it would serve as a showcase for research-technology's achievements and promise. Control engineering would demonstrate by dint of concrete practices how research-technologists and narrow-niche instrument engineers can work together.

The experience of the Eastman Kodak Company in the field of control engineering is an often cited instance of cooperative interaction between research-technology and niche instrument requirements. In 1946 the Eastman Kodak Company set out to make control engineering the hub of all of the company's processing and production operations. The giant firm initially purchased commercially available instrumentation. Eastman Kodak set up a program of in-house training to enable its engineers to utilize the new equipment. In an attempt to profit from control engineering systems to a maximum, the company also founded an instrument division that was to provide additional properly adapted control engineering devices. Soon, however, company management began to doubt the effectiveness of a policy which based control engineering instrument research on criteria of expediency and short term demand. Some technical managers inside the firm believed that if control engineering was to advance technically and become profitable, it would be necessary to generate new species of baseline devices from which narrower application apparatus tailored specifically to Eastman Kodak's requirements could then be derived. Now leery of the company's capacity to develop narrow-niche devices independent of recourse to upstream instrument design, Eastman Kodak turned to outside instrument sources, and more significantly, to the sponsorship of generic devices. The giant company sought out designers of template devices who worked in universities, state metrology centers, or in consultancy firms. The goal was to promote generalist scientific instrument research that might eventually contribute to better control engineering apparatus at Eastman Kodak. This policy quickly bore fruit. Template devices were located that narrow-niche instrument men converted to satisfy in-house control engineering demand. The two instrument currents – template device engineering and narrow-niche instrument engineering – proved complimentary, and the groups worked together harmoniously. It was in anticipation of reaping further gain through research-technology that Eastman Kodak encouraged the generalist instrumentation efforts of the new Instrument Society of America.

Demand for instrument theoreticians, theoretical metrologists, and generalist instrument designers grew appreciably after World War II. The *Review of Scientific Instruments* and *Instrument Magazine* printed job offers for this category of personnel, as did the *Instrument Society of America Journal* after 1954. There was a rise in advertising for research-technologists as the U.S. Air Force and Navy, the National Bureau of Standards, and companies, such as AT&T, expressed a need for instrument theoreticians, theoretical metrologists, and generalist instrument designers. Demand similarly increased for narrow-niche instrument men who were expert in the design, installation and use of highly specific devices for a restricted environment. Many firms sought specialists in narrow-niche devices as well as research-technologists.

STRANGE COOPERATIONS

The first years of the *Instrument Society of America Journal* were beset by the fear of dissension between research-technologists and narrow-niche instrumentation specialists. Articles by the journal editor, Richard Rimbach and his associates insisted on the need for generalist instrument theoreticians and designers and specialized instrument experts to cooperate (Akins 1954; Batcher 1954). The nation's industries and armed forces were dependent on both kinds of instrument men, and there was a legitimate place for both inside the Society and the Journal. Moreover, instrument and metrology theoreticians and narrow-niche instrument personnel needed one another. The apparatus designed by narrow-niche instrument experts was often based on generic instruments that had been designed and constructed by instrument generalists. Conversely, the technical input, ideas and demand incorporated in generic equipment were frequently tied to narrow-niche instrumentation and its practitioners.

The Journal called for mutual good relations between the two sometimes discordant instrument currents. Rimbach likened the activities of research-technologists and narrow-niche instrumentalists to a grid, with an x-axis and a y-axis. Specialized instrument design experts master the intricacies of their devices which incorporate the specifications necessary to carry out a predetermined task. These instrument men must know the detailed requirements of the workplace where the device will operate, and must know exactly how to make a device perform effectively. In contrast, research-technologists operate "transversely." They master general metrological and instrument theory which enables them to design multipurpose generalist equipment. The diffusion of this equipment then spreads horizontally outward to a variety of potential users, whose narrow-niche instrument engineers modify it to address specific needs. The Journal's editorial staff insisted that the Society's local and regional chapters could accommodate both activities. Some chapters were indeed so big that if called for, sub-chapters might be opened to cater to the demands of generalists and theoreticians or narrow-niche designers. Nevertheless, Rimbach considered that when possible, it was advisable to maintain mixed chapters, which would promote contact and communication between the two mutually interdependent currents.

Rimbach's views were echoed and amplified in six articles that appeared in the *Instrument Society of America Journal* in 1954. The articles sought to establish a taxonomy of the functional purviews of research-technologists and narrow-niche instrument designers. The texts introduced a vocabulary that labeled the instrument groups, and labeled their operations and institutional space. These articles were authored by industrial managers, scientists, engineers, and by both categories of instrument men.

According to all of the authors, the instrument landscape can be divided into two spheres. One sphere consists of specialized devices; the other is composed of generalist instruments. Specialty equipment belongs to engineers and technicians, and also to most academic scientists. These factions develop tailored apparatus or use existing equipment to solve the detailed, local technical problems associated with the detection, measurement and control of specific processes or physical phenomena. The design and specifications of equipment is determined by the particulars of well defined tasks. Inside industry, the "specialty instrumentalist" tailors equipment for the needs of production; inside academia "instrument specialists" devise apparatus for use in a given experiment.[8] In these cases new, albeit restrictive use instruments are developed by the specialty instrument expert. For example, in the industrial setting, this often takes place in "instrument development bureaus." These bureaus perform a research function for user-guided specialty apparatus. This research is upstream from designated applications, but the research agenda is nevertheless dictated by precisely formulated client need (Brand 1954; Fletch 1954; Lee 1954; Web 1954).

In contrast, in these articles research-technologists received the labels "generalists," "instrument theoreticians," and "basic instrument designers" (Brombacher 1954; Draper 1954; Lucks 1954). It was suggested that through the mastery of metrological principles, instrument generalists design multipurpose equipment which is later adopted and adapted by specialized instrument designers located in instrument development bureaus and university research laboratories. A single generic instrument, designed by an instrument theorist, can ultimately find its way into innumerable universities and firms where it performs many different kinds of work. Its movement is transverse.

The authors of these texts stressed that the work trajectories of metrological and instrument theorists and generalists were by necessity similarly transverse. The authors claimed, for example, that at Bell and at AT&T one can find the usual instrument development bureaus (of the type described above) which are connected to specific military applications, civilian telecommunication applications, radio frequency power amplification, and production. These bureaus (sometimes also known as "instrument divisions") spawn devices for their particular industrial division. But operating alongside the bureaus, or between them, one can often find a section called "instrument operations." In these articles, the work and philosophy of instrument operations are likened to the activities and outlook of systems engineering – but in

---

[8] In some instances, the dual function of experimental benchwork and specialty instrument design/construction is performed by scientists.

the realm of generalist equipment. From this instrument operations vantage-point, instrument theoreticians pass freely back and forth between the numerous instrument development bureaus, thereby gleaning strategic information about categories of demand and collecting ideas and technical information. During these forays, research-technologists indicate to instrument specialists why a given generic device should be adopted, and suggest the lines along which it can be appropriately redesigned.

Several of these articles specify three additional sites of template research where intellectual and occupational transverse movement are considerable. The National Bureau of Standards undertakes instrument research in generalist devices as well as specialist equipment. By dint of its dealings with industry, government services, the military and academia, some categories of Bureau of Standards personnel necessarily conceive instruments design along sweeping lines, and this personnel is well placed to ensure that generalist equipment finds a multi-disciplinary and multi-institutional outlet. Consulting firms and private instrument-makers occasionally work along generalist and theoretical metrological lines as well. Research-technologists are often found in America's universities. While theory and experimentation are the mainstays of academia, work associated with metrological theory and template apparatus is tolerated, and it sometimes even receives active support. For academics who are involved with generalist instrumentation, their contacts with industry, with state technical services and with the military are just as important as the contacts they have with their university colleagues. Here, the university functions as a platform for transverse movement in the same way that instrument operations laboratories perform this function inside industry.

CROSSING BOUNDARIES

In an article that examines the career characteristics of U.S. instrument men, Daniel Shimshoni points to a high rate of job mobility. He states that instrument men change firms far more frequently than other categories of industrial personnel with the same amount of training and occupying similar positions in company hierarchy (v. Hippel 1988; Shimshoni 1970). The career data from my prosopography of 137 American research-technologists also indicates considerable mobility, but the boundaries crossed by research-technologists are not uniquely, or even principally, inside industry. To the contrary, research-technologists' boundary crossings are multi-sector. They circulate freely and frequently from university to business, to the military or state metrology services, to consultancy firms, and often back to university. The preceding list should not be interpreted by the reader as an indication that all U.S. research-technologists are located inside academia. While academia constituted the main institutional attachment for 55% of the total sample, 30% never worked in a university. At least at one point in their career, all of the research-technologists for whom data has been collected were connected to industry: 45% spent time in state technical services; 85% dealt with the military; 35% worked with instrument design or manufacturing companies; and 20% had links to instrument consulting firms.

Boundary crossing was a fundamental component of research-technology on numerous counts. Of course, multiple audiences allowed the diffusion of research-technologists' generic instruments, and multiple audiences constituted an information

channel for new technical data and ideas. Equally significant, working at the interface of many boundaries established a career space which enabled generic device
designers and constructors to escape impinging demands exacted by the particular
body (university, firm, state agency) paying their salary at the time. Through the multiplication of their commitments, research-technologists were committed to no one.
They thereby managed to safeguard their primary commitment to genericity. Boundary crossing thus allowed research-technologists to create, occupy and configure an
interstitial arena – a neutral space between the spheres which came into existence in
association with specific tasks.

Men out of Academia

For a long time, the military's link with American universities, like that of industry,
was unfavorable to research-technology, because the attachments choked off the
freedom of maneuver that was so indispensable to interstitial action. The eventual
expansion of the playing field for academic instrument-minded men, with its enhanced opportunity for maneuver, was in large part one result of post 1945 growth in
instrument demand of all sorts, and was similarly a consequence of the post war
strengthening of the American instrument community. The instrument work of Jesse
Beams at the University of Virginia and W.W. Hansen at Stanford exemplify how
academia-associated U.S. research-technologists have operated.

Jesse Beams (1898-1977) was one of America's foremost research-technologists
(Elzen 1986; Gordy 1983). He combined expertise in mechanics, thermodynamics,
magnetism, metallurgy, fluid physics and control engineering in his work on the ultracentrifuge, which he developed as a generic instrument between 1930 and 1950.
Beams' breadth of knowledge and skills went beyond that of most scientists and engineers. His instrument incorporated numerous cognitive and technical fields, and it
served many areas of application which stretched from aviation to nucleonics, from
biology and medicine to metallurgy and thin film physics, and included quantum
physics and metrology. While Beams' institutional base was a university, the potential of his generic ultracentrifuge propelled him toward military circles, the National
Institute of Health (NIH) and the Rockefeller Foundation, the Atomic Energy Commission (AEC), the General Electric Company, a host of small high technology
firms, and numerous academic physics departments. He spent more time at these
sites than he did at the University of Virginia Physics Department. Although recognized by scientists as one of their own and often heralded by them for his work,
Beams' endeavors and trajectory do not coincide with the norms of either an academic or an engineering career.

Beams undertook his pre-doctoral studies at Wichita State University and Auburn
University where he specialized in physics. He did his doctoral work at the University of Virginia, completing it in 1924. His dissertation was intended to deal with the
time interval between absorption and emission in the photoelectric effect. In order to
measure this brief interval, Beams engineered a rapidly rotating mirror, using a
steam-driven centrifuge. He failed to measure the absorption/emission interval; his
thesis dealt mainly with metrology, problems of measurement, and the principles
associated with rotating devices. Based on his outstanding instrument skills, he won a

National Research Council grant for postdoctoral investigations. He spent 1926 at Yale, where he worked on theories of rapid rotation and on how to use rotation in the accurate measurement of time. While there, he temporarily joined forces with Ernest Lawrence, who was also investigating how circular trajectories could be used to raise the energy levels of sub-atomic particles. In 1927 Beams returned to the University of Virginia, where he conducted more work on rotary devices.

The difficulties encountered by Beams in the accurate measurement of time came from insufficient rotational speeds. Beams judged that a fuller grasp of the basic mechanisms and principles of rotation could lead to a multipurpose device which, among other things, could provide more precise time measurement. His technical research entailed four steps which were carried out over a span of twenty years:

1) Based on what he was able to learn from earlier European studies, between 1926 and 1931 Beams made slight improvements to existing devices that incorporated flexible drive shafts and steam-driven propulsion systems, and that operated in a relatively reduced-atmosphere, low-friction environment.

2) Beams soon drew the conclusion that friction continued to constitute a key obstacle to achieving rotation beyond 2,000 rps. He consequently designed and built a vacuum capsule, in which he placed his rotating capsule. But it never proved possible to sustain a valid seal as air seeped into the centrifuge at the point of the drive shaft connection, thereby raising pressure, increasing friction, and slowing rotation. Friction-induced heating sometimes jeopardized the physical or biological materials that were being centrifuged.

3) From 1933 to 1937, Beams built a high vacuum ultracentrifuge. The principles of magnetism were central to this endeavor, as were techniques of control engineering. Beams built his rotor from magnetic metals. By applying an exogenous magnetic field to this magnetized rotor, he suspended the rotor inside the centrifuge's gas-free, vacuum capsule. Of course, during rotation the center of gravity of the spinning rotor shifted. To compensate for this, Beams adopted a servomechanism, of the kind frequently used in control engineering, to stabilize the position of the rotor.

4) Extremely high rotating speeds continued to be limited by the mechanical techniques then employed for spinning the rotor. In 1931 Beams had developed a new approach to rotation, which consisted of using piano wire to induce circular movement, and while higher speeds were attained, the number of rps nevertheless remained far below what he calculated was physically feasible. To overcome this deficiency, Beams again turned to magnets. He set up a series of high performance electromagnets around the outside of his vacuum capsule and applied alternating current to them to spin the magnetic rotor without the intervention of any mechanical apparatus. This setup gave rise to rotating values in excess of three million rps, generating pressures of one billion gravities.

Throughout this instrument research, Beams was a professor at the University of Virginia, and for much of the time he was head of the physics department. But can one say that Beams was a "scientist"? He was not an experimenter in the normal sense of the term; his experiments dealt not with physical systems, but instead with devices. Neither was he a theoretician in the standard sense of the term; for the theories that Beams advanced related to metrology and not the regularities of a physical world. Inside the University, Beams' home was the vast and well equipped workshop, for which he was himself the principal architect (Brown 1967). While Beams certainly had a place in science, science occupied a moderate place in his personal agenda. This can be seen by looking at the spectrum of his institutional and personal interlocutors and by viewing the extent of the technical networks into which his generic ultracentrifuge diffused.

Between 1930 and 1960 Beams participated in the re-embedding of his generic device in no fewer than seven distinct spheres: experimental biology, medicine (see the chapter by Gaudillière in this volume), isotope separation, jet propulsion research, research on thin metal films, and related to that, the development of test procedures and new industrial norms, research on photon pressure, and the determination of a high precision gravitational constant. Jesse Beams worked with or for the Manhattan Project, the AEC, the NIH, the NSF, the Rockefeller Foundation, the U.S. Air Force and Navy, the Spinco Company that built ultracentrifuges, and the Beckman Instrument Company. Beams' work involved designing a device open to disembedding, and it involved participating in the task of re-embedding that device in different environments. Beams was in constant motion; his operational base was interstitial, situated between multiple institutions and employers and between innumerable channels of output and input.

A comparison of the instrument work of the research-technologists Jesse Beams and Edward Pickles (1911–67) underlines some of the differences between generic instrument research and narrow-niche instrument development (Elzen 1986; see Gaudillière's chapter in this volume). Pickles, one of Beams' students, is emblematic of a narrow-niche instrument designer. He was trained at the University of Virginia, where he received his B.S., M.S., and in 1936 his Ph.D. in physics. Beams was Pickles' dissertation supervisor, and they worked together on advanced models of the high speed ultracentrifuge. Pickles served as an instrument specialist at the Health Division of the Rockefeller Foundation between 1934 and 1946, where he was first a laboratory assistant and then a member of the permanent staff. He collaborated with biologists in their research on bacteria and viruses. His ultracentrifuge proved crucial in separating and isolating particular strains of biological materials. Pickles acquired a high level of training in biology during this period, but he remained an instrument specialist, which he viewed as a sub-division of physics. Regardless of his self-identification with physics and his commitment to that discipline, Pickles' instrument agenda was nevertheless strictly determined by the momentary demands of the current research project of his biology colleagues. Pickles' instrument endeavors were in effect biology driven.

In 1946–7 Pickles founded Spinco Instruments Inc., a small high technology company specializing in the design and manufacture of ultracentrifuges. Pharmaceutical companies and biology research laboratories were the firm's principal clients, and Pickles' instrument perspectives continued to be circumscribed by the specifications set by clients. In 1955 Spinco was sold to Beckman Instruments Inc., and Pickles became laboratory director at Beckman. While continuing to develop ultracentrifuges that were adapted to biological programs, he also developed instrumentation for particle identification and classification, and he worked on a new species of electrophoresis.

Pickles' trajectory differs from that of Beams in two fundamental respects: First, Pickles' activities were far removed from the generic instrumentation commitment of Beams. Pickles re-embedded his apparatus, custom-adapting it to satisfy specialized demand. In contrast, Beams explored the underlying principles associated with high speed rotational systems, and by this procedure, he designed general instrument systems which were susceptible to dis-embedding. A part of Beam's activities entailed accompanying his generic device across boundaries, as he assisted in re-embedding it in various environments. Second, Pickles' career trajectory was linear; he moved from the Rockefeller Foundation to Spinco and then through industrial merger to Beckman. Beams, however, worked for no fewer than sixteen employers in the public and private sector, and he frequently worked simultaneously for several. He worked in an interstitial setting; that is to say, he occupied the nooks and crannies that exist between the masonry of established institutions. Working out of this set of coordinates facilitated Beams' commitment to generic instrumentation as opposed to specialized equipment imposed by employer demands.

The research-technology path of another of America's most gifted generic instrument men, W.W. Hansen (1900–1949) (Bloch 1952), parallels the trajectory of Jesse Beams in important ways. Hansen, a specialist in electronics, worked in the San Francisco economic, institutional and intellectual environment for most of his short life. From an early age, Hansen's interest lay primarily with technology and engineering rather than science. He viewed science as a tool through which to look at engineering, and unlike many of his later associates, he did not abandon engineering in favor of science. Hansen received his training at Stanford University, where he began to work in the laboratory of the physics department at the tender age of sixteen. He obtained his MS in physics in 1926 and his Ph.D. in 1929. Hansen was made an instructor in the Stanford University physics department in 1926, and he was later given a professorship. His early research dealt with electron emission, and later with the excitation mechanism of cathode ray emission. His dissertation dealt with cathode ionization of silver. Hansen's research focused on technical systems, and while his official position was in the physics department, during the 1920s and 1930s the department did not distinguish rigorously between engineering and science. The playing field remained open, thereby leaving ample space for those committed to technology and to fundamental research into instrumentation.

Hansen's interests lay at the crossroads of technology, experimentation and theory. His talent lay in the theory of instruments. In 1938 he received an invitation from MIT to do research there. Hansen accepted, and for the initial months of his brief stay he sharpened his mathematical and theoretical skills in the domain of quantum mechanics. He believed that therein lay the solution to the thorny technical problem of vacuum tube microwave power amplification. Failure to obtain satisfactory technical results in this field, he argued, stemmed directly from a poor understanding of the relations between the electro-magnetic and magnetic dipoles and quadrapoles operating inside amplifier tubes. Hansen further believed that a general, theoretical analysis of dipole and quadrapole interactions would prove crucial to the improvement of numerous existing technical systems and to the discovery of new ones. In the past, scientists had depicted pole interaction in the framework of a two-dimensional plane. Hansen, however, succeeded in representing them in a spherical frame of reference. By dint of this, Hansen proposed a new model of dipole/quadrapole electromagnetic/magnetic dynamics. While his approach garnered some attention at MIT, the audience was chiefly in theoretical physics; but Hansen felt that the most exciting field of relevance was in technology and engineering.

In late 1938 he returned to the Stanford physics unit and to a circle of bay area instrument-minded men whom he had befriended in the 1930s. This circle included the Varian brothers who were already instrument entrepreneurs (Varian 1983). Russel Varian's instrument experience was in proto-television and low power microwave amplification, and before traveling east, Hansen had assisted him with projects on several occasions. Hansen resumed this collaboration, while at the same time at the Stanford physics department he began to construct a vacuum tube designed to depict the complex interaction that occurs during microwave amplification. Hansen considered this work as investigatory experiments to uncover basic theoretical principles involved in a family of technical systems. He dubbed his new vacuum tube setup the "rumbatron," because the geometry of the interaction between the different poles which the setup depicted swung about in a pattern and rhythm reminiscent of the rumba dance-step. Hansen quickly realized that the technical principles demonstrated by the rumbatron had vast technical implications.

In 1940 he established, together with the Varian brothers, a new scientific instrument company, Varian Instruments. Hansen was the firm's principal investor. Working with Russel Varian, Hansen struggled to dis-embed relevant components of the generic electronic distributions revealed through the rumbatron in a manner consistent with their re-embedding in a microwave amplifier of the sort used in radar. Brought quickly to fruition, this technology became the basis for the Varian klystron, which emerged as the centerpiece of America's wartime radar detection system. Perceiving the generic qualities of his technical discovery in dipole/quadrapole electromagnetic/magnetic dynamics, Hansen next showed how this understanding pertains to high frequency radio communication. Based on a hypothesis of electromagnetic behavior which was derived from instrument theory grounded in his rumbatron, Hansen wrote a series of technical reports for the Office of Navel Research in which he argued for a new species of radio antenna design. As the war neared its end, some of his recommendations were incorporated into radio antennas for transcontinental and transoceanic communications. Between 1946 and his untimely death in 1949, Hansen, the consummate research-technologist, re-embedded technical aspects of his rumbatron into the Stanford University linear accelerator. Once again, Hansen saw his rumbatron as an instrument whose principal strength lay in its capacity for multiple dis-embedding. Hansen's triumph also lay in his capacity to see precisely where this dis-embedding potential could be translated into effective technical re-embeddings.

Men out of Industry

The instrument career of Winfield Salisbury (b. 1903) indicates how research-technology functioned in the U.S. during the middle decades of the twentieth century when a practitioner's main employment and institutional referent was industry.[9] Salisbury's achievement in research-technology lay in methodology as much as in the design and construction of artefacts. Between 1925 and 1970 he engineered extremely high voltage systems, which entailed very huge transformers, rectifiers, capacitors, and transmission cables. Salisbury received his B.A. at the University of Iowa in 1926, and he was thereafter self-taught in engineering. Like many industry-based American research-technologists, his career trajectory was characterized by innumerable moves – moves from industry to academia, from industry to state metrology agencies, and moves from firm to firm. Over a span of thirty-five years, Salisbury changed research sites no fewer than sixteen times.

In 1926-27 he remained at the University of Iowa where he taught introductory courses in physics. In 1928 he left academia and for the next ten years he worked as a private consultant. Salisbury's real interest lay in engineering and instrumentation, especially in high voltage systems. He designed technical circuitry which was remarkable not only for high voltage, but was also remarkable for its great electrical stability and for its capacity to discharge huge quantities of power over a very short time interval. One of his setups became known as the "Salisbury engine." As opportunities for consulting diminished during the years of the great depression, in 1937 Salisbury took work at the U.S. Department of Labor, where he worked as a statistical analyst. Fortunately for him though, because of his already strong reputation in the realm of discharge electricity, he received an offer to work on the big new cyclotron at the University of California Berkeley Radiation Laboratory. He remained there until 1941. He then moved to the MIT Radiation Laboratory in 1942, and later he became director of the high power division of the Harvard Radiation Research Laboratory, where he remained until 1945. Between 1946 and 1951 Salisbury directed research operations at the world famous Collins Radio Company. During the academic year 1951-52, he was appointed McKay Professor of Engineering at the University of California, Berkeley. He became chief physicist at the California Research and Development Company from 1952 to 1954, and then Director of the Gray Science Division of Remler Company Inc. in 1954-55. By this time Salisbury had become a much sought after research-technologist. In acknowledgment of this, and in order to bring him to the firm, the Zenith Radio Corporation established a new laboratory in high voltage generic instrumentation which was named after Salisbury. Salisbury temporarily directed the unit from 1956 to 1958. He next worked at the Varo Company in the capacity of senior researcher from 1959 to 1965. Between 1966 and 1970 he ended his career as a consultant with the Harvard Observatory.

Although career mobility of the sort exhibited by Salisbury suggests that he might be a research-technologist, it is rather Salisbury's generic instrument endeavors that clinch the matter. This engineer designed and built a fundamental apparatus that stabilized and stored an immense electrostatic charge. The setup's principles were disembedded and re-embedded in a variety of domains. Re-embedding extended to the production of electron-induced generation of relatively pure light. This had both civilian and military applications. The underlying principles of the Salisbury engine also proved central to the operation of the Berkeley cyclotron. Radio communica-

---

[9] 'Salisbury' entry in *Modern Scientists and Engineers* 1980.

tions firms found high voltage stability useful in some of their high power commercial transmitter units. In the 1960s, the electronic devices of astronomical observatories similarly found uses for the Salisbury engineered system. Salisbury was never the master of specific applications, but through his generalist instrumentation he facilitated the development of new technologies.

One could point to scores of other research-technologists who primarily worked out of an industrial setting. Candidates for this would include Arnold Beckman and Russell Varian. In this volume Hans-Joerg Rheinberger analyses Packard's subtle movements between industry and academia as he went about engineering and re-embedding his liquid scintillation counters.

EXIT

Due to its in-between character research-technology was, and remains, an unstable activity. On the basis of this one has to question its long-term prospects. This final section will examine statistical career data of research-technologists as a basis for raising theoretical questions about research-technology and its future.

How do practitioners of research-technology finish their careers in the field? Research-technology is by dint of its vocation and the social and occupational arena that it occupies a remarkably unstable realm. But does this instability mean that research-technology is a precarious activity per se and in danger of dissolution? (See the chapters of Nevers et al., and Johnston in this volume.)

There are three likely scenarios: 1) Research-technology is often a relatively stable domain, and many practitioners continue to design, dis-embed and re-embed generic instruments until retirement. 2) Practitioners design a generic device early in their careers. They re-embed it in a sphere or two; they try to maintain research-technology activities, but are forced out of generic instrumentation by a combination of professional and market forces. 3) Research-technology work constitutes a brief chapter in a career. Individuals profit from the many professional opportunities resulting from their design of a generic instrument. In the process of re-embedding, they often move away from generic design and into the development of narrow-niche devices.

My sampling technique focuses on individuals who published frequently and over a long period in the *Review of Scientific Instruments* during the decades 1930 to 1960. It is weighted in favor of men who spent a sizable portion of their career in the research-technology field. It is not surprising that my data show that many practitioners were successful in finding ways to continue generic instrument work up to their retirement. This was clearly so in the cases of Beams and Salisbury, and Hansen's continuation in research-technology was assured had he not died prematurely In my research-technology population only 15% of the individuals left the field permanently. This generally happened at a relatively early age, and individuals gravitated to industry in every instance. Another 25% of the people in my prosopography had careers where they initially developed a generic artifact or methodology. They then became involved in re-embedding for a protracted period, before once again taking up generic instrument work, and then resuming re-embedding for a considerable time. While movement away from research-technology mainly occurred in the direction of industry, a sizable number of practitioners also moved to military re-embedding. The important point here is that almost two thirds of the research-technologists who comprise my prosopography managed to sustain their generic instrument commitment without interruption, sometimes in the face of adverse institutional and market conditions.

The internal evolution of technical and scientific fields, and the willingness of instrument end-users to re-embed and absorb new generic devices influence the career options of research-technologists. In my prosopography there is a marked over-representation of durable research-technologists in the areas of control engineering and cybernetics, electronics, and the new physics. For example, opportunities for generic work abounded in nucleonics immediately after World War II, as witnessed by events in liquid scintillation counting (see Rheinberger in this volume). In contrast, other areas which for a time proved propitious to research-technology eventually lost their promise, thus forcing many practitioners to leave research-technology altogether, to circulate in and out of the domain, or to shift to alternative spheres of generic instrumentation. This happened to some extent in plant genetics (see Nevers et al. in this book), and in Fourier transform spectroscopy (see Johnston in this volume).

Individuals who stayed only briefly in research-technology were also concentrated in particular fields, for example in mechanics. This observation coincides with the drift away from mechanics in instrument reviews (see above).

The stability of research-technology thus appears highly contingent on the structures and performance of its intellectual, institutional, and economic environment. Enhanced differentiations within industry, science and the state, and growing differentiations between the three, probably reinforce the demand for research-technology. It is research-technology's "in between" position that permits it to service heterogeneous audiences, and it is research-technology's transverse boundary-crossing, "universalizing" products that help actors inside the spheres of science, industry and the state to speak effectively to one another across the boundaries of their separate interests, conventions, and institutions.

GEMAS/CNRS, Paris, France

REFERENCES

Akins, G.S. (1954). Principles of automatic control. *Instrument Society of America Journal* 1 (January): 40–4.

Batcher, R. (1954). Electronics instrumentation. *Instrument Society of America Journal* 1 (January): 46–7.

Ben-David, Joseph (1991). *Scientific growth. Essays on the social organization and ethos of science.* Berkeley, CA: University of California Press.

Bloch, Felix (1952). William Webster Hansen 1909–1949. *Biographical Memoirs* 27 (National Academy of Sciences of the United States of America: 121–37.

Brand, W. (1954). Your responsibilities. *Instrument Society of America Journal* 1 (February): 6-7.

Braverman, Harry (1974). *Labor and monopoly capital: The degradation of work in the twentieth century.* New York: Monthly Review Press.

Brombacher, W. (1954). Problems in research. *Instrument Society of America Journal* 1 (March): 8–10.

Brown, Frederick L. (1967). *A brief history of the physics department of the University of Virginia, 1922–1961*. Charlottesville, VA: University of Virginia Press.

Bruce, Robert (1973). *A.G. Bell and the conquest of solitude*. Boston, MA: Little Brown.

Chandler, Alfred D. (1977). *The visible hand: The managerial revolution in American business*. Cambridge, MA: The Belknap Press of Harvard University Press.

Chemical Abstracts of the American Chemical Society 26 (1933).

Chemical Abstracts of the American Chemical Society 28 (1935).

Chemical Abstracts of the American Chemical Society 29 (1936).

Draper, C. (1954). Definitions of research development and instrumentation. *Instrument Society of America Journal* 1 (March): 12–14.

Elzen, Boelie (1986). Two ultracentrifuges: A comparative study of the social construction of artefacts. *Social Studies of Science* 16(4): 621–62.

F.K.R. (1930a). Editorial. *Review of Scientific Instrument* 1(8): 427–8.

F.K.R. (1930b). The national industrial and standardization movement. *Review of Scientific Instrument* 1(12): 705–10.

Fletch, E. (1954). Instrumentation. *Instrument Society of America Journal* 1 (February): 12.

Gordy, W. (1983). Jesse Wakefield Beams. *Biographical Memoirs* LIV (National Academy of Sciences of the United States of America): 3–49.

Hippel, Eric von (1988). *The sources of innovation*. Oxford: Oxford University Press.

Hughes, Thomas (1976). *Thomas Edison, professional inventor*. London: HMSO.

Journal of the Optical Society of America 1(1). (1917). Resistance to instrumentation.

Kevles, Daniel (1971). *The physicists: The history of a scientific community in modern America*. New York: A. Knopf.

König, Wolfgang (1980). Technical education and industrial performance in Germany: A triumph of heterogeneity, in R. Fox and A. Guanini (eds.), *Education, technology and industrial performance, 1850–1939* (pp. 65–87). Cambridge: Cambridge University Press, 1980.

Lee, E. (1954). The fundamentals. *Instrument Society of America Journal* 1 (February): 10.

Lucks, C. (1954). Quality of research. *Instrument Society of America Journal* 1 (March): 7.

Mindell, David (1996). *"Datum for its own annihilation": Feedback, control and computing, 1916–1945*. Ph.D. dissertation. Cambridge, MA: Massachusetts Institute of Technology.

Modern Scientists and Engineers (1980). Salisbury entry in 1: 621–3. New York: McGraw-Hill Book Company.

Noble, David (1977). *America by design: Science, technology and the rise of corporate capitalism*. New York: A. Knopf.

Pasachoff, Naomi (1996). *A.G. Bell*. New York: Oxford University Press.

Rimbach, Richard (1954). Position of the technical press in relation to industry. *Instrument Society of America Journal* 1 (January): 15–18.

Servos, John (1980). The industrial relations of science: Chemical engineering at MIT, 1900–1939. *ISIS* 71: 531–49.

Shimshoni, Daniel (1970). The mobile scientist on the American instrument industry. *Minerva* 8(1): 58–89.

Varian, Dorothy (1983). *The inventor and the pilot: Russel and Sigurd Varian*. Palo Alto, CA: Pacific Books.

Web, R. (1954). Features of industry and instrumentation systems. *Instrument Society of America Journal* 1 (January) 12–18.

CHAPTER 6

PATRICIA NEVERS[a], RAIMUND HASSE[b],
RAINER HOHLFELD[c], and WALTHER ZIMMERLI[d]

MEDIATING BETWEEN PLANT SCIENCE AND PLANT BREEDING: THE ROLE OF RESEARCH-TECHNOLOGY

> Strategically speaking it probably would
> have been better if I had been trained in a
> proper craft.
>
> *A plant gene technologist in the 90s*

The statement cited above was made by a scientist interviewed in the course of an empirical study aimed at identifying certain types of research in the area of plant genetics and plant breeding (Hasse et al.1994). The research activities of this scientist could not be readily equated with those typical of either basic research or applied research since they seemed to be neither theory-oriented nor product-oriented. Thus this scientist was neither primarily interested in expanding a theory designed to explain a phenomenon of plant biology nor was he concerned with developing new plant varieties to meet certain needs in agriculture and industry. Instead his work seemed to represent a distinct type of research that is basically aimed at improving and proliferating a new kind of research-technology, plant-gene technology. The statement is appropriate as an introductory statement since it points to a salient feature of research-technology in general and gene technology in particular, their emphasis on experimental craftsmanship. Moreover, it raises the critical question of the future of gene technology and of researchers trained exclusively in this field, a topic discussed at the end of the chapter.

In order to better understand the features of this type of research, we shall begin with a brief description of how research-technology emerged in the course of history in one particular institution involved in plant genetics and plant breeding. This will be followed by a summary of the results of an empirical study we carried out in 1995–1996 at three universities, three non-university research institutes and in the research divisions of three commercial enterprises involved in plant breeding in both the old and new states of Germany. Readers will hopefully indulge our reluctance to reveal the identity of these research establishments. All together fifty-four semi-

B. Joerges and T. Shinn (eds.), Instrumentation: Between Science, State and Industry, 97–118.

standardized interviews were conducted, each one and a half to three and a half hours long. The interviews were analyzed by three different people in parallel using some analytical categories that had been postulated before the interviews were conducted and others that were derived inductively from the interview material.[1]

The characteristics of a distinct type of research corresponding to what has been termed "research-technology" will then be described and compared with those of theory-oriented and product-oriented research in terms of three analytical dimensions – organization, practice and interpretive framework.

THE EMERGENCE OF RESEARCH-TECHNOLOGY IN PLANT GENETICS AND PLANT BREEDING IN GERMANY (1900–1990): A CASE STUDY

Prior to the rediscovery of Mendel's rules of heredity in 1900, botany and plant breeding were largely separate domains with diverging goals and interests. Many researchers in botany were interested in classifying plants and clarifying questions of species constancy and plant sexuality, while breeders focused on developing new varieties for practical purposes. However, when the significance of Mendel's results was recognized by C. Correns, E. Tschermak und H. de Vries at the turn of the century, various kinds of exchange between theory-oriented biological research and the practice of plant breeding began to emerge, accompanied by the transfer to breeding of theoretical knowledge and experimental techniques from biological research, in particular from the newly established discipline of genetics. Not long afterwards, in 1911, the Kaiser-Wilhelm-Gesellschaft (KWG) was founded in Germany for the express purpose of encouraging such exchange. Modelled after the *Physikalisch-Technische-Reichsanstalt*, the institutes of the KWG were established to promote a new type of research intermediate between basic research and applied industrial research (Hohn and Schimank 1990). One of these was the Kaiser-Wilhelm-Institute (KWI) for Breeding Research established in 1928.

The KWI for Breeding Research and its first director, Erwin Baur, can be regarded as exemplary for the aims of the newly founded KWG. Baur was actively involved in basic research on evolution and heredity, but he was also very interested in practical plant breeding. He required the same dual orientation from his students (Stubbe 1959: 4). The focal point of Baur's research was the snapdragon, *Antirrhinum majus*, and Bauer collected from all over the world specimens of different species of snapdragon as well as varieties of a given species with different forms of growth, flowers and leaves. He studied the heredity of their traits by performing crosses between them. In addition, for more practically oriented breeding projects he promoted expeditions for collecting the wild forms of many cultivated plants such as potato, wheat and legumes (Straub 1986: 12). He helped to establish both the genetically oriented *Zeitschrift für induktive Abstammungs- und Vererbungslehre* and the more practically oriented journal *Der Züchter*. He was a member of both the *Gesellschaft für Vererbungslehre* and the *Deutsche Landwirtschaftsgesellschaft*. The inter-

[1] For a full report see "Die Technologisierung der Biologie am Beispiel der Pflanzengenetik und Pflanzenzüchtungsforschung," a project conducted by Raimund Hasse, Rainer Hohlfeld, Patricia Nevers, and Walther Ch. Zimmerli from 1990-1993 and funded by the German Ministry of Science and Technology (Hasse et al. 1994).

dependence of genetic research and breeding practice was reinforced by the financial structure of the institute. Originally the institute was expected to cover its own research expenses at least in part through the sale of seeds and varieties it developed, but in the long run it was forced to attract funds from agricultural sources as well. However, while the structure of the institute was designed to encourage exchange between research and breeding, a type of technology comparable to modern research-technology had not yet evolved. The techniques used for developing new varieties of plants were still quite simple, consisting of manually performed crosses, and these had been developed by breeders outside of biological research.

After Baur's death in 1933, the institute was taken over by a new director, W. Rudorf, whose background and training were more closely connected to agriculture. He dealt with the theoretical problems of vernalization and photoperiodism and their significance for breeding. Another branch of his research was directed toward breeding frost- and drought-resistant varieties of wheat as well as virus-resistant varieties of beans and wheat. He also was interested in testing varieties under various different climatic conditions and therefore leased land for this purpose near Heidelberg and Klagenfurt. Thus the research he favored was similar to what we have termed product-oriented research, and this orientation continued until well after World War II.

In view of the atrocities that resulted from the misuse of genetic research during World War II, the successor organization to the *Kaiser-Wilhelm-Gesellschaft*, the *Max-Planck-Gesellschaft* (MPG), decided to abandon applied research in favor of basic research. The key to achieving this kind of strict separation was to be an independent system of financial support subsidized by the state. However, this was not realized until 1964. By this time various different kinds of sophisticated research-technology had emerged in biological science, and some of these were subsequently exported to breeding. These included techniques for inducing mutations and polyploidy, hybrid breeding techniques, and tissue culture techniques as well as techniques for the in vitro propagation of plants. With the help of the new finance program of the MPG, the development of such costly technology became more feasible. And J. Straub, who became the director of the *Max-Planck-Institut* (MPI) for Plant Breeding Research in Cologne (successor to the KWI for Breeding Research) in 1961, took advantage of this opportunity to encourage the development of "unconventional breeding techniques" in breeding research, in particular cell-biology and tissue-culture techniques. For example, during his directorship these techniques were used to generate haploid lines of normally tetraploid potato plants. Thus instead of promoting the separation of basic research from applied research the new financial program eventually served the original concept of the KWG by furthering the proliferation of new research-technology.

Another significant turn of events occurred in 1978 when J. Schell succeeded Straub as director of the MPI for Plant Breeding Research. Before his appointment the institute had still been actively involved in developing, registering and selling new varieties of crop plants and had maintained close contacts to smaller breeding companies in Germany. Scheibe (1987) lists thirty-nine different varieties of grain, clover, legumes, potatoes and grass that were bred by the *Max-Planck-Institut* between 1946 and 1978. Schell, however, had been trained as a bacterial geneticist and was interested in exploiting the natural gene transfer system of *Agrobacterium tumefaciens* for developing a general system of gene transfer in plants. His appointment as director was accompanied by a new phase of research policy at the MPI for Plant Breeding Research. The institute ceased developing varieties for practical purposes and began to concentrate its efforts on producing prototypic organisms such as genetically modified tobacco plants and research-technology such as gene-transfer techniques.

PLANT GENETICS AND PLANT BREEDING IN THE 1990s

According to certain theories of technological change the once separate domains of science and technology are currently beginning to approach each other and intermingle in certain areas (see, for instance, Weingart 1978, 1982; and, more recently, Zimmerli 1988, 1990). This rapprochement is supposedly mediated by the growing potential of science for producing technology and technologically useful results and by the increasingly scientific orientation of technology itself. A number of indicators of this development bear a striking resemblance to features attributed to research-technology. One indication of technological change is that the boundaries of basic research, applied research, development and production become fuzzy, suggesting a greater degree of fluidity between these areas. Secondly, a form of hybridization between cognitive and technical elements becomes increasingly evident. Science no longer generates only "pure" theory but technical elements as well. Zimmerli refers to these elements as "coagulated" forms of theoretical knowledge (Zimmerli 1990), and Hacking maintains that they encode "pre-established knowledge which is implicit in the outcome of an experiment" (Hacking 1992: 42). Scientific knowledge is increasingly gained through technical construction processes in the course of experimentation so that a distinction between scientific discovery and technological construction becomes more and more difficult (Hohlfeld 1988). The material results of such processes appear to be what Joerges and Shinn (this volume) refer to as core generic devices: open, prototypic devices that mediate the flow of knowledge and expertise between research and technology.

The intermingling of science and technology contributes to a hightened societal interest in science, manifested in the form of increased expectations regarding both economic utility and ethical reflection. Due to the latter, instances of confrontation with and moral pressure from the public occur more frequently, and a shift of public protest can be observed from the final production sites to the scientific laboratories where the technology originates. This, in turn, results in the need for new coping strategies, a subject not taken up in this chapter (for a discussion see Hasse and Gill 1994). Finally, technological change may lead to a shift in the criteria by which scientific results are deemed acceptable. *Utility* rather than truth may become the most important measure of acceptability (Bayertz and Nevers 1998: 127).

Genetic engineering is a field of biology in which these general trends in scientific and technological change can be readily observed. In contrast to various studies indicating that science and technology are converging in this area and forming a unified, homogeneous type of research (see, for example, Hack and Hack 1985; Krimsky, Ennis, and Weissman 1991; Yoxen 1982), we postulate in the following that changes in the relationship between science and technology can result in a redistribution of previously established types of research in institutions.

At the beginning of the 1990s, when we began our investigation, we felt that the field of plant genetics and plant-breeding research might be in a state of transition due to the recent introduction of gene technology. While genetic engineering of less complex microorganisms began in the 1970s and had already become a standard instrument of microbiology-based industry, plant-gene technology was initiated more than ten years later when the first gene-transfer systems for plants were developed. At the time the project reviewed here was conceived, the genetic engineering of animals was still in a preliminary state. How did this transition affect research styles and, to introduce a more inclusive term, "types of research" at the interface between genetics and breeding?

THREE TYPES OF RESEARCH

We use the general term "type of research" to refer to a conceptual composite encompassing three different dimensions: the prevailing organizational structures and processes, the practice of research and the interpretational framework in which it is embedded. These dimensions overlap more or less with those of various other such terms currently circulating in the history, philosophy and sociology of science, in particular concepts such as "style of thought," "thought collective," or "epistemic style." The term "style of thought," introduced by Mannheim (1986) and interpreted by Nelson (1992), is restricted for the most part to cognitive dimensions shared by a particular group and thus overlaps only partially with the term "type of research" we have selected (see Mannheim 1986; Nelson 1992).[2] Our terminology has more in common with that of L. Fleck, who developed the concepts of a "style of thought" and a "thought collective" in describing how scientific facts are generated (Fleck 1980). According to Harwood, Fleck uses the term "style of thought" to refer to a "tradition of shared assumptions" which are largely invisible to and unquestioned by those who bear them (Harwood 1986: 177). A "thought collective," as defined by Fleck, refers to a community of people who regularly exchange thoughts with one another. Thoughts compatible with the basic assumptions of the collective are readily integrated, while incompatible ideas may be rejected.[3] The term "epistemic style" was employed by Maienschein in her comparison of German and American embryologists around the turn of the century (Maienschein 1991). According to Maienschein, the epistemic style of a group of researchers embraces the particular goals

---

[2] According to Nelson, Mannheim's definition of a style of thought includes both the theoretically refined style of a group as manifested in its texts and the "diffuse cognitive and emotive worldview" underlying this style. While we have considered the explicit and implicit theories which guide researchers in their work, we have not attempted to assay as broad a dimension as "worldviews." However, some of the traits of a style of thought as outlined by Nelson are very similar to the analytical categories we have used to characterize the interpretational framework of a type of research.

[3] Membership in a thought collective is established through education, and experience and craftsmanship play an important part in the transmission of a thought style (Harwood 1986: 177). Thus Fleck's terminology has a definite sociological skew, in that it includes the bearers of a style of thought and their scientific activities in addition to cognitive components.

deemed worthy of pursuit, the processes of investigation and the standards of evidence considered acceptable. She suggests that this term overlaps with Laudan's definition of traditions (Laudan 1981: 151), although she maintains that Laudan is more interested in theories and theory change while her emphasis is on the actual practice of performing science.

Based on his experience in modern cell biology Grinnell has expanded the concept of a thought collective (see Grinell 1992). He maintains, in keeping with Mannheim and Fleck, that a thought collective has common goals and beliefs but adds that members of a thought collective share knowledge and assumptions about methodological approaches, observations, accepted hypotheses and important problems requiring investigation. This concept of a thought collective is similar to that inherent in Darden and Maull's concept of a field of research, Burian's concept of a discipline and Hacking's concept of closed systems of research practice (see Burian 1992; Darden and Maull 1977; Hacking 1992). In addition, however, Grinnell describes the sociological structure of a thought collective as having an older, established core and a younger, more mobile group of marginal members. Another very interesting perspective of Grinnell's interpretation is the idea that the boundaries of a thought collective usually remain invisible until a conflict arises (Grinnell 1992: 48). These are aspects that we, too, have considered in identifying types of research.

In the past, the most important category used to distinguish different types of research has been the degree to which research is oriented towards the solution of either theoretical or socio-technical problems. If this category is applied in the present investigation, two extremely variant types can be identified that correspond to what has been traditionally referred to as basic research and applied research. At one end of the spectrum one finds theory-oriented research, concerned, for example, with elucidating the genetic basis of plant development. The other end of the spectrum is occupied by product-oriented research, which in our case deals with breeding new plant varieties to meet certain needs in agriculture and industry.

However, other data in our investigation did not seem to fit this simple pattern. For example, many of those we interviewed claimed to have an intermediate or dual orientation with interests in advancing theory as well as in improving plant varieties. They described their research as "*anwendungsorientierte Grundlagenforschung*," which literally translates as "application-oriented basic research." When we examined exactly what such researchers do that would permit them to claim a dual orientation, we found that they basically develop and improve experimental methods that could be used either in theory-oriented or in product-oriented research, even though they are not directly involved in either of these. In our investigation these were mostly researchers working in cell biology or gene technology. This led us to consider the researchers attitudes towards methods as a second analytical category, that is, the extent to which methods are regarded simply as means for attaining certain goals or as legitimate ends in themselves. By including this category in our analysis, we were able to distinguish a third, distinctly different type of research which we preliminarily termed "methods-oriented research" and which exhibits many of the general features associated with "research-technology" in this volume. In determining types of research, we assessed both research practice and the interpretational framework in which it is embedded. However, by examining a third dimension, the organizational structures and processes characteristic of different types of research, we have included a perspective not covered in depth by any of the investigations referred to above.[4] In addition we also attempted to evaluate social factors that influence research such as public opinion, legal structures and political processes.

RESEARCH-TECHNOLOGY IN THE FORM OF PLANT-GENE TECHNOLOGY AND PLANT-CELL BIOLOGY

The main goal of research-technology at the interface between plant genetics and plant breeding is the perfection and proliferation of experimental techniques that can be used by others in either theory or product-oriented research. This was summarized by the director of an institute that focuses on research-technology of this kind[5]:

Participant:

> Our institute can never develop products but at the most technology. That is the maximum that it can achieve. It can attempt to protect this technology and thus make itself attractive for other, external partners . . . .

Interviewer:

> By technology you mean certain procedures?

Participant:

> Yes, exactly. Certain procedures. It might also mean certain plants, but for the most part it will be procedures for the generation of certain plants. Therefore, the most we can do is to produce prototypes, but by no means the final products that are truly of value for commercialization. That's not our job.

The technology referred to in this passage corresponds to what Joerges and Shinn (this volume) call "core generic devices." The term "devices" suggests some form of machinery. However, in modern molecular biology such devices may include not only machines but also procedures such as gene-transfer methods or methods of marker-dependent selection, research commodities such as DNA probes, clones or sequence data, or even organisms such as potato plants with a modified metabolic pathway for producing amylopectin instead of starch. Rheinberger refers to these materials as "epistemic things" (see Rheinberger 1992: 67–86).

Organizational Structures and Processes – Managing Heterogeneous Networks

Research-technology exhibits a unique profile with respect to organizational structures and processes. One very significant aspect of this profile is determined by the fact that sophisticated new equipment (centrifuges, electrophoresis apparatus, sequencing machines, etc.) and novel research commodities (clones, primers, probes, etc.) are constantly being designed and modified, and in turn are also constantly in demand. The equipment and materials that are designed or modified in the context of

---

[4] For further details on the importance of organizational variables in mediating between social contexts and knowledge production see Hasse (1996: 9–16).

[5] Most of the interviews were conducted in German. Translations were made by P. Nevers.

research-technology are mass produced in specialized branches of industry and then offered to other areas of science and technology. Because of this interdependence, research-technologists operate in a broad and dense network of exchange. Partnership (and competition) is not restricted to actors of a particular scientific community but extended to members of other areas of science and technology as well.

Compared to other areas of biological or plant-breeding research of more modest dimensions, the prevailing organizational structure of research-technology is one of "big science." Research-technology is mainly employed and developed in large departments with numerous teams and many researchers since a critical mass of specialists is required to generate research-technology. Research-technology departments may develop high degrees of specialization associated with particular methods or techniques, and sometimes such specialization even occurs within individual research teams. Thus one group may be involved with tissue-culture techniques while another works on cloning or sequencing. An important by-product of this organizational structure is that competitive effects are weakened as long as the cooperation within a department or research team is based on specializations with clear cut domains of competence. On the other hand, however, specialization tends to reduce incentives to collaborate. Thus successful cooperation requires a balance between common tasks and competition on the one hand and specialization on the other.

In general, growth and proliferation are prominent organizational characteristics of the kind of research-technology examined in our case study. Both factors make the acquisition of funds and other resources crucial to this type of research, but since it is an area in which the interests of science and technology coincide, it is usually generously funded. However, horizontal differentiation and corresponding growth also increase the span of control required in research-technology. Consequently the heads of research-technology departments may find that too much of their energy is absorbed by administrative duties. They usually deal with this problem by delegating numerous managerial duties to the leaders of their research teams. Therefore in research-technology more often than in other areas of research the elaboration of research proposals, the supervision of subordinate scientists, educational tasks and other duties are placed on the shoulders of team leaders. This in turn leads to the emergence of an intermediate management level uncommon in the two-tiered organizational structure of more traditional research units.

This structural change has far-reaching consequences. On the one hand, establishing an intermediate managerial "caste" allows the heads of research-technology departments to expand their task profile by engaging more actively in "foreign affairs." They may influence the course of research in other fields of science by setting methodological standards as referees for journals and as advisors for funding agencies. They may become actively involved in shaping scientific policy or formulating legislation relevant to their research such as federal laws concerning gene technology. Or they may function as technical experts and consultants to industry. On the other hand, by expanding their tasks in this manner scientific managers run the risk of losing contact with the everyday life of laboratory research. Thus it may become increasingly difficult for them to coordinate in great detail the research activities of their departments.

The networking characteristics of research-technologists result from the fact that experimental products and procedures developed in research-technology are transferred to other areas of science and technology. Research-technologists have to be actively involved in both of these social contexts. Compared to more theoretically oriented research units, the networks of research-technologists are quite heterogeneous, including contacts with politicians, managers and administrators. Since the generic devices developed in research-technology are designed for end-of-the-pipeline use in a larger socio-technical context – in our case in agriculture – one might expect to find contacts with representatives of this larger context, more specifically with farmers and agricultural organizations. However, although scientists in research-technology may claim to be concerned with socio-technical problems such as those of agriculture, this is not reflected in their network structures. Compared to scientists in research more directly connected with agriculture, who are embedded in a broader technological trajectory, the networks of research-technologists tend to be dominated by strategic ties based on immediate reciprocity. This probably is due to their increased need to acquire resources and to proliferate and justify the results of their work. In contrast, theoretically oriented research and research more closely attuned to agriculture are characterized by strong commitments by the respective participants to a common goal so that strategic considerations are less important.

In addition to contacts with potential consumers of research-technology in other areas of science and technology, informational exchange and cooperation with scientists in other specialized research units is central to research-technology. Interactions of this kind often occur between subordinate researchers. They foster mutual recognition as research-technologists and representatives of a rational avant-garde that promotes breakthroughs in science and technology. However, we suggest that this loose association of research-technologists is not comparable to that of a community since important institutional qualities such as a formal structure and group characteristics are only weakly developed.

Research Practice – Developing Generic Devices

In general, laboratories centered around research-technology resemble workshops rather than "think tanks." Work is often distributed on the basis of different experimental techniques, e.g. one group grows cell cultures, another clones the DNA and a third sequences it. Scientists are hired because of their expertise in certain techniques, and novices can gain merit points by introducing new techniques from elsewhere. This is reflected in the explanation one researcher gave us for how he got his job:

> Well, in principle it happened this way. Over there I learned a few techniques, how to work with protoplasts, maize protoplasts, and then the technique of direct gene transfer. And, you see, X was very interested in that, and at that time I guess there was no one here who could work with cereal protoplasts . . . .

Since research-technology is centered around experimental skills and craftsmanship, the success of this type of research depends not only on the presence of "master craftsmen" capable of conveying their experience to others but also on the availability of opportunities for practical learning experiences. Trial and error learning is apparently more important in some areas of research-technology than in others, in particular in the area of cell and tissue culture. A researcher in this field summarized his experiences in the following manner:

> Yeah, I mean, it took me a long time and a lot of frustration to learn that, because for two years it didn't work, and then somehow after a while, somehow I figured it out, and that's the thing now. I can't simply teach other people how it works. I can show them and tell them exactly how they should do things, but there are some things . . . – even if it's just the color of the culture or the cycle, or how you inoculate – you can't describe that with words, explain things verbally. They simply have to make their own experiences, and that takes a half a year or a whole year until they can manage things on their own.

Tacit knowledge based on experience and tradition, such as that required for tissue culture represents a valuable form of scientific capital that can give a scientist an edge on others. For example, a researcher in industry who works with cell cultures remarked that he didn't have to worry about competition from scientists in other companies since his practical knowledge was unique and could not be readily conveyed to others. However, he did have to be careful about whom he trained – only people who were certain to remain in his company.[6]

Another important factor for the success of the laboratory/workshop is obviously the availability of good material resources. The material research capital of a laboratory includes not only apparatus and instruments but also other commodities such as clones, probes and sequence data. As one researcher in methods oriented research surmised:

> We work here experimentally, and that's the same as if you want to repair a car or some such thing. The most important thing is the equipment.

Scientists in research-technology select the laboratories in which they want to work not so much because of the intellectual prowess of the director but rather on the basis of a matter of fact evaluation of his material research capital:

> Why I came to X is because Y had a relatively large collection of genes for . . . , and I saw that as a good possibility for further work in this area.

In research-technology, few people are involved with the highly creative act of actually *inventing* new technology from scratch. The majority of scientists in this type of research works on modifying and testing the applicability of certain aspects of a particular kind of research-technology. Since the outcome of such research is usually visible and predictable, this kind of work is characterized by low task uncertainty (Whitley 1982, 1984). However, significant differences can be observed, for example, between cell biology and gene technology. In research involving gene technology, technical task uncertainty is low since the work process consists of many steps

---

[6] The significance of the element of secrecy surrounding artisanal knowledge has also been demonstrated for Fraunhofer's research on optical lenses, see Jackson, this volume.

with a well-defined time allotment and predetermined sequence. One researcher described work in gene technology in the following manner:[7]

> . . . yeah, it's more like what we'd call cookbook work. There's a certain recipe. If you do this and this, then you get these results. And it's quite reproducible . . .

In research involving tissue culture, on the other hand, more task uncertainty is evident. This is expressed in the following statement:

> In cell biology you always have to deal with plants that either grow or don't grow, and if you repeat an experiment three times, you get three different sets of results. . . . The molecular biologist is used to seeing results after two days. An experiment is one step and never lasts longer than one or two days, and then you see whether you've got something, and then you go on to the next step. But when we (cell biologists, ed.) perform a certain step, it sometimes takes six months before you see whether you performed the right step.

In both cell biology and genetic engineering, research seems to be wrought together by strong functional dependence. People depend upon one another for research materials such as cell lines, sequence data or recipes, although the exchange of such materials is becoming more and more restricted due to patenting. And in some cases dependence is caused by the need to use certain pieces of equipment that are very costly and thus available in only a limited number (e.g. automatic sequencing machines or scanning electron microscopes). This kind of dependence differs in two respects from Whitley's notion of functional dependence (Whitley 1984). Firstly, it is not restricted to the exchange of knowledge, because, as noted above, it may involve access to material products and research equipment as well. Secondly, dependence is not restricted to other researchers in the same field since it may be extended to products and processes from other research fields (e.g. information technology).

In research-technology there seems to be strong agreement on what problems are important and how they should be investigated. The standards are set by a well defined group of journals with a tradition of restrictive acceptance, in which results from this type of research are published. In the present case these include journals of molecular biology such as "Cell" or "Proceedings of the National Academy of Science (USA)." Publication in these journals is important for securing and maintaining a reputation in technology-oriented research. However, if research-technology experts become engaged as specialists in theory or product-oriented research, they may occasionally be co-authors of papers submitted to other research communities, for example, to journals of plant physiology or applied plant-breeding journals.

In addition to publications in journals, other forms of communication can be observed in plant-gene technology. Patents represent one such option. However, some scientists are skeptical about their value as a form of communication. Thus one researcher surmised:

> I would guess that patent registrations are for the most part less reliable than scientific publications are supposed to be and usually are. You see, malicious tongues maintain that a patent registration is something in between truth and fantasy.

---

[7] The terms "molecular biology" and "gene technology" were used synonymously by most of our interviewees. We interpret this as an indication of the "genetic fix" that prevails in current molecular biology.

Another obstacle is the necessity to maintain secrecy until a patent has been registered. This restricts reporting on ongoing research.

An additional change in communication processes observed in research-technology involves a shift from written to oral communication, whereby informal communication is considered more important than lectures at meetings, at which often only published results are presented. Thus the telephone and e-mail seem to be the most important means of keeping up with the ongoing research of fellow research-technologists.

The low task uncertainty and high mutual dependence of current molecular biology with its focus on gene technology is a prerequisite for employing the large, semi-skilled student labor force that forms its foundation. At the same time, packaging research in a series of certain tasks also has a problematic feed-back effect on scientific tradition. Students learn to *expect* certain results and shy away from topics with a less certain outcome. They *expect* to be able to publish one or two papers during the time of their Ph.D. work. As one senior researcher commented:

> They want, actually they really want – and I've already had quite a bit of trouble in this respect that has resulted in dissatisfaction – they want a project that is such that there will be certain results that they can then publish in a good journal with no difficulty. But when you begin a new project, you can't always tell what will come out of it. So sometimes it can be really tricky to start a new project. Sometimes it seems to be easiest – it depends – to work with undergraduates. Because they don't have such definite expectations. They have to demonstrate that they know how to work in a laboratory, but they are more willing to risk working on uncertain topics.

Contrary to research oriented toward needs in agriculture (product-oriented research, see below), in which research problems are defined by a process of negotiation with others outside of science, the technological problems under investigation in research-technology are for the most part generated by the internal dynamics of the field. Scientists in this type of research are accustomed to offering their technology to others, but they rarely ask beforehand what might be needed. As formulated by Joerges and Shinn in this volume, research-technology offers technological answers to questions that have not yet been raised. A classic example of this situation is the current development of technology for inducing resistance to herbicides in plants.

In research-technology situated between plant genetics and plant breeding the objects under investigation consist primarily of plant cells and subcellular components of plant cells, in particular DNA, and modified plants derived from cells or manipulation of DNA. In principle, this type of research could be directed towards other objects such as populations of organisms or interactions between plants and their environment. However, a different kind of technology would be required for research with such objects.

The objects of research-technology serve to facilitate the flow of knowledge and expertise between science and technology. In this function as mediators between different worlds with different viewpoints and agendas they resemble what Star and Griesemer refer to as boundary objects (see Fujimura 1992, Star and Griesemer 1989; see also Gaudillière, this volume). An example derived from our investigation that illustrates the concept of a boundary organism is that of the weed *Arabidopsis*. Once exclusively a model organism of theoretically oriented biology and currently a popular object of developmental genetics, it is now rising to new fame in gene technology-research because of its small genome. It is relatively easy to identify and clone genes from this plant and transfer them to others, and researchers are currently in the process of sequencing its entire genome. Since all flowering plants, including *Arabidopsis*, are closely related, researchers assume that their genes are also similar so that an *Arabidopsis* gene can be substituted for a missing or defective gene in any other flowering plant (including crop plants). As a result, genes from *Arabidopsis* can be used in theoretical studies and in breeding projects as well. Gene technology-research has thus converted *Arabidopsis* to a decontextualized source of genes for both scientific and technical purposes.

Interpretational Framework – Competing Repertoires

Among scientists working in plant genetics and plant-breeding research two basic interpretational frameworks can be observed that resemble those of the two feuding groups that once existed in the debate about the driving force of evolution described by E. Mayr (1982). One group, comprising experimental geneticists and termed "essentialist," was convinced that discontinuous variation caused by sudden mutations in the genetic material of an organism is the most important factor in evolution. Their opponents, supporters of Darwin's theory of selection, regarded continuous variation in response to changes in the environment as the decisive factor. In plant-breeding research a similar constellation can be observed. Many gene technologists are or were convinced that breeding success depends primarily upon generating new plant variants with the help of gene manipulation, while researchers in conventional plant breeding tend or tended to emphasize the importance of field trials and selection. This stems from the very different experiences that have molded the two groups. While researchers in conventional plant breeding spend a great deal of time with intact plants in the field and have in the past had little experience with molecular biology, gene technologists are absorbed with manipulating DNA in laboratory experiments and often lack a perspective of the whole plant. As one gene technologist put it:

> Because I work at the level of DNA, what I do with plants I could, in principle, also do with humans or whatever, even elephants.

At one point, some gene technologists even seemed to believe that conventional plant-breeding techniques would eventually become superfluous, as the following statement by an influential gene technologist indicates: "Once a useful gene has been isolated, it can in principle be transferred to many different crops without altering the other properties of the recipient plants and without a lengthy breeding program."[8] Of course, this attitude heightened the mutual animosity of both groups and served to sharpen the contours of the respective thought collectives.[9]

---

[8] Passage derived from a text from 1988; the source is not cited to maintain anonymity.

[9] Contrary to the view expressed here, most of our interview partners, both gene technologists and researchers in conventional plant breeding, were of the opinion that gene technology will neither reduce the time required for plant breeding nor replace conventional techniques.

That these two groups represent two different thought collectives is also indicated by differences in their concept of the term "gene." This is summarized in the following statement of a researcher in conventional plant breeding concerned with measuring and statistically evaluating the multiple properties of plants in field trials:

> Somebody who comes from biology (as opposed to agriculture, ed.) can't even imagine that a gene is a statistical concept. And I couldn't at all imagine that you can wrap a gene around your finger.

The head of one plant-breeding company was aware of these discrepancies when he first set up a section for molecular biological research in his firm and wanted to make sure that the two groups rectified their differences and communicated with each other. Thus he wisely decided to intersperse the gene-technology groups among those of the researchers working with conventional techniques and even arranged for them to have their coffee breaks and lunch breaks at the same time.

If research-technology is to become reputable and influential – and proliferation is a major goal of this type of research – it must be accepted by, and incorporated into, other kinds of research. Thus finding a common basis for communication is essential. The following example from our study demonstrates one strategy employed by some plant-gene technologists in dealing with conventional plant breeders. Instead of promoting *gene transfer* as a means of generating new plant variants, a process quite foreign to conventional plant-breeding practice, this group of technologists chose to emphasize the use of gene technology for *diagnostic* purposes in a process known as *marker-dependent selection*. By shifting emphasis from prototype production to selection they were able to mollify the suspicions of conventional plant breeders by drawing on their previous experience with classical genetics and selection:

> The most important point in RFLP (method of marker-dependent selection, ed.) today . . . is to be able to integrate your field breeder with this technology. – It's a yes or no answer, and it has the beauty to, as they say, bring back all the mysteries of basic genetics to Mendelian units. And you understand, you'll start talking about the art of chromosomes, and linkages and breaking linkages. It is . . . very amenable to the way of thinking of the breeder, you know. – Of course, for them, for the school of plant breeding, this is a kind of resurrection.

On the other hand, the style of thought of a particular field of research-technology may become so influential, that the "language" of a different thought collective simply goes out of fashion. Thus one geneticist interested in theoretical problems maintains that methods of proof based on classical genetics once generally accepted in the field are no longer convincing whereas methods based on gene technology are "in."

> You see, for most people – although there are exceptions – a band in a gel is more convincing than the results of a cross . . . . A band in a gel is somehow easier to grasp, and what it means is unimportant – people believe the band. Numbers in a table are too abstract, I think, more difficult to grasp. Actually, I see things the other way around. A band in a gel can result from hundreds of different artifacts.

BETWEEN THEORY-ORIENTED AND PRODUCT-ORIENTED PLANT BIOLOGY

Our results indicate that research-technology in the form of plant-gene technology and plant-cell biology is a novel type of research that has emerged in a space between basic research and applied research in the conventional meaning of these terms. In order to understand how research-technology differs from these other types of research, we will briefly point out some their characteristics.

Theory-Oriented Research

Theory-oriented biological research, i.e. research concerned with finding explanations for biological phenomena, is found primarily at universities and some non-university research institutes. Research on the developmental genetics of *Arabidopsis* is an example of this type of research from our case studies in plant genetics and plant breeding. Its organizational structure is best characterized as decentralized and flexible. The head of the department or division is still actively involved in research practice and has an expressed interest in maintaining this involvement. Thus he tends to try to keep his group small and to delegate managerial duties that would detract from direct involvement in research practice. An important function of the head of such a department is the intellectual integration and coordination of the activities of his group, and his intellectual superiority is an important source of authority in the group.

Topics and experimental approaches are selected that can be successfully dealt with by a small group. The resources required for this kind of research are therefore comparatively modest. Since both resources and manpower in theory-oriented research are limited, it cannot successfully compete with more expansive types of research such as research-technology. Therefore the "strategy of survival" for this type of research consists of avoiding competition by finding and forming an exclusive "research niche."

Because of the specific orientation and size of theory-oriented research, the number and intensity of contacts that scientists in this type of research maintain are also of limited dimensions. Contacts are cultivated that are conducive to exchanging ideas, establishing a reputation in the scientific community and securing sufficient research funds. But other developments relevant to research such as special programs for technology transfer, changing interests in business and industry or new legislation on gene technology are followed with as little active participation as possible.

Since theory-oriented research is concerned with finding explanations for biological phenomena exhibited by organisms, the problems under investigation will at least theoretically embrace objects corresponding to a wide range of organizational levels, from the subcellular level to that of ecosystems. However, the actual objects of research practice may be limited to only one or a few organizational levels. This may be problematic if a level is selected that is the current focus of a flourishing field of research-technology. If, for example, research activities in developmental genetics are restricted to the examining of genes or cells, then this particular kind of theory-oriented research runs the risk of being transformed into technology-oriented research. Scientists originally interested in theoretical problems may become absorbed by details of experimental technology and be forced to succumb involuntarily to the pull of research-technology. However, sometimes research-technology is courted deliberately in hopes of being able to profit from the ample funding capacities associated with it. The idea is to secure funds for a technology-oriented project and redirect some of them to more theoretically oriented projects, a practice commonly known as "bootlegging" (Greenberg 1966). For example, a researcher interested in the developmental genetics of *Arabidopsis* may agree to sequence any genes he isolates, regardless of whether or not this serves his particular theoretical interests. This permits him to tap the attractive funding capacities of a large, technologically oriented EU-sequencing project aimed at determining the entire sequence of the plant's genome.

Methods in theory-oriented research are generally selected as means for solving theoretical problems. Since researchers in this type of research are more interested in theory than in details of experimental technology, they tend to select methods that will not dominate the research process. Alternatively, specialists for certain experimental techniques may be hired to deal with phases of the investigation that require a stronger methodological orientation. If several different organizational levels of an organism are included in theory-oriented investigations, a wide variety of methodological options is available, which can be combined in various different ways. The results of such research are not always predictable and may exhibit a significant amount of variation.

Since the problems of theory-oriented research are generally of a broader scope, and since the objects usually are more complex than in research-technology focused on gene technology, a greater amount of uncertainty prevails regarding the best way to solve a particular problem. Because of the broader scope of their research problems, scientists in theory-oriented research may be active in more than one scientific community or discipline. People working on the developmental genetics of *Arabidopsis*, for example, may publish in traditional journals of developmental biology, in journals of genetics or in journals of molecular biology. Thus they are not dependent upon a single scientific community for reputational gains, so mutual dependence may be lower than in research-technology.

Even though this type of research is oriented towards solving theoretical rather than socio-technical problems, results or data may be produced as spin-off products that are of socio-technical value. The probability of this kind of coincidence increases when methods are employed that are the focus of an active field of research-technology, and when research is directed towards appropriate boundary objects. Thus when research on the developmental genetics of *Arabidopsis* employs gene-technology techniques, there is an increased likelihood that genes useful for technical purposes in plant breeding will turn up by chance.

Product-Oriented Research

The ultimate goal of product-oriented research is the fabrication of a marketable product, and the problems investigated are an integral part of a long-range trajectory aimed at producing such a product. In the area of plant genetics and plant breeding the goal of product-oriented research is a new plant variety for use in agriculture, and the traditional sites of this research are either the plant-breeding research departments of agricultural faculties at universities or research teams in private breeding companies.

The trajectory of breeding a new variety begins with either the discovery of a natural plant variant or the fabrication of a plant prototype with some interesting new property, e.g. a nematode-resistant line of sugar beets or a line of rape lacking erucic acid. If constructing the prototype requires the use of gene technology, it may take place in a research-technology laboratory. If less "fashionable" research-technology is required, the task may be assumed by researchers involved in product-oriented research. The next step is to establish the new property in a line with other important agronomic traits such as high yield or pathogen resistance. Then years of field trials and backcrossing are needed to generate a marketable variety. These tasks may also be included in product-oriented research. Other parts of the trajectory such as production and marketing are usually conducted in non-scientific institutions.

Problems in product-oriented research are selected by a process of negotiation between interested parties outside of science who express certain needs and scientists capable of offering certain goods and services that may be useful for satisfying these needs. In our study, partners in such cooperative efforts may include farmers, seed growers and breeders with insufficient research facilities of their own or representatives of industries that process agricultural products. In one case, for example, the head of the department of plant breeding in the agricultural faculty of a university negotiated with seed growers and farmers interested in having rape bred that lacks erucic acid so that it could be used as feed for animals. He also dealt with representatives from a soap industry interested in securing the production of rape *with* erucic acid since it is a component of certain oils used in soap production. In addition he also provided technical training for plant breeders in methods for identifying erucic acid in extracts from rape.

In the cases we have investigated, the internal organization of product-oriented research, like that of theory-oriented research, was generally found to be decentralized. Different teams work on different problems of practical relevance such as breeding male sterility in rape or elucidating the genetics and biochemistry of plants with unusual kinds of oil. Each group may even work on a different crop plant (e.g. maize, rape, sugar beet). Since the work is highly diversified and organized in separate projects that are not directly related to one another, there is little internal pressure due to competition.

The role of the head of a department in product-oriented research is defined by the main goal of this type of research, which is to generate results that can be fed directly into a production process. Thus he cultivates many contacts with cooperation partners outside of science who represent potential consumers of the research results from his department. In the case of a department at a university or some other non-commercial research institution contacts are maintained with small plant-breeding companies that have no research departments of their own or with partners in industry. A research department within a large company, on the other hand, will cultivate contacts to the other parts of the concern involved in development, production and marketing. In addition, the head of a department in product-oriented research often functions as a kind of scientific foreign minister by formulating and promoting policies that are favorable to the research in his department. In the case of plant-breeding research he may participate in agricultural policy making in order to guarantee the use of plant varieties developed in his research department. He may be involved in organizing funding programs for special breeding projects to be conducted in his department. Or he may participate in planning programs for the release of genetically modified plants. An example encompassing a number of such activities is the breeding of genetically engineered rape with a modified composition of fatty acids as part of a federal program designed to promote the development of renewable resources.

As in almost all areas of science, product-oriented research is usually parceled out to individual projects that last one to three years. However, since these projects are embedded in a long-term production process with a time projection that may be much greater than 1-3 years, an atmosphere of security and relative composure prevails that is less common in other types of research.

The objects of product-oriented research are specified by its socio-technical orientation. In the present case, therefore, crop plants are the objects of investigation. Since the long-range goal is a new plant variety for use in agriculture, work with whole plants and populations of plants in the field and in greenhouses as well as in the laboratory is an integral part of this research, although lower levels of organismic organization usually are also included. The final choice of object is reached by negotiation with others involved in the breeding process.

A particularly important feature of the kind of product-oriented research we investigated is its organismic orientation. Scientists in this kind of research usually work with whole plants during some stage of their research, and field trials are often a routine part of their investigations. Thus their work involves objects of a different organizational level than those of a plant-gene technologist and the experiences they gain in dealing with these objects differ as well.

Because different levels of organismic complexity are involved, researchers in product-oriented research, like those in theory-oriented research, could in principle select from a wide variety of methodological options. However, since topics and methods must be negotiated with partners outside of science, and since economic considerations play an important part in this type of research, a more conservative choice is usually made that excludes high risks. Thus methods that are the focus of an expanding field of research-technology are generally employed very sparingly. The need to negotiate topics and methods results in greater mutual dependence and the more conservative choice of methods results in lower task uncertainty.

The results of product-oriented research are designed to be channeled into a long-range production process. In the case of plant-breeding research they may include cell lines, plants or even registered varieties. They may also consist of statistical data gained in field trials. With these results researchers in this type of research are not only interested in making a contribution to science. They also hope to make a valuable contribution to a production process. Consequently they are integrated in a research community that is distinct from those of theory-oriented research or research-technology and publish in different journals. Moreover, they are not as exclusively and strongly integrated in their community as scientists in other types of research, since they have other sources of reputational gains outside of science. This leads to lower mutual dependence on the research community, which is counterbalanced by greater mutual dependence on non-scientific arenas.

THE ROLE OF RESEARCH-TECHNOLOGY IN BIOLOGY AND ITS FUTURE

The focus of modern molecular biology on the development and proliferation of methods has been pointed out by various other authors. For example, Hack and Hack noted that basic research in molecular biology "often consists of a basic alteration of methodological and technical-instrumental knowledge" (Hack and Hack 1985). In her history of molecular biology at Caltech, Kay (1993) describes how the designers of the research programs of the Rockefeller Foundation, with which molecular biology became established throughout the world, drew upon their experiences in chemistry and physics in organizing projects on the basis of certain instruments and technical procedures. Busch et al. (1991: 179) remarked on the central meaning of methods and instruments in plant molecular biology and on their significance for future developments in agriculture and breeding. Rheinberger has clearly documented the importance of "epistemic things" for one area of molecular biology, the field of protein synthesis (Rheinberger 1997). And Magner maintains that "all too often, the history of molecular biology is treated as a grand epic of charismatic and eccentric personalities and brilliant guesses and ideas, while the importance of new physical instruments, techniques, and experimental craftsmanship is neglected" (Magner 1994: 440). Thus our observations seem to be well supported by those of others and are in keeping with a general trend in the history, philosophy and sociology of science towards directing more attention to the significance of research methods and technology in addition to theory (Bayertz and Nevers 1998).

The type of research we have termed research-technology seems to be comparable to what Darden (1991) has referred to as a field in which techniques rather than ideas are of central importance, or what Burian has called a "techniques-based discipline" as opposed to "problem-based disciplines" (Burian 1992) It also conforms well with Bechtel's description of the development of cell biology between 1940 and 1965 as that of a techniques-based discipline (Bechtel 1993). Bechtel suggests that biochemistry and molecular biology may be examples of other such disciplines.

The objects of all three areas – biochemistry, cell biology and molecular biology – are cellular or sub-cellular components, and the genesis of all three areas of research involved departing from traditional objects of biological research at a higher organizational level such as entire organisms or systems of organisms. Darden regards changes in the organizational level of research objects as an effective strategy for generating theories in biology; she notes, however, that such changes may require the development of new techniques (Darden 1991: 221).

In some cases techniques-based disciplines seem to function as mediators between other, more traditional disciplines. For example, Bechtel maintains that with the help of electron microscopy and histology modern cell biology was able to link together the fields of cytology and biochemistry (Bechtel 1993). And through gene technology modern molecular biology seems to have reunited the fields of genetics and embryology. Burian has listed a number of papers dealing with how classical embryology has been rejuvenated by modern methods of molecular biology (Burian 1993). Our own investigation of research on the developmental genetics of *Arabidopsis* provided evidence for a similar effect of molecular biology in establishing connections between classical genetics and plant developmental biology. As Bechtel has noted, these observations lend support to the model of theory change in science developed by Darden and Maull (see Bechtel 1993; Darden and Maull 1977), in which neither the reduction of theories nor the displacement of one theory by another is considered the main driving force. Instead, theoretical development is seen as a dynamic process in which the knowledge of different fields of research becomes interwoven. Thus processes by which separate fields of research become interconnected are considered important forces in theory formation. In this sense research-technology could be regarded as a linking element in the epistemological construction of theory.[10]

As techniques are perfected, they also become increasingly routine and eventually can be assumed by unskilled technicians or even machines. Since novelty and originality are important reputational criteria in the scientific community (Whitley 1984: 184), research focused on a particular kind of experimental technology eventually becomes obsolete. Although it may be highly novel and highly expansive in the beginning, as the technology it produces becomes more and more standard, its novelty and consequently also the momentum of the field diminish. In modern science areas of research-technology such as plant-gene technology may rise and fall within a period of 30-50 years. The current status of DNA sequencing suggests a development of this kind. As one scientist commented:

> I think in the past – perhaps about five years ago – if you cloned and sequenced the gene, you could publish it anywhere. Whereas now people would want you to ask biological questions about what the gene is doing at least.

What happens to a particular area of research-technology when the entire field reaches this stage? What happens to a techniques-based discipline once it has formed a bridge between two other disciplines in science? And what happens to researchers during such a climax phase? Do they persevere in the kind of work in which they have been trained, or do they migrate into one of the fields they have helped to unite?

Our case study suggests that there are different strategies for surviving such a crisis. When a particular field of research-technology wanes, a few researchers may manage to obtain permanent positions as specialists for a certain kind of technology and remain immune to changes in their scientific environment. These people may then build up a reputation on the basis of excellent craftsmanship, which leaves them relatively free from publication pressure or pressure to attend congresses. In the present investigation, they were found in the form of a rare biochemist here or an isolated electron microscopist there, relics of former phases of methodological proliferation. However, the number of positions in research of this kind is extremely limited.

---

[10] Support for this idea is found in the statements of several of our interviewees who maintain that molecular biology is an instrument for approaching new research problems with objects at a higher level of organization. However, these remarks must be viewed with caution due to the unique position of scientists in research-technology which requires them to justify their research by demonstrating possibilities for application in other fields.

Thus the great mass of trained research-technologists has no other option than to seek employment in either theory- or product-oriented research. Accordingly, a number of those we interviewed expressed interest in applying their technological knowledge to problems of plant physiology, developmental genetics or plant breeding. However, gaining a foothold in a new field of research is only possible if one acquires additional knowledge and experience in the field, and this is no easy task for a researcher who has been trained and socialized exclusively in a particular field of research-technology. Thus restricting one's training in science to a single field of research-technology is precarious, and many research-technologists may eventually face unemployment. Furthermore, trying to streamline education to meet the needs of a particular kind of research-technology is obviously a hopeless venture. By the time the lugubrious educational system has changed its course to meet the demands of one area of research-technology, a new one may be on the rise.

*a*Universität Hamburg, Germany
*b*Rheinisch-Westfälische Technische Hochschule Aachen, Germany
*c*Berlin-Brandenburgische Akademie der Wissenschaften, Berlin, Germany
*d*Universität Witten/Herdecke, Germany

REFERENCES

Bayertz, Kurt and Nevers, Patricia (1998). Biology as technology, in R. Porter and K. Bayertz (eds.), *From Physico-Theology to Bio-Technology* (pp. 108–32). Amsterdam: Rodopi.

Bechtel, William (1993). Integrating sciences by creating new disciplines: The case of cell biology. *Biology and Philosophy* 8(3): 277-99.

Burian, Richard (1992). How the choice of experimental organism matters: Biological practices and discipline boundaries. *Synthese* 92: 151–6.

Burian, Richard (1993). How the choice of experimental organism matters: Epistemological reflections on an aspect of biological practice. *Journal of the History of Biology* 26(2): 362–4.

Busch, Lawrence, Lacey, William B., Burkhardt, Jeffrey and Lacey, Laura L. (1991). *Plants, power and profit*. Cambridge, MA: Basil Blackwell.

Darden, Lindley (1991). *Theory change in science. Strategies from Mendelian genetics*. Oxford: Oxford University Press.

Darden, Lindley and Maull, Nancy (1977). Interfield theories. *Philosophy of Science* 44: 43–64.

Fleck, Ludwik (1980). *Entstehung und Entwicklung einer wissenschaftlichen Tatsache*. Frankfurt/Main: Suhrkamp.

Fujimura, Joan (1992). Crafting science: Standardized packages, boundary objects and "translation", in A. Pickering (ed.), *Science as practice and culture* (pp. 168-211). Chicago: The University of Chicago Press.

Greenberg, D.S. (1966). "Bootlegging": It holds a firm place in conduct of research. *Science* 153(3738, August): 848–9.

Grinnell, Frederick (1992). *The scientific attitude*. New York and London: Guilford Press.

Hack, Lothar and Hack, Irmgard (1985). "Kritische Massen " Zum akademisch industriellen Komplex im Bereich der Mikrobiologie/Gentechnologie, in W. Rammert, G. Bechmann, and H. Nowotny (eds.), *Technik und Gesellschaft*, Vol. 3 (pp. 132-58). Frankfurt/Main and New York: Campus.

Hacking, Ian (1992). The self-vindication of the laboratory sciences, in A. Pickering (ed.), *Science as practice and culture* (pp. 29–64). Chicago: The University of Chicago Press.

Harwood, Jonathan (1986). Ludwik Fleck and the sociology of knowledge. *Social Studies of Science* 16: 173–87.

Hasse, Raimund (1996). *Organisierte Forschung. Arbeitsteilung, Wettbeweb und Networking in Wissenschaft und Technik*. Berlin: edition sigma.

Hasse, Raimund and Gill, Bernhard (1994). Biotechnological research in Germany: Problems of political regulation and public acceptance, in U. Schimank and A. Stucke (eds.), *Coping with trouble. How science reacts to political disturbances of research conditions* (pp. 253–92). Frankfurt/Main and New York: Campus.

Hasse, Raimund, Hohlfeld, Rainer, Nevers, Patricia, and Zimmerli, Walther Ch. (1994). *Die Technologisierung der Biologie am Beispiel der Pflanzengenetik und Pflanzenzüchtungsforschung*. Abschlussbericht zum BMFT-Projekt "Die Technologisierung der Biologie: Zur Durchsetzung eines neuen Wissenstyps in der Forschung". Bonn: Bundesministerium für Forschung und Technologie.

Hohlfeld, Rainer (1988). Biologie als Ingenieurskunst. Zur Dialektik von Naturbeherrschung und synthetischer Biologie. *Ästhetik und Kommunikation* 69: 61–7.

Hohn, Hans-Willy and Schimank, Uwe (1990). *Konflikte und Gleichgewichte im Forschungssystem*. Frankfurt/Main und New York: Campus.

Kay, Lily (1993). *The molecular vision of life*. Oxford: Oxford University Press.

Krimsky, Sheldon, Ennis, James G., and Weissman, Robert (1991). Academic corporate ties in biotechnology: A quantitative study. *Science, Technology and Human Values* 16(3): 275–87.

Laudan, Larry (1981). A problem-solving approach to scientific progress, in I. Hacking (ed.), *Scientific revolutions* (pp. 144–55). Oxford, New York, etc.: Oxford University Press.

Magner, Lois N. (1994). *A history of the life sciences*. New York, Basel, Hong Kong: Marcell Dekker, Inc.

Maienschein, Jane (1991). Epistemic styles in German and American embryology. *Science in Context* 4(2): 407–27.

Mannheim, Karl (1986). *Conservatism: A contribution to the sociology of knowledge*. London and New York: Routledge and Kegan Paul.

Mayr, Ernst (1982). *The growth of biological thought*. Cambridge, MA: Cambridge University Press.

Nelson, Rodney D. (1992). The sociology of styles of thought. *British Journal of Sociology* 43(1): 25–54.

Rheinberger, Hans-Jörg (1992). *Experiment, Differenz, Schrift*. Marburg: Basilisken Presse.

Rheinberger, Hans-Jörg (1997). *Toward a history of epistemic things. Synthesizing proteins in the test tube*. Stanford, CA: Stanford University Press.

Scheibe, Arnold (1987). *Bedeutung der wissenschaftlichen Institute für die private Pflanzenzüchtung*. Hamburg und Berlin: Verlag Paul Parey.

Star, Susan Leigh and Griesemer, James R. (1989). Institutional ecology, "translations," and boundary objects: Amateurs and professionals in Berkeley's Museum of Vertebrate Zoology, 1907-39. *Social Studies of Science* 19: 387–420.

Straub, Joseph (1986). Aus der Geschichte des Kaiser-Wilhelm-/Max-Planck-Instituts für Züchtungsforschung. *Berichte und Mitteilungen der Max-Planck-Gesellschaft* 2: 11–36.

Stubbe, Hans (1959). Gedächtnisrede auf Erwin Baur, gehalten am 25. Todestag (2. Dezember 1958) in Müncheberg/Mark. *Der Züchter* 29(1): 1–6.

Weingart, Peter (1978). The relationship between science and technology – A sociological explanation, in W. Krohn, E.T. Layton, and P. Weingart (eds.), *The dynamics of science and technology* (pp. 251–86). Dordrecht, The Netherlands, and Boston: D. Riedel.

Weingart, Peter (1982). Strukturen technologischen Wandels. Zu einer soziologischen Analyse der Technik, in R. Jokisch (ed.), *Techniksoziologie* (pp. 112–41). Frankfurt/Main: Suhrkamp

Whitley, Richard (1982). The establishment and structure of the sciences as reputational organizations, in N. Elias, H. Martins, and R. Whitley (eds.), *Scientific establishment and hierarchies* (pp. 313–53). Dordrecht, Boston and London: D. Riedel.

Whitley, Richard (1984). *The intellectual and social organization of the sciences*. Oxford: Clarendon Press.

Yoxen, Edward (1982). Giving life a new meaning: The rise of the molecular biology establishment, in N. Elias, H. Martins, and R. Whitley (eds.), *Scientific establishment and hierarchies* (pp. 313-57). Dordrecht, Boston, London: D. Riedel.

Zimmerli, Walther Ch. (1988). Ethik der Wissenschaft als Ethik der Technologie. Zur wachsenden Bedeutsamkeit der Ethik in der gegenwärtigen Wissenschaftsforschung, in P. Hoyningen-Huene and G. Hirsch (eds.), *Wozu Wissenschaftsphilosophie?* (pp. 391–418). Berlin: Walter de Gruyter.

Zimmerli, Walther Ch. (1990). Handelt es sich bei der gegenwärtigen Technik noch um Technik? Elf Thesen zu "langen Wellen" der Technikentwicklung, in B. Bievert and K. Monse (eds.), *Wandel durch Technik? Institution, Organisation, Alltag* (pp. 289–310). Opladen: Westdeutscher Verlag.

PART III

PURVIEWS OF GENERIC INSTRUMENTS

CHAPTER 7

SEAN F. JOHNSTON

IN SEARCH OF SPACE: FOURIER SPECTROSCOPY, 1950-1970

In the large grey area between science and technology, specialisms emerge with associated specialists. But some specialisms remain 'peripheral sciences,' never attaining the status of 'disciplines' ensconced in universities, and their specialists do not become recognized 'professionals.'[1] A major social component of such side-lined sciences – one important grouping of technoscientific workers – is the 'research-technology community' (e.g. Shinn 1997).[2] An important question concerning research-technology is to explain how the grouping survives without specialized disciplinary and professional affiliations. The case to be discussed below illustrates the dynamics of one such community.

The specialists surrounding the technology of scientific instruments have been cited as major contributors to research-technology activities (Joerges and Shinn, this volume). Their professional posts, social status and monetary support are more uncertain than is the case for either academic scientists or qualified engineers. Instrument specialists and their products are less established and often more transitory than their counterparts on either side of the science/technology divide.

The volatility and survival tactics of such interstitial communities can be illustrated by the case of Fourier spectroscopy (also known as Fourier transform spectroscopy, FT spectroscopy, FTS or FTIR). This technique, described briefly below, was developed after the Second World War. It attracted a fairly robust coterie of participants for about two decades before becoming successfully commercialized and supplanting older technology in the 1970s and 1980s.

To a great extent, all spectroscopies (and, more generally, instrumental techniques and indeed measurement technologies) are 'peripheral' to academic science, which they supply (see, for example, the contributions by Jackson and Mallard in this volume). Fourier spectroscopy, though, is in some respects an extreme case. It existed for a very long time without convincingly 'serving' academics, and without generating uncontentious results for all of its proponents. As a result, this small community employed several tactics to enhance its credibility. These were truly survival

---

[1] For a more lengthy discussion and examples of other cases of peripheral science, see Johnston 1996a,b.

[2] Other communities associated with such sciences in the twentieth century include those of customers, salespersons and publishers.

B. Joerges and T. Shinn (eds.), Instrumentation: Between Science, State and Industry, 121–141.
© 2001 *Kluwer Academic Publishers.*

tactics rather than *strategies*, because they consisted of what can be seen as disciplinary border skirmishes without any overarching plan of attack.

The result was a transformation of the *discipline* of spectroscopy, but a continued *professional* invisibility of the community which achieved it. I contend that this is the norm rather than the exception for research-technologies. Moreover, this fluid, transdisciplinary behavior is increasingly common in modern technoscience; a fluid community, indeed, can increase its enrolment by drawing upon practitioners in cognate fields.[3]

In the following sections, the principal players – including the concept, its proponents, sponsors and opponents – will be introduced. Their various attempts to influence the adoption of the technique will then be discussed.

THE TECHNOLOGY AND ITS PROPONENTS

Many of the technical details of this technology can be by-passed for the social features that I want to stress, but it is necessary to sketch a few distinctive features to set the context and to make subsequent events understandable.

Devised in large part by Albert Michelson in the 1890s, Fourier spectroscopy was developed after the Second World War by researchers in Britain, France and America. The name Michelson produces a 'knee jerk' response by philosophers and historians of science as the man who tested the hypothesis of aether drift, or perhaps for his precise measurement of the speed of light. But he was much more than that: Michelson was in many respects a prototypical research-technologist.

A professor of physics at a time in America when physics was subservient to engineering, Michelson spent most of his career perfecting and applying his version of an optical device dubbed an *interferometer* to numerous problems. In adapting his generic device to a myriad of uses, he found himself participating with astronomers, chemists and naturalists; interacting with standards institutions, universities and government – but never entirely embraced by any community. Through his career, Michelson stressed the multifarious applications of his instruments rather than any scientific objective. With his interferometers, Michelson measured differences in the speed of light; the separation of binary stars; the length of metrological standards; and the spectral components of colored light.[4] This latter spectroscopic application was analytically the most complex.

An interferometer is a simple but unintuitive instrument (see Figure 7.1). It consists of a detector of light intensity and three mirrors, one of which typically moves,

---

[3] This account, in at least its early stages, fits well with Bruno Latour's conceptions (e.g. Latour 1987, esp. 146–53) of how science gets done. As he describes, proponents enrol support by marshalling a variety of resources, both technical and social. There are, however, some distinct divergences. For example, even when the 'new science' had replaced the old, its proponents remained a shadowy, peripheral group who merely passed 'power' (that is, subsequent control of technical development and application) to analytical chemists. Similarly, while the discussion of 'communities' developed here has resonances with Kuhn, Merton and others, it can most usefully be associated with recent work on the emergence of technical professions. See, for example, Abbott 1988 and MacDonald 1995.

[4] For an autobiographical account aimed at general audiences, see Michelson 1902.

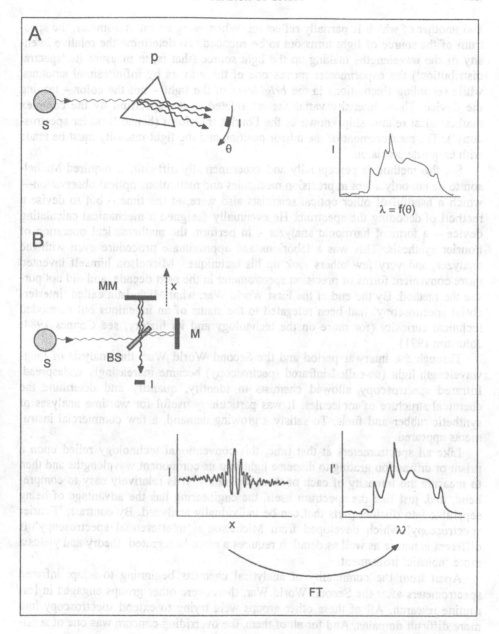

Figure 7.1. Distinguishing features of optical spectroscopies. A. Dispersive spectroscopy:
S, *light source;* P, *prism or diffraction grating;* I, *intensity measured by detector swept*
through angle θ. *In the resulting spectrum* I(λ), *the wavelength* λ *is a function of* θ.
B. *Fourier spectroscopy:* S, *light source;* BS, *beam-splitting mirror;* M, *fixed mirror;*
MM, *moving mirror, translated through distance* x; I, *intensity measured by detector. The*
resulting record of I *versus x, or 'interferogram,' is related to the spectrum* I'(λ) *(or,*
more accurately, I'(1/λ)) *by the Fourier transformation* FT. *Both techniques involve spe-*
cific complications to calibrate the wavelength scale and instrumental response.

and another of which is partially reflecting. When using an interferometer, the spectrum of the source of light turns out to be encoded. To determine the relative intensity of the wavelengths making up the light source (that is, to measure its 'spectral distribution') the experimenter moves one of the mirrors by infinitesimal amounts, while recording fluctuations in the *brightness* of the light – not the color – leaving the device. These intensity variations are related to the spectrum by the complex mathematical relationship known as the Fourier transform (hence 'Fourier spectroscopy'). The measurement of the mirror position and the light intensity must be made with exquisite precision.

So, this method is perceptually and experimentally difficult. It required Michelson to be not only adept at precision mechanics and meticulous optical observation—which a handful of other optical scientists also were, at the time – but to devise a method of decoding the spectrum. He eventually designed a mechanical calculating device – a form of harmonic analyzer – to perform the mathematical operation of Fourier synthesis. This was a laborious and approximate procedure even with the analyzer, and very few others took up his technique.[5] Michelson himself invented more convenient forms of precision spectrometer in the next decade, and did not pursue the method. By the end of the First World War, what Michelson called 'interferential spectroscopy' had been relegated to the status of an ingenious but outmoded technical curiosity (for more on the technology and its history, see Connes 1984; Johnston 1991).

Through the interwar period and the Second World War, the analysis of long-wavelength light (so-called infrared spectroscopy) became increasingly widespread. Infrared spectroscopy allowed chemists to identify, quantify and determine the chemical structure of molecules. It was particularly useful for wartime analyses of synthetic rubber and fuels. To satisfy a growing demand, a few commercial instruments appeared.

Like all spectrometers at that time, this conventional technology relied upon a prism or diffraction grating to disperse light into its component wavelengths and then to measure the intensity of each of them separately. It is relatively easy to comprehend, and, just like the spectrum itself, the engineering has the advantage of being separable into distinct parts that can be individually analyzed. By contrast, 'Fourier spectroscopy' which developed from Michelson's 'interferential spectroscopy' is different in nature as well as detail: it requires a more 'integrated' theory and yields a more 'holistic' instrument.

Apart from the community of analytical chemists beginning to adopt infrared spectrometers after the Second World War, there were other groups engaged in less routine research. All of these other groups were trying to extend spectroscopy into more difficult domains. And for all of them, the overriding concern was one of sensi-

---

[5] Replication of Michelson's results was difficult even for experts, and later was to dog his claims concerning the absence of an aether drift (see Swenson 1972). The tacit knowledge required to perform interferometry, and Michelson's own reticence to support close collaborators or graduate students (as opposed to associates from other fields), militated against its development in his hands as a viable research-technology.

tivity – to measure weak signals in difficult conditions. This concern led some to explore the nature of the instrument itself.

This new cognitive basis for spectroscopic design infused a new generation of investigator. According to one of them, Peter Fellgett, 'the real trigger' for development was *instrument science*. "The basic idea of 'instrument physics' is to understand in a full scientific sense why an instrument has a particular performance," he argued:

> In most cases, a scientific instrument is devised in the first place as a means to the end of making some physical phenomenon or quantity susceptible to observation or measurement, and once it has served this purpose nobody thinks very deeply about it again. Consequently, it is often tacitly accepted that 'in theory' an instrument should have a particular performance, but 'in practice' it does not. This however is not good science, which demands that if theory and practice differ, then one or both must be improved. Had Adams and Le Verrier been content to say that 'in theory' Uranus moves in a particular orbit but 'in practice' in a slightly different one, the planet Neptune would never have been discovered (Fellgett 1984).

This was hardly a new idea in Germany, where instrument science had been on the agenda since the 1870s.[6] It was, however, an epistemic basis little employed in Britain (except, perhaps, at a handful of sites such as the Cambridge Scientific Instrument Company, founded in 1878, see Cattermole and Wolf 1987). Such a hybrid was unappealing both to academics, who frequently disdained engineering sciences, and to British engineers who, in most branches of the subject, were still trained by apprenticeship more often than by academic studies.[7]

The need to concentrate on instrument design was a theme repeatedly broached by post-war spectroscopists, because all were seeking to make measurements at the limits of practicality. Several of them, though, were diverted from their original research to instrument science itself. I will briefly give a narrative of the emergence of the subject through these individuals, and then relate this to a more analytical description of their activities. A prosopography of the group that emerged, which will be termed the 'Fourier community,' reveals some distinct career features that, I would argue, extend to other collectives of research-technologists.

During the Second World War, Peter Fellgett had worked as part of a group analyzing aviation fuels, using prism spectrometers. At that time, he says,

---

[6] For the fortunes of German *Instrumentenkunde*, see Terry Shinn (this volume). On the other hand, such an emphasis was sufficiently unusual even in 1960s Germany that another Fourier pioneer, Günther R. Laukien, a co-founder of the successful Bruker Instrument company (specializing in Fourier and NMR spectrometers from the 1970s) was praised by his company upon his death for having "recognized the need to apply scientific knowledge to commercial products in order to foster research and development," a "highly unusual" idea "only achievable through confrontation and reluctance within the academic community" (Laukien 1997).

[7] An extreme case is chemical engineering, arguably the most 'academicized' engineering occupation in Britain. By the end of the Second World War, there were departments in only two British universities, both with technological connections (London and Glasgow). About half of the qualified practitioners held degrees, mostly in chemistry or mechanical engineering. Departments and degree-holders increased rapidly after the war, a department and professorship being created at Cambridge, for example.

I had become interested in why infra-red detectors . . . were so relatively insensitive; or, more generally, whether it were possible to identify some physical limit of radiation detectors (Fellgett 1991).

As a PhD student from 1948 at Cambridge, Fellgett realized that most of the radiation passing through a prism spectrometer is wasted because only a narrow band of wavelengths passes through a slit to be measured by the detector at any one time. He consequently looked for some means of measuring all wavelengths simultaneously. This would provide a stronger detector signal that was less obscured by sources of noise. His initial thoughts were for a rather complicated arrangement of rotating optical disks that would 'multiplex' different infrared wavelengths on a single detector, combined with an undefined scheme to somehow separate the encoded wavelengths from the detector signal afterwards.[8]

Within a year, however, Fellgett concluded that a Michelson interferometer could quite elegantly achieve what he wanted: to encode the entire spectrum of light into a single time-varying electrical signal. This would yield a much more sensitive form of spectrometer. This principle of 'multiplexing' eventually became known as the 'Fellgett advantage.' The drawback for Fellgett, as for Michelson, was that the signal was encoded in the form of a Fourier transform. Fellgett adopted a manual calculation aid developed before the war for analyzing X-ray diffraction patterns. This was sufficient to demonstrate the principle.

Fellgett did not publish his results, but publicized them by word of mouth. John Strong, an infrared spectroscopist at Johns Hopkins University in Baltimore, who had been designing infrared instruments since the early 1930s, visited Cambridge in 1949 and saw some of Fellgett's apparatus and results. He misunderstood the principle and the multiplex advantage, however, and decided to build his own version simply as a means of improving conventional spectrometers (Strong 1984). Strong's postgraduate students and technicians at Johns Hopkins constructed and used the device as a sort of optical filter for a grating spectrometer. He was thus adapting this quite novel technology to an old problem. When Peter Fellgett visited Johns Hopkins the following year, however, one of Strong's students, George Vanasse, understood and rekindled interest in Fellgett's original ideas (Vanasse 1982).

Fellgett also presented some of his results in a 1952 meeting of the Optical Society of America (Fellgett 1952). A few commercial designers were immediately intrigued. Indeed, designers in the engineering department of Perkin-Elmer Corporation in New Jersey, producers of one of the few commercial prism-based infrared spectrometers, for a time investigated the practicality of interferometer designs (Scott and Scott 1953; the design they later patented was not commercialized).

Strong invited a British physicist, H. Alistair Gebbie, who shared his interest in atmospheric measurements, to join his group on an Air Force contract in 1954. Gebbie proposed using a digital computer to transform the measurements into spectra (Gebbie 1984). Strong arranged for an IBM 605 computer at Binghampton, New York, to be programmed for the Fourier transform, and a single spectrum was calcu-

---

[8] The purpose of encoding the signal was to create the largest possible signal so that it was stronger than the relatively constant electrical noise, thus improving sensitivity. See Fellgett 1951. Another encoding scheme, with a less articulated theoretical justification, was Golay 1951.

lated. Owing to the high cost of computer calculation (some $25,000 per spectrum), it remained the only published transformed spectrum until 1956, and by 1959 it had appeared in no fewer than four papers and a book by Strong.

In France, Pierre Jacquinot, who had employed large and high resolution spectrographs for his research, found that he needed a cheaper form of instrument when he moved to a new French university post in 1942 under the Vichy government. Through the mid-1940s he experimented with other forms of instrument and eventually discovered that the Michelson interferometer had another advantage: unlike dispersive instruments, it did not require a slit which inevitably limited the amount of light that could pass through it (and which thereby reduced the so-called optical 'throughput' or 'étendue'). This attribute later became known as the 'Jacquinot advantage.' Jacquinot judged, however, that 'the precision requirements to obtain good spectra were so severe that is was almost unbelievable that they could ever be met' (Jacquinot 1984). He did not publish his work until 1954, and even then focused on other kinds of instrument (Jacquinot 1954). Like Strong, Jacquinot directed his students to instrument research. Two of them, Janine and Pierre Connes, began studying the theory and practice of Fourier spectroscopy in 1954. Pierre Connes later wrote of Jacquinot's "will and ability to treat menial instrumental matters as parts of physics just like more high-flown subjects" (Connes 1992).

In 1954 another American, Lawrence Mertz at Baird Associates, was working on a contract for the Army Signal Corps because two of his superiors had heard of Fellgett's work. Mertz developed a Fourier spectrometer of his own design, but again encountered the problem of calculation, and did not obtain a spectrum from his original measurements for over a decade.

By the late 1950s, then, there was a collection of perhaps a dozen persons in three countries pursuing this new instrument technology. But how self-conscious was the collectivity of these groups? Commonality was not immediately obvious: according to Mertz "various clans coagulated to develop and exploit" the techniques, centerd on the concepts of either Fellgett and Jacquinot (Mertz 1971). Fellgett later remarked that Jacquinot's work was largely irrelevant to his own studies of stars because the 'Jacquinot advantage' related only to large light sources (Fellgett 1991). If, then, the group interests were not strong, by what mechanisms did this new subject survive?

FINDING A COMMON LINE: THE 1957 BELLEVUE CONFERENCE

In 1957, Pierre Jacquinot promoted a merging of interests by organizing a conference entitled 'interferometric spectroscopy' to cover the non-traditional but expanding list of techniques being investigated by his laboratory at Aimé Cotton. Only five of the papers dealt with Fourier spectroscopy. The conference "served to bring all Fourier spectroscopists . . . together and to provide an arena for exchanging their ideas" and to share their meagre results (Sakai 1991). Fellgett showed some extensions of his thesis work; Strong, Vanasse, and Gebbie, their single spectrum calculated the previous year; Mertz, his still undecoded data of a star; and Janine Connes, a theoretical analysis of the underlying mathematics. The conference was nevertheless seen by contemporaries as important. This first joint meeting provided encouragingly complementary information. Within the jumble of scanty results, the proponents saw a pattern that seemed to support what had been separate theoretical derivations: Fourier spectroscopy appeared able to give recognizable spectra, and promised to be orders of magnitude more sensitive than conventional methods. The ensemble of papers also provided a critical mass that gained attention for them in the specialist press (Conference Proceedings 1958). This potential for opening hitherto impracticable domains such as far infrared, atmospheric and stellar spectroscopy was reason enough to tackle the serious problems of computation and demanding mechanics.

NEW COMMUNITIES AND THEIR PATRONS

After the Bellevue conference, these individuals were linked collectively by a growing sense of community – or at least an awareness of common goals. The groups and the claims they made expanded, and as they did so, began to attract both allies and foes. The directions of expansion were quite distinct in each country.

Fellgett remained in academia, and from the end of the 1950s onward was relatively inactive in Fourier spectroscopy.

In France, Jacquinot continued to supervise workers, particularly the Connes, at the Laboratoire Aimé Cotton (LAC) until 1964 (see Connes 1995).[9] This research site was arguably the most stable one for the emerging subject. There, a tradition and acceptance of such research-technology activities had already been established (Shinn 1993). Moreover, they were able to extend their influence. The bargaining chip, in negotiating recognition, was the promise of dramatically improved instrumental sensitivity. In late 1963, the Connes took a sabbatical year at the Jet Propulsion Laboratory in Pasadena, California, where a development project for a Mars spacecraft was being planned. There they developed a very successful Fourier spectrometer to be used on an earth-based telescope to study the atmospheres of planets. After using it at an Arizona observatory to observe Venus, they brought the instrument back to France in 1965 for studies of Mars. The number of investigators at the LAC grew and developed a widening range of instruments. A number of French spectroscopists were trained at the LAC and have subsequently populated other laboratories at the CNRS and elsewhere.[10]

Vanasse, and Strong's other students, made Fourier spectroscopy the focus of their PhD dissertations. Strong, however, was not closely involved in the research. With the waning of his personal influence came a decline of the Baltimore research program. Vanasse eventually went to the Air Force Cambridge Research Laboratory (AFCRL) in Massachusetts, with which Strong had developed links through earlier consulting and research contracts. Along with him went a number of other former Johns Hopkins students, forming a development team in the 1960s concerned with

---

[9] Jacquinot subsequently moved to another division of the CNRS.

[10] See, for example, a special issue of *Spectrochimica Acta*, Vol. 51A (1995) on 'High resolution Fourier transform spectrometry in France.' The authors, several former members of the LAC, are affiliated with the *Institut d'Astrophysique de Paris*, CNRS; *Laboratoire de Physique Moléculaire et Applications*, Université de Paris Sud; *Groupe de Spectrométrie Moléculaire et Instrumentation Laser*, CNRS. Connes himself left the Laboratoire Aimé Cotton for the *Service d'Aeronomie* of the CNRS in 1974. Note, however, that these sometimes academic off-shoots were concerned primarily with *application* of the technology rather than its development as such.

atmospheric research. Thus a second generation of research-technologists – the students of Strong and Jacquinot – were populating this interstitial science outside universities but within government-sponsored laboratories.

The second wave of practitioners also filled another occupational niche: product-oriented research and development in commercial firms. Alistair Gebbie was largely responsible for starting this. He joined the British National Physical Laboratory (NPL) in 1957. By 1960 he had developed his own design of Fourier spectrometer, and benefited from a faster (and effectively free) 'in-house' computer to calculate the spectra. His design was truly a 'generic device,' in that the prototype was reworked and essentially resold to three companies over a period of twenty years.[11]

Lawrence Mertz, too, promoted commercial applications. In 1960, he became Vice President of the fledgling Block Associates (later Block Engineering), a Cambridge, Massachusetts instrument consulting company. The company provided special purpose Fourier spectrometers to the Air Force, including one for a 1962 satellite (Block and Zachor 1964).[12] Here, again, the promise of highly sensitive instruments was promoted, but on rugged air-borne or space-borne platforms rather than on the ground. Both Gebbie and Mertz thus indoctrinated new and experienced instrument designers in Fourier technology while encouraging its transformation.

PROVOKING OPPOSITION

The first advocates of the technology thus found allies at the AFCRL, CNRS and NPL – large, well-funded organizations that allowed considerable leeway in research and which were relatively undemanding of immediate results.[13] Each of these sponsors was government-supported and had academic and industrial links.

But the advocates of Fourier spectroscopy found rising opposition closer to home. From the late 1950s onward, their increasing visibility among colleagues provoked disapprobation. Their critics were spectroscopists and optical physicists employing conventional prism and grating spectrometers. The ostensible technical issues were that (a) Fourier spectroscopy was no better, and probably worse, than existing techniques; (b) it was prone to poorly understood errors which were difficult to discern or resolve because (c) the technique was unintuitive; and, (d) impracticable because of computational difficulties and unrealistic demands for mechanical precision.

---

[11] Sir Howard Grubb-Parsons & Co. c1960–62; Research and Industrial Instrument Company (RIIC) c1963–8; Lloyd Instruments c1983–8. Gebbie also approached manufacturers such as Perkin-Elmer, where he was informed that "no-one will ever depend on computers to get their spectra" [H. Alistair Gebbie, personal communication, 21 Nov 1990]. Grubb-Parsons and RIIC ceased trading in the late 1970s; Perkin-Elmer began manufacturing Fourier spectrometers then, and by the 1990s was the chief commercial producer.

[12] Note that Mertz's instrument – a simple fist-sized device – was very dissimilar to the Connes' large and multiply-compensated instrument intended for telescopic use, but incorporated similar underlying principles.

[13] The more demanding, poorer-funded exceptions were Grubb-Parsons and RIIC in England. Their first commercial products were, however, acceptable to the small number of versatile physicists purchasing them, who were able and willing to engage in their own 'instrument science' activity.

As Latour, Galison and others have shown, however, scientific disputes are seldom resolved by objectively agreed cognitive evidence (Galison 1987; Latour 1987). Many of the critics, for example, were to be found working in similar environments: under government contracts, in national laboratories, or as associated academic staff at universities. Indeed, the 'Fourier community' was constructed by the system of alliances and oppositions which developed. In a very real sense, the nascent Fourier collective was threatening conventional spectroscopists and instrument designers on both intellectual and social territory, by vying for publication space, development contracts and employment openings. Intellectual disagreements consequently were accompanied by other, non-cognitive, arguments. The skirmishes between the old guard and new took place in various locales and employed a variety of tactics. For over a decade, they failed to be decisive.

TACTICS OF THE FOURIER COMMUNITY

For the first young Fourier spectroscopists, gaining acceptance by their peers for the technology went hand in hand with their professional accreditation. Fellgett found his PhD advisor unreceptive to his plans to test his notions of the multiplex advantage experimentally, and was advised to downplay his Fourier spectroscopy research (Fellgett 1991). Similarly Mertz, pursuing a PhD at Harvard after his time at Baird Associates, found his work in Fourier spectroscopy rejected as a subject unsuitable for a thesis topic. He later complained that "the academic community would have nothing to do" with the technique.[14]

On the other hand, those students with an influential academic patron fared rather better. John Strong had a firm reputation as a designer and spectroscopist, as well as influence in industry and government as a consultant (see Vanasse 1982: 190–1). The AFCRL, which had supported research contracts for Gebbie and others, also employed Strong's former students.[15] In the same way, Janine and Pierre Connes found guardianship under Jacquinot both as students and research workers. Both Strong and Jacquinot were securely placed instrument designers able to shepherd their students across the bridge from personal study to academic certification to employment.

The ostensible criterion of scientific validity is to demonstrate convincing experimental evidence. The criteria for convincing evidence, however, are frequently difficult to negotiate between advocates and their opponents. For Fourier spectroscopy, direct confrontation with the prevailing technology proved ineffective.

Not only the nature of evidence, but its manner and sequence of presentation, can be crucial in enrolling support. At the 1957 Bellevue conference Fellgett, and Strong's group, presented only scanty data demonstrating that the method could yield a spectrum of trivial light sources. Such evidence in no way threatened conventional instruments which had been producing copious results routinely for a quarter-century, and did little to convince skeptics either of its potential or justification. The

---

[14] Mertz received a PhD some eight years later: "eventually, after Pierre Connes so convincingly demonstrated the capability of FTIR, Harvard conceded that I might receive a PhD for my part" (Mertz 1991, personal communication).

[15] George Vanasse, Ernest Loewenstein, and Hajime Sakai, by 1963.

presentation by Mertz of other, untransformed data only underlined the suffocating burden of calculation carried by the technique. Strong himself was biased against the technique by "an experimentalist's natural distrust of anything involving such a prodigious calculation" (Strong 1984).

Other investigators entertained more serious doubts. In 1959 Franz Kahn published a note in *Astrophysics Journal* concluding that Fellgett's 'advantage' was in fact a serious disadvantage when the light source was not extremely stable (Kahn 1959). Another notable early critic was Gerard Kuiper of the Lunar and Planetary Laboratory at the University of Arizona. A prominent planetary astronomer, Kuiper published an analysis in 1962 which, he argued, proved conclusively that Fourier spectroscopy could not deliver on its promise even in principle (Kuiper 1962: 83). Mertz suggests that Kuiper was converted to Fourier technology during a planetary observing experiment by the substitution of a Fourier spectrometer for Kuiper's malfunctioning dispersive instrument, although this change of faith probably involved other factors (Mertz 1991. Kuiper subsequently made links with the Connes group at the CNRS; they initiated one of the members of his group in Fourier spectroscopy, and gave him the instrument they had developed at JPL to begin his own program back in Arizona (Connes 1995: 1102).

Among the few vocal critics were designers of optimized conventional instruments, some of whom undertook their own comparisons, such as Fritz Kneubühl at the Swiss Federal Institute of Technology (see, for example, Loewenstein, Moser, and Steffan 1996). Kneubühl's conclusion that the Fourier community had made exaggerated claims were countered by their criticisms of his expertise in operating the new technology.

But even direct comparison could prove remarkably unpersuasive. In 1966 Janine and Pierre Connes demonstrated remarkable spectra of planets demonstrably better than prior results which, they admitted, had been inferior or at best comparable to those obtained using conventional instruments (Connes and Connes 1966). One sympathetic commentator wrote that the results were "far superior to that attained by conventional means, and all doubts about the importance of Fourier spectroscopy should be laid to rest by this work" (Loewenstein 1966).[16] Such evidence, deemed 'conclusive' by proponents, was nevertheless ignored by many spectroscopists still mistrustful of the indirectness of the technique. The opponents disputed the very definition of 'evidence,' some of them arguing that the claimed improvement of spectral resolution was in fact due to instrumental artefacts. Gebbie later reported that he, too, had long worried over unexplained spectral features in his first spectrum of the atmosphere, suppressing the data while trying to rule out the possibility of some unexplained instrumental effect (Gebbie 1984). Thus the influence of the public demonstration of the technique was tempered by a seeming morass of tacit considerations.

Experimental evidence arguably weaves a convincing tapestry only with the supporting thread of theoretical justification. Here too, however, obtaining consensus on what constituted adequate and convincing evidence was difficult.

---

[16] Loewenstein later worked in ophthalmic optics.

Janine Connes published her PhD thesis in 1961. This was, for several years, the most careful and complete analysis of Fourier spectroscopy in print. Published in French, it was translated by the AFCRL for its own use (J. Connes 1961). Rather than attracting praise by advocates, however, it was simply ignored by critics of the technique. Those that did take note called attention to another worrying 'artefact': the appearance of a transformed spectrum could be altered dramatically by the kind of 'apodization,' or mathematical fine-tuning, employed with the Fourier transformation. The Fourierists countered that such 'filtering' was an inevitable feature of any optical instrument, but their opponents, largely unversed in, and mistrustful of, such mathematical niceties, rejected such 'sophisticated' arguments.[17]

A thematic issue of *Applied Optics* in 1969 rehearsed the disputes concerning the superiority of conventional versus Fourier spectroscopy. Jacquinot, no longer active in the subject and cast as impartial arbiter, strove to "objectively survey both methods (i.e. interferometry and grating spectroscopy)" (Jacquinot 1969). Gebbie recast the mathematics to highlight similarities with conventional spectroscopy, but still failed to convince skeptics, who now cited experimental evidence to back up their claims (Gebbie 1969). Critics and proponents, experimentalists and theorists, were speaking largely incomprehensible languages.

Part of the problem was that there was no truly 'neutral' venue for discussing claims. The publication of results was thus of mixed benefit to the early Fourier community. On the one hand, it communicated the members' work to a broader audience and promoted the technology. On the other, it provoked attacks of the still-contentious results.

For at least a decade, too, publications were incoherent and sparse. Fellgett's thesis remained unpublished, and most of his papers were difficult to obtain. After Fellgett's 1952 Optical Society abstract, the next publication was a brief note by John Strong in 1954. He reached a much wider audience with a lengthy description of the technique as an appendix to a popular optics textbook in 1958 (Strong 1954; Vanasse and Strong 1958).

As a means of articulating group interests, such publications were unsuccessful. The community was small and already better served by direct communication through meetings or correspondence. Moreover, its members found their interests poorly served by existing journals. A major vehicle for papers on spectroscopic instrumentation had, since 1919, been the *Journal of the Optical Society of America* (*JOSA*). The Fourier community, apart from John Strong's group, found the *JOSA* editor, Wallace Brode, reticent to publish papers on the technology. Mertz, for example, whose papers were repeatedly rejected, subsequently employed the tactic of submitting a series of advertisements to *JOSA* consisting of technical abstracts.[18] As a result, he said, "there were no hassles with referees, they were inexpensive . . . and

---

[17] The central issue was the subjective interpretation and judgement of distortion of the spectrum that was caused by applying a Fourier transform to a limited quantity of data instead of an infinite range of frequencies. Proponents 'solved' the problem by applying various versions of mathematical smoothing, or apodization (literally 'foot removal'). Their opponents argued that such 'artificial' and 'cosmetic' manipulation was disturbingly sensitive, and consequently labeled the entire technology as *ad hoc* and scientifically unsound.

[18] E.g. *Journal of the Optical Society of America*, Vol. 50, No. 3 (1960) and Vol. 51, No. 8 (1961).

publication was swift" (Mertz 1991). Such non-traditional beacons of communication may revert to historical invisibility, however: advertisements are frequently stripped from archived journals. *JOSA* was not unique in opposing publication. Other optics journals, too, were unenthusiastic about a technology that relied so intimately on me-chanical design, electronic components and mathematical manipulation. The Mertz episode has nevertheless been credited with fostering, in 1960, the launch of a new OSA journal, *Applied Optics*. Significantly, the editor was employed at AFCRL.[19] This proved a more welcoming repository for papers on the subject as did, from 1961 onward, the new British journal *Infrared Physics*. Thus, the 1969 'debate' was con-ducted on the home ground of the Fourierists under a sympathetic editor who had already published a plethora of papers on the technique. On the other hand, tradi-tional designers' periodicals such as the British *Journal of Scientific Instruments* and the American *Review of Scientific Instruments* were infrequent vehicles for publica-tion.[20]

Books, which necessarily presented more accomplished results, also were insig-nificant for the first decade and were devoted to the technology only since about 1970.[21] Those books that did appear, such as Strong's *Fundamentals of Optics* (1959) and Mertz's *Transformations in Optics* (1966), referred to Fourier spectros-copy as a side topic.

The seminal Bellevue conference has been mentioned above. Other important stages for uniting the Fourier community were a conference at Orsay, France, in 1966, again organized by the CNRS, and at Aspen, USA, in 1970. At Orsay 23 pa-pers on Fourier spectroscopy were presented, including the widely cited Connes' work on planetary spectra.[22]

Aspen, four years later, was sponsored by the AFCRL and was the first full con-ference on the technique.[23] If Bellevue had given a sense of collectivity to the work-ers, Aspen provided evidence of a burgeoning community having autonomy over an instrumental technology. Attendance swelled to over 400. Members of the commu-nity achieved a transient visibility: photographs of prominent participants were pub-lished over the legend "some pioneers of Fourier transform spectroscopy" (*Applied Optics* 11: 1673 (1972)). A sense of collective history also developed. Twelve years later, a session of another conference at Durham, England, was devoted to historical reminiscences (published in *Infrared Physics* 24 (1983)).

Such conferences were successful largely because they preached to the converted, attracting audiences already disposed towards the technique or in fact already study-ing or using it. Joint conferences, sponsored by optical societies and others, were considerably less successful during this period for precisely the same reason. Such conferences, such as the annual Optical Society of America meetings, brought to-

---

[19] See, for example, John N. Howard, editor of *Applied Optics*, quoted in R.A. Hanel 1970.

[20] These journals both stressed non-mathematical, non-precision metrologies. The papers that were ac-cepted, and their readership, emphasized relatively unsophisticated mathematical analysis.

[21] This first wave of books, notably Chantry (1971) and Bell (1972), came from physicists, and stressed applications in far infrared and solid state research.

[22] For conference proceedings see *Journal de Physique* 28 (1967).

[23] See *AFCRL Special Report* 114 (Jan. 1971). At least seven AFCRL designer/researchers were there, several of them former students of John Strong.

gether experts on conventional techniques including a large number involved with applying conventional technology. This was the worst of all possible worlds: young Fourier advocates, employing relatively exotic but unexplored devices, faced older practitioners who had decades earlier moved beyond 'proof of concept' to applications. A secondary problem was that such conferences, while covering a broad range of topics, still generally limited those topics to optics or spectroscopy. Their attendees had less familiarity with either electronics or computation, both of which were integral to the new technology. The reasons for the relative failure to communicate at annual conferences, then, were similar to the problems with publishing in established journals: Fourier spectroscopy was seen by organizers and a majority of participants as too far removed from the traditional audiences, and too much a 'hybrid' of disciplines to be comprehensible to them.

Yet communication at non-specialist conferences *did* finally improve after the 1970s for two reasons. First, the number of practising Fourier spectroscopists had by then risen sufficiently to support sessions devoted to the technology; the phenomenon of a 'conference within a conference' appeared. Secondly, a prominent advocate of Fourier spectroscopy papers and conferences emerged in the guise of the Society of Photo-optical Instrumentation Engineers (SPIE). The SPIE had begun in the 1940s as the Society of *Photographic* Instrumentation Engineers, a specialist organization devoted mainly to the design of still and motion picture cameras. These devices involved precision mechanics, optics (although for imaging, rather than high-precision interferometric optics) and, increasingly, electrical and electronic components. The SPIE had also become a vehicle for the work of American military- and civil government-funded investigators, the very contractors that had become prominent advocates of American Fourier spectroscopy research. By the mid-1970s the SPIE had recast itself as an organization devoted to optical engineering, which itself was understood to include expertise in mechanics, electronics and computing. On the other hand, the conferences devoted entirely to Fourier spectroscopy after the 1970 Aspen conference were populated substantially by spectroscopists and analytical chemists, not the designer-researchers of the original Fourier community.[24]

Beyond meeting at conferences, there was a very limited amount of assimilation of new ideas by the sharing of researchers and equipment, but not post-doctoral workers or technicians, between groups. Pierre Connes notes, for example, that the research students at the Laboratoire Aimé Cotton were "sadly innocent of digital techniques" and unaware of any practical details of other research, during their first years there (Connes 1992). The Connes' sojourn at JPL, and their subsequent loan of their instrument to the University of Arizona, are exceptions to the general isolation of workers.

While a disciplinary presence of sorts thus emerged in America, academic posts for the Fourier community failed, on the whole, to materialize. Instrumentation development was seen as too application-driven to serve as a suitable subject for a per-

---

[24] These conferences were at Columbia, SC, 1977 and 1981; Durham, England, 1983; Ottawa, Canada, 1985, and Vienna, Austria, 1987.

manent post. Those few who did settle in academia in the 1960s gained posts in electrical engineering departments more often than in physics departments.[25]

The situation changed in the late 1970s for two reasons. First, 'optical engineering' became an increasingly recognized academic specialism at a few centers, notably the University of Rochester (New York) and the University of Utah.[26] Second, analytical chemists were gaining interest in Fourier spectroscopy as a viable and sensitive technique (see, for example, Finch 1970). Academic-based chemists such as Peter Griffiths began to make the development of new Fourier-based measurement techniques the basis for their work. Even Griffiths, though, had had an earlier career in instrument development.[27]

Acquiring sponsors proved to be the most socially effective tactic for these research-technologists. Early British and American companies have already been mentioned. According to one participant, British spectrometer manufacturers "were interested rather than wildly excited by the development because there was not much market," and mechanical accuracy problems "made it look like an uneconomic proposition" (Threlfall 1990). The first American companies survived by development contracts for government departments intrigued by the promise of measurements in difficult circumstances: to observe aircraft, to scan terrain or to probe the atmosphere from satellites and space probes. Indeed, the region of Massachusetts surrounding the Air Force Cambridge Research Laboratory (later the Air Force Geophysical Laboratory at Bedford) and populated by its contractors has subsequently become the principal American center of so-called 'electro-optical' technology. It is noteworthy that acquiring this sponsorship required the instrument to mutate, to become more 'generic': it had to become portable, more robust and capable of employment with different kinds of light source.[28]

As late as 1978, about one-third of commercial spectrometers were still sold to government (Dunn 1978). The AFCRL was a major early patron, but was joined by other significant research and development sites in America: the Jet Propulsion Laboratory in Houston, and the NASA Goddard Space Flight Center in Maryland.[29] The Aerospace Corporation, in Los Angeles, also began developing and purchasing

---

[25] H. Alistair Gebbie, at Imperial College, and Günther R. Laukien, at Karlsruhe, were both trained in physics and became professors of electrical engineering. Similarly, until the early 1970s, most practitioners of the subject had backgrounds either in physics or electrical engineering.

[26] Rochester had long associations with optics through its links with Eastman Kodak. The Utah connection grew from its links with planetary astronomy. On the other hand, technological universities such as the Massachusetts Instiute of Technology and the California Institute of Technology developed no particular connection with Fourier spectroscopy.

[27] Griffiths worked as a designer at Digilab, a successor to Block Engineering, in the early 1970s.

[28] The physical transformation of Fourier spectrometers, based on local experience and particular applications, was profound. At the NPL, smooth operation relied on precision-ground mechanical guides; at Block Engineering, the small and simple instruments were scanned using converted loudspeaker voice coils; at the LAC, complex optical and servo-mechanically controlled devices evolved. By the late 1970s, however, all such instruments incorporated a laser as a precise reference of wavelength, and employed the fast Fourier transform (FFT, first publicized in 1965 by a mathematician but exploited by the Fourier community a year later), which sped computation by orders of magnitude.

[29] GSFC developed Fourier spectrometers that flew on Nimbus satellites and the Voyager and Mariner spacecraft, and JPL applied the technology to atmospheric remote sensing.

Fourier spectrometers in the early 1960s, maintaining informal communications with John Strong and his former students at AFCRL (Randall 1991, personal communication).[30]

As at the CNRS and NPL, groups at these American government institutions enjoyed the luxury of relatively good funding, an absence of commercial pressures and considerable freedom to pursue engineering innovations. Its designers maintained close links with industry.[31] Moreover, further funding for both the Connes and Mertz came from the American Air Force, primarily through George Vanasse and Alistair Stair, an associate (Mertz 1991). This financing of fundamental research was both possible and common in the USA until the Mansfield Amendment prevented such military spending.

More companies were founded from the late 1960s onward, generally with Fourier spectrometer designers as instigators or consultants. Such individuals were often quite anonymous; the fledgling companies strove to market innovations and were consequently reticent to publish. Moreover, the commercialization of Fourier spectroscopy has been marked by numerous small and often short-lived companies (Johnston 1991: chap. 15).[32] A small number of persons broadcast the 'seeds' of the technology widely. Besides Alistair Gebbie, for example, Ray Milward, a Briton who did postdoctoral research at MIT in the late 1950s and then went to the Royal Radar Establishment in England, where he borrowed Gebbie's instruments, was a proselytizer. Milward subsequently joined the RIIC company in England, then the French company Coderg in 1973, and then another French company, Polytec, at each of which he created a product line of Fourier spectrometers.[33] He afterwards founded the British branch of an American spectrometer company (Mattson) and subsequently continued to hop between manufacturers of the instruments as a manager or consultant (Milward 1990, for one of his more widespread designs, see Milward 1969).

Those publications that did appear fulfilled the role of publicity and marketing as much as technical content. In the company brochures, just as in the literature and conference papers, there was an awkward coexistence of publicity hype and reticence about technical details.[34] Moreover, the skills demanded of such instrument designers were unusually broad (optics, electronics, mechanics, and computing expertise). Most had been trained as physicists or electronics engineers. Owing to the small size

---

[30] The Aerospace Corporation was founded in 1960 specifically as a contractor for the Air Force.

[31] Companies such as Block Engineering (founded 1960), Idealab (c1962), Midac and Bomem (1973) were maintained by such contracts in their early years.

[32] The companies evolved rapidly in an increasingly competitive market. Some seeded new companies or product lines [e.g. Digilab (later Bio-Rad), and Nicolet from Block Engineering; Mattson Instruments from Nicolet]; others merged with larger firms (RIIC with Beckman; Analect with Laser Precision Analytical; Bomem with Hartmann & Braun and then Elsag Bailey; Mattson with Philips); still others failed or left the market (Beckman; Coderg; IBM Instruments; Lloyd Instruments; Spectrotherm). During the 1960s, a good sales volume was of the order of two dozen instruments per year. By the mid-1980s, this had risen to several hundred per year for the largest companies.

[33] All three firms ceased manufacturing Fourier spectrometers by the mid 1970s.

[34] This was, in fact, a replay of the public demonstration/tacit knowledge conundrum that had mired the first presentations of 'convincing evidence.'

and financial fragility of the small instrument manufacturers and the uncertainty of the market, the spectrometer designers in industry not infrequently added marketing and administrative skills to their already broad technical backgrounds.

The technology was made more marketable in the late 1960s by the publication of a much more efficient calculating algorithm (the Fast Fourier transform, or FFT), the development by Mertz and associates of a rapid-scanning form of the instrument, and the commercial availability of minicomputers. These increased the speed and economy of the instruments dramatically. Companies also provided an important vector for change. Attempting to increase their markets, spectrometer companies tried in the late 1960s to interest chemists in the technology. These attempts largely failed at the time because Fourier spectrometers were less reliable and more demanding of technical knowledge than the, by then, highly automated dispersive instruments.[35] A growing trend towards computerizing measuring instruments from the 1970s onward, however, allowed its advocates to present Fourier spectroscopy in a better light. The technology had the disadvantage of demanding computers, but this could be portrayed as an advantage in itself: digital manipulation of data from a Fourier instrument could be much more informative than that from a dispersive machine because of the precision of the scale of wavelength – which became known as the 'Connes advantage.' What physicists had identified as the serious drawbacks of the technology (its reliance on unintuitive and expensive equipment, which necessitated waiting to obtain measurements rather than observing them in 'real time') were increasingly down-played for its putative advantages (the ability to measure more 'difficult' samples with less preparation, and to undertake more elaborate analyses of the data).[36] This recasting of the criteria of judgement by chemists substantially disregarded the criticisms of earlier critics. Not surprisingly, they renamed their appropriated and remoulded technology (FTIR, for Fourier transform infra-red). It is significant, too, that these instruments became more generic – more versatile – when applied to the wide variety of new applications that chemists, and their commercial markets, provided.[37] The instruments were taken up in routine testing labs and then the shop floor in the process industries. The instruments only became 'black-boxed' in this way in the early 1980s when the technology had been rendered reliable and automated, and when adequately powerful microcomputers rendered the technology less expensive

---

[35] Through the 1960s, physicists were the primary purchasers of Fourier spectrometers, which were applied to problems where the intensity of optical radiation was impracticably low for conventional instruments. Investigations included far infrared spectroscopy, atmospheric studies such as airglow and auroral emission, telescope-based observations of stars and planets, and studies of very weak optical absorption such as by impurities in semiconductors or dilute liquids. Somewhat later, the inherent sensitivity was used for measurement of short-lived phenomena such as transient chemical species and magnetically-confined plasmas.

[36] Chemists and industry were attracted to the opportunities for energy-wasting optical sampling techniques provided by this efficient form of spectroscopy. Techniques such as diffuse reflectance, gas chromatography/infrared (GC/IR) and Fourier transform-Raman spectroscopy were made practicable by Fourier techniques, and considerably extended the precision, speed and versatility of analytical chemistry.

[37] The evolution of Fourier spectrometers along a distinctly nonlinear path as they were appropriated, refashioned and extended by different groups, supports what Michel Callon and others have termed a 'model of translation.' See, e.g., Callon and Law 1982; Latour (1987: 133–44).

and thus competitive. Chemists, inexperienced in such instrumentation, neither wanted nor needed to take on the peculiar culture of the research-technologists: it was now embedded in their instruments.

The enrolment of support from this new community was not universally applauded, however. Connes later complained that "if you are an instrument builder, your viewpoint differs greatly from that of the person who buys a ready-made interferometer, and I personally have some doubts about Fourier spectrometers being used properly even when producing indisputably fine results" (Connes 1978). An American contemporary emphasized the distinction and desire for continued independence between communities:

> The ideal operation of an FT-IR system should be a closed shop with one key operator. Furthermore, the key operator should be electronically oriented with a background in both machine language and high order programming. This type of key operator can easily be trained in infrared sample handling and would provide an ideal interface between the analytical chemist and the system . . . . (Dunn 1978).

The original opponents of Fourier spectroscopy were not widely converted to the new technology; they continued to employ conventional instruments in proven applications. Eventually, however, Fourier technology and its development community became more numerous than conventional spectroscopists. This was due both to the inevitable retirement of older practitioners and to the Fourier community's success in recruiting new adherents, particularly in the form of contract sponsors and analytical spectroscopists. The old guard was merely superseded. The Fourier community itself remained stable but not sizeable: by the early 1970s it included perhaps 500 investigators, with certainly fewer than 100 making it a full-time occupation.

FATE OF THE COMMUNITY

The Fourier community thus nurtured its nascent technology for more than two decades, before introducing an academic community, the analytical chemists, to it. In this time, the Fourier community managed not only to survive, but to find influential patrons in the American military and space programs, at national laboratories and in industry. Many of them (e.g. Gebbie, Mertz and other less familiar names) were funded by government contract and a succession of companies.

The members of this fluid community were disciplinary hybrids who had difficulty in establishing academic homes. Their visibility to colleagues in science and engineering was mixed. Those in industry published relatively little; they skirted recognition just as the community itself fell between the domains of academic science and commercial engineering. Much of their expertise remained tacit knowledge, embodied in a handful of practitioners fertilizing new companies. The sense of community was fading, too, by the mid-1970s, when the number of designers employed by commercial firms began to dominate those in relatively open and large institutions. Moreover, the 'external threat' of conventional technology was ebbing as Fourier spectroscopy became commercially established, which also diminished the strength of collective identity. While the interstitial specialism persisted and indeed prospered, its technological shepherds lost their social coherency.

CONCLUSION

The early survival success of the Fourier community was achieved principally through two tactics: affiliation with generous sponsors in the government and military, and by association with companies, which then largely proselytized the separate chemistry community. It is noteworthy that the early success was largely contingent on a particular political context: the uncritical cold-war funding for high-status and militarily promising projects. The community's employment of technical literature relied on new forms of journal that stressed 'applied science' or hybrid specialisms such as 'optical engineering' or 'electro-optics.' New modes of communication such as advertisements became persuasive. All these features deviate markedly from models of scientific development that emphasize the importance of university research and open publication in disciplinary journals.

What is distinct from many of the other contributions in this volume is the existence of a true *community* forged around a specific device. The members of this fragile skill collective enhanced their own chances of professional survival by cooperating professionally. They socialized at conferences, freely exchanged and debated ideas on engineering approaches, shared equipment and, most importantly, presented a united front to sway conventional spectroscopists. Their collectivism and tactics were an evolutionary necessity to enable each of them to cling to their untenured and vulnerable posts.

The Fourier community was 'emergent' in the sense of appearing only above of certain scale of activity, and not being predictable from smaller-scale events. This group became coherent only when defined by the acceptance of the technology by its sponsors and by its opposition to other groups such as conventional spectroscopists. The members were defined, and 'emerged,' not only through the technology they employed, but in relation to their social alliances and oppositions. The 'old guard' is, in fact, difficult to characterize in this episode because the principal form of opposition 'they' employed was to ignore the Fourier community. Thus, the small group of advocates for the new technology became more visible than the much more numerous supporters of the conventional technologies. This is intriguing because the Fourier community continued to survive in the cultural interstices: *between* science and engineering, *between* academia and industry, *between* design and application. The incongruity of this 'holistic' subject, like other research-technology specialisms that rely upon integration of multiple disciplines, is that its specialists have remained socially dispersed through industry, government and the fringes of academia.

University of Glasgow, Scotland

REFERENCES

Abbott, Andrew (1988). *The system of the professions*. Chicago: University of Chicago Press.
Bell, Robert J. (1972). *Introductory Fourier transform spectroscopy*. New York: Academic Press.
Callon, Michael and Law, John (1982). On interests and their transformation: Enrolment and counter-enrolment. *Social Studies of Science* 12: 615–26.

Cattermole, Michael J.G. and Wolf, A.F. (1987). *Horace Darwin's shop: A history of the Cambridge Scientific Instrument Company 1978-1968*. Bristol: Hilger.

Chantry, George W. (1971). *Submillimetre spectroscopy*. London: Academic Press.

Connes, Janine (1961). Recherches sur la spectroscopie par transformation de Fourier. *Revue d'Optique* 40: 45, 116, 171, 231 [Transl. by C.A. Flanagan as AD 409 869 (Defense Documentation Center, 1963)].

Connes, Janine and Connes, Pierre (1966). Near-infrared planetary spectra by Fourier spectroscopy. I. Instruments and results. *Journal of the Optical Society of America* 56: 896–910.

Connes, Pierre (1978). Of Fourier, Pasteur, and sundry others. *Applied Optics* 17: 1318–21.

Connes, Pierre (1984). Early history of Fourier transform spectroscopy. *Infrared Physics* 24: 69–93.

Connes, Pierre (1992). Pierre Jacquinot and the beginnings of Fourier transform spectrometry. *Journal de Physique II* 2: 565–71.

Connes, Pierre (1995). Fourier transform spectrometry at the Laboratoire Aimé Cotton 1964–1974. *Spectrochimica Acta* 51A: 1097–104.

Dunn, S.T. (1978). Fourier transform infrared spectrometers: Their recent history, current status, and commercial future. *Applied Optics* 17: 1367–73.

Fellgett, Peter (1951). *Theory of infra-red sensitivities and its application to investigations of stellar radiation in the near infra-red*. Unpublished PhD Thesis, University of Cambridge, UK.

Fellgett, Peter (1952). Multi-channel spectrometry. *Journal of the Optical Society of America* 42: 872.

Fellgett, Peter (1984). Three concepts make a million points. *Infrared Physics* 24: 95–8.

Fellgett, Peter (1991, January 18–30). Personal communication.

Finch, Arthur (1970). *Chemical applications of far infrared spectroscopy*. New York: Academic Press.

Galison, Peter (1987). *How experiments end*. Chicago, IL: Chicago University Press.

Gebbie, H. Alistair (1969). Fourier transform versus grating spectroscopy. *Applied Optics* 8: 501–4.

Gebbie, H. Alistair (1984). Fourier transform spectroscopy – recollections of the period 1955-1960. *Infrared Physics* 24: 105–9.

Golay, Marcel J.E. (1951). Multislit spectrometry and its application to the panoramic display of infrared spectra. *Journal of the Optical Society of America* 41: 468–72.

Hanel, R.A. (1970). International conference on Fourier spectroscopy, Aspen, 16–20 March 1970. *Applied Optics* 9: 2212-15.

Jacquinot, Pierre (1954). The luminosity of spectrometers with prisms, gratings or Fabry-Pérot etalons. *Journal of the Optical Society of America* 44: 761.

Jacquinot, Pierre (1969). Interferometry and grating spectroscopy: An introductory survey. *Applied Optics* 8: 497–9.

Johnston, Sean F. (1991). *Fourier Transform infrared: A constantly evolving technology*. London: Ellis Horwood.

Johnston, Sean F. (1996a). Making light work: Practices and practitioners of photometry. *History of Science* 34: 273–302.

Johnston, Sean F. (1996b). The construction of colorimetry by committee. *Science in Context* 9: 387–420.

Kahn, Franz D. (1959). The signal: Noise ratio of a suggested spectral analyzer. *Astrophysical Journal* 129: 518.

Kneubühl, Fritz K., Moser, J.-F., and Steffan, H. (1966). High-resolution grating spectrometer for the far infrared. *Journal of the Optical Society of America* 56: 760–4.

Kuiper, Gerard (1962). *Communications of the lunar & planetary laboratory of the University of Arizona*, Vol. 1: 83.

Latour, Bruno (1987). *Science in action*. Cambridge, MA: Harvard University Press.

Laukien, Günther R. (1997). Obituary. *Spectroscopy Europe* 9: 6.

Loewenstein, Ernest V. (1966). The history and current status of Fourier transform spectroscopy. *Applied Optics* 5: 845–54.

MacDonald, Keith M. (1995). *The sociology of the professions*. London: Sage Publications.

Mertz, Lawrence (1971). Fourier spectroscopy, past, present, and future. *Applied Optics* 10: 386–9.

Mertz, Lawrence (1991, 29January and 19 February). Personal communications.

Michelson, Albert A. (1902). *Light waves and their uses*. Chicago, IL: University of Chicago Press.

Milward, Ray C. (1969). A small lamellar grating interferometer for the very far-infrared. *Infrared Physics* 9: 59–74.

Milward, Ray C. (1990, 15 November). Personal communication.

Proceedings of the 1957 Bellevue Conference (1958). Les progrès recents en spectroscopie interférentielle. *Journal de Physique et Radium* 19: 185.

Proceedings of the 1966 Orsay Conference (1967). Méthodes nouvelles de spectroscopie instrumentale. *Journal de Physique* 28: suppl. C2.

Randall, Charles M. (1991, 22 April). Personal communication.

Sakai, Hajime (1991). *Unpublished notes*. Deptartment of Physics and Astronomy, University of Massachusetts at Amherst.

Scott, L.B. and Scott, R.M. (1953). A new arrangement for an interferometer. *Perkin-Elmer Corporation Engineering Report* No. 246 (May 22).

Shinn, Terry (1993). The Bellevue grand électroaimant, 1900–1940: Birth of a research-technology community. *Historical Studies in the Physical and Biological Sciences* 24: 157–87.

Shinn, Terry (1997). Crossing boundaries: The emergence of research-technology communities, in H. Etzkowitz and L.A. Leydesdorff (eds.), *Universities and the global knowledge economy. A triple helix of university-industry-government relations* (pp. 85–96). London: Cassell Academic.

Shinn, Terry (forthcoming). *Operation of the triple helix*.

Strong, John (1954). Interferometric modulator. *Journal of the Optical Society of America* 44(A): 352.

Strong, John (1958). *Concepts of classical optics*. San Francisco: W.H. Freeman.

Strong, John (1984). Fourier transform spectroscopy reminiscences. *Infrared Physics* 24: 103.

Swenson, Loyd S. (1972). *Ethereal aether: A history of the Michelson-Morley-Miller aether-drift experiments, 1880-1930*, Austin, TX: University of Texas Press.

Threlfall, Terry (1990, 25 September). Personal communication.

Vanasse, George A. (1982). Infrared spectrometry. *Applied Optics* 21: 189–95.

Vanasse, George A. and Strong, John (1958). Applications of Fourier transformation in optics: Interferometric spectroscopy, in J. Strong (ed.), *Concepts of classical optics*, Appendix F (pp. 419–34). San Francisco: W.H. Freeman.

CHAPTER 8

HANS-JÖRG RHEINBERGER

PUTTING ISOTOPES TO WORK:
LIQUID SCINTILLATION COUNTERS, 1950-1970

Perhaps no other instrument has symbolized the techno-myth of an avant-garde science, so widespread in the expanding community of molecular biology and radio-medicine in the 1960s and 1970s, more powerfully than the liquid scintillation counter. It was an apparatus that effectively came to represent three key technologies of this century: mechanical automation, electronics, and radioactive tracing. Yet in contrast to other instruments and techniques characteristic of the rising new biology and medicine such as electrophoresis, ultracentrifugation, electron microscopy, NMR (Nuclear Magnetic Resonance), and PCR (Polymerase Chain Reaction) (Elzen 1986; Kay 1988; Lenoir 1997: chapt. 9 [in collaboration with Christope Lécuyer]; Rabinow 1996; Rasmussen 1997) liquid scintillation counting has so far received no attention from historians of science and technology. This paper intends to exemplify the coming into being and the trajectory of a research enabling instrument. As I will show below, within twenty years liquid scintillation counting (LSC) developed from a clumsy technology for special purposes of radiation measurement into a generic technology that became ubiquitous in molecular biology and medicine laboratories in the 1970s. Among a few other early models, it was particularly Packard's Tri-Carb® Liquid Scintillation Spectrometer that made its way into university institutes, national laboratories, hospitals, and research departments of companies. A Packard Tri-Carb with its calculator data output connected to an IBM typewriter-printer became a signpost of an up-to-date modern biomedical laboratory in the 1960s and 1970s (Figure 8.1).

In this study, I briefly discuss the introduction of emitters of low energy β-particles (electrons) such as ^{35}S (sulfur), ^{14}C (carbon), and ^3H (tritium) into bio-medical research during the 1940s and early 1950s. The technologies initially available to follow these biological tracer elements in biochemical reactions were of very limited efficiency. I then show how several epistemological, technological, and cultural factors came to interact in the aftermath of World War II, which finally happened to establish liquid scintillation counting as an alternative to the traditional methods of solid sample counting or gas counting based on ionization. The first commercial liquid scintillation counter that became the prototype of a continuous production series was built for the University of Chicago by Lyle E. Packard in 1953.

B. Joerges and T. Shinn (eds.), Instrumentation: Between Science, State and Industry, 143–174.

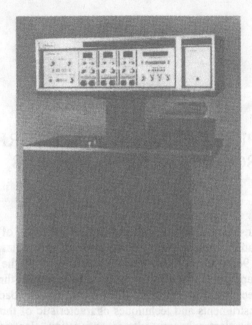

Figure 8.1.   First Liquid Scintillation Counter with an Absolute Activity Analyzer (Collection of Edward F. Polic, 702 Glenn Court, Milpitas, CA 95035-3330).

I have chosen the story of this prototype to describe the trajectory of a piece of research-technology, without aiming at an encompassing history of LSC in all its bewildering technical details and scientific ramifications. Between 1953 and 1970 the design of the instrument underwent a cascade of technological mutations that, after much exploratory tinkering, made its generic application possible. This, in turn, opened new epistemic dimensions for radioactive experimentation in biology and medicine. I follow the main events in the development of sample preparation, data processing, and the instrument's circuitry, a development that exemplifies the commercialization of nuclear energy in postwar America and beyond. Finally, I look at how the relations between producer and customers took shape as Packard Instrument grew from a one-man home business into an international corporation.

Radioactive tracing is an example of what Gerhard Kremer, a former president and now retired executive of Packard Instrument at the International Bureau of the Company in Zurich, insists on calling a "research enabling technology," i.e. a technology opening new fields of investigation.[1] According to Kremer, it is characteristic of such technologies, to put it paradoxically, that they permit questions to be answered that have not yet been posed. Basically, radioactive tracing consists of three components. The first is the production of suitable radioactive isotopes and the incorporation of these isotopes into a variety of organic molecules. It is characteristic of radioactive tracers that they do not noticeably alter the chemical or biological characteristics and functions of the compounds into which they are incorporated. The

---

[1] Interview with Dr. Gerhard Kremer, Zurich, 2 April, 1996.

second is the development of experimental systems in which particular metabolic reactions can be visualized and represented, preferably in vitro, through the addition of radioactive molecules as tracers. The third is the development of corresponding measuring devices. This paper will be largely confined to the third aspect of radioactive tracing.[2]

RADIOLABELS IN BIOLOGICAL AND MEDICAL RESEARCH

Biological and medical research began to change from the 1920s to the 1930s, and effectively was revolutionized in the course of World War II and immediately thereafter through the advent of artificial radioactive isotopes.[3] Of special importance in the context of this paper are those elements that represent major constituents of biological molecules such as hydrogen, carbon, phosphorus, and sulfur. Radioactive phosphorus (^{32}P) was one of the first cyclotron-produced radioactive isotopes. In 1939 Luis Alvarez and Robert Cornog, using the cyclotron at the Radiation Laboratory of the University of California at Berkeley's Department of Physics, obtained radioactive hydrogen (^3H) by bombarding deuterium gas with deuterons (Alvarez and Cornog 1939). Ernest Lawrence immediately realized the potential of the finding: "Radioactively labelled hydrogen opens up a tremendously wide and fruitful field of investigation in all biology and chemistry" (quoted in Heilbron and Seidel 1989: 272). He began negotiations with the Rockefeller Foundation over a large grant. Less than a year later, Samuel Ruben and Martin Kamen, using Lawrence's facility, recovered radioactive carbon (^{14}C) by bombarding graphite with deuterons (Kamen and Ruben 1940; Ruben and Kamen 1940; see also Kamen 1963[4]). The construction of powerful particle accelerators, and later, controlled fission, were crucial in the production of a large variety of new isotopes (Whitehouse and Putman 1953, esp. chapt. IV on the production of radioactive isotopes). In the aftermath of World War II, ^3H, ^{14}C, ^{32}P, and ^{35}S, derived in particular from reactor production, became widely available for biological and medical experimentation from the Isotope Distribution Program of the Atomic Energy Commission.

Like phosphorus and sulfur, hydrogen and carbon are ubiquitous constituents of organic matter. And like phosphorus-32 and sulfur-35, the radioactive isotopes hydrogen-3 and carbon-14 emit β-particles upon decay and have a half-life long enough to offer the prospect of being used as convenient tracers in metabolic studies. Typically, in such experiments the in vivo distribution or the uptake of these atoms into biological molecules is monitored. Alternatively, the metabolic fate of molecules isotopically labeled prior to their application is followed in vitro. With the exception of phosphorus, however, their radiation energy was not high enough to be measured

---

[2] Elsewhere, I have given an example of the second aspect, i.e. the construction of appropriate experimental systems. See Rheinberger 1997.

[3] For an early overview see Hevesy (1948); for the production of biologically and medically relevant istotopes in the first particle accelerators see, for example, Heilbron and Seidel (1989, esp. chapt. VIII); for the revolution of biomedicine in the wake of the Manhattan Project see Lenoir and Hays (2000).

[4] This paper was widely circulated; it appeared also in *The Journal of Chemical Education* (1963), and as an introduction to Rothchild (1965).

reliably by the conventional Geiger-Müller counting tubes then in use, which recognized carbon-14 only poorly and tritium not at all because its low energy β-particles could not penetrate the walls of the tubes.

The huge war efforts of the United States, epitomized by the Manhattan Project, resulted in an unprecedented expansion of radiation and radiation research, as well as in its diagnostic and therapeutic use in nuclear medicine, including human experimentation (Heilbron and Seidel 1989, chapt. VIII; Lenoir and Hays 2000). As a by-product of reactor development, radioisotopes came to abound. When the war was over, the United States concentrated atomic research in a network of national laboratories, among them Los Alamos, Berkeley's Radiation Lab, Oak Ridge, and Argonne. They were supervised by the Atomic Energy Commission (AEC) that had been set up in 1947 as a civilian, governmental agency to coordinate the military, economic, political, and scientific interests in atomic energy. Promoting the production of fissionable material and of atomic devices for military use was top priority, but the mission of the commission was also to succeed in "giving atomic energy a peaceful, civilian image" and, therefore, to promote research in such areas as radiobiology and radiomedicine (for the changing images of nuclear energy, see Weart 1988). In the first year of its existence, a division of biology and medicine was added to the AEC. Within the first few postwar years, radioisotopes flooded the laboratories and hospitals. In 1947 the isotopes produced in the reactor at Oak Ridge alone were "the equivalent of thousands of years of cyclotron production" (Hewlett and Anderson, Jr. 1962, Vols. I and II; Vol. III was added later, Hewelett and Holl 1989, see footnote[5]). In the summer of 1946, the Oak Ridge laboratory began delivering its radioisotopes to hospitals and universities nationwide as part of the Isotope Distribution Program of the AEC under its director Paul Aebersold who had done his PhD with Ernest Lawrence in Berkeley (Lenoir and Hays 2000). In 1947 AEC sold phosphorus-32 for $ 1.10 per millicurie; iodine-131 for $ 1.70, sulfur-35, $ 35.00, and carbon-14, $ 50.00 (AEC 1947). In 1948 the supply of isotopes required for biomedical research as well as for cancer diagnostics and therapy, as part of the atoms-for-peace campaign, even became free of charge (Hewlett and Anderson 1962). Phosphorus-32 and iodine-131 had already been tried in cancer diagnostics and therapy for a decade (Heilbron and Seidel 1989, chapt. VIII; Lenoir and Hays 2000), sulfur-35 and carbon-14 held promise to become ideal tracers for biochemical assays. The impact was massive. Between 1945 and 1956 the fraction of the total number of studies which used radioactive isotopes increased from 1 to 39% in the American *Journal of Biological Chemistry* (Broda 1960: 2). A brief sampling indicates that with this percentage, a saturation level had been reached for the years to come.[6] In 1966, more than 5,000 shipments with altogether 2.5 million curies left Oak Ridge (Cohn 1968).

The virtually ubiquitous presence of radiation in military as well as in civilian, environmental, and medical contexts called for new, sensitive and reliable detection, monitoring, and measurement devices. Postwar declassification of investigations in radiation instruments additionally triggered the search for alternative counting methods (Hewlett and Anderson Vol. II 1962: 247). Companies such as Radiation Counter

---

[5] The quotes are from Vol. II, pp. 96 and 109.
[6] In the first quarter of 1959, the number was 39%, in 1961, 43%, and in 1963, 33%.

Laboratories (Chicago), Instrument Development Laboratories (Chicago), North American Philips Company (New York), Victoreen Instrument Company (Cleveland), General Radio Company (Cambridge), Cyclotron Specialties Company (Moraga, CA), Engineering Laboratories (Tulsa), Geophysical Instrument Company (Arlington), soon produced counters of all sorts and sizes which were widely advertised in the scientific and technical literature.[7] The beginnings of a nuclear industry enhanced the trend (Balogh 1991).

EARLY STEPS IN RADIATION MEASUREMENT

In the early years of radiation research around the turn of this century one of the first methods for quantification of the activity of radioactive samples was based on the phenomenon of scintillation. Sir William Crookes in London developed a method for counting what he called the "emanations" of radium based on the scintillation, or light flashes, that these emanations provoked on a screen of zinc sulfide (Crookes 1903). Crookes' observation was immediately confirmed by Julius Elster and Hans Geitel from Wolfenbüttel (Elster and Geitel 1903). Five years later, Erich Regener in Berlin recorded the α-particles of polonium by using the scintillation method (Regener 1908). The light flashes were counted visually by using a simple microscope. This method gained wide acceptance in nuclear physics for about two decades, although it had major disadvantages grounded in the fact that the 'counters' were humans:

> Rapid fatigue of the observer and subjective influences require a frequent change of observers. They can only observe for half a minute up to a minute, and need long intervals in between. During a whole week, the time in which they can reliably observe amounts to two hours at best. The net effect is poor; good and useful computations can only be expected upon 20 to 40 scintillations per minute (Krebs 1953: 362).[8]

The scintillation method gradually fell into oblivion when Geiger-Müller counters came into use in the late 1920s (more details on the early history of radioactivity research and measurement can be found in Hughes 1993; Rheingans 1988; Trenn 1976, 1986). These instruments were based on the ionizing capacity of the emitted particles and the ensuing discharges produced in an electrical field in a gas-filled tube. Geiger-Müller counters proved useful for the detection of β-particles of higher energy. γ-rays could be measured, albeit with low efficiency, through the secondary electrons which they produced when penetrating the walls of the tube. Later versions of the Geiger-Müller counting tubes were supplied with a thin mica end window in front of which a solid sample could be mounted after plating it directly on aluminum planchets. Using this device, which remained in use well into the 1950s, β-particles emitted by radioactive carbon could be measured with an efficiency of about 10%. The weak β-emissions of tritium, however, remained beyond the scope of this tech-

---

[7] These companies are a sample taken from the first issues of *Nucleonics* in 1947.

[8] "Schnelle Ermüdbarkeit des Beobachters und subjektive Einflüsse erfordern häufigen Wechsel der Beobachter, die nur, mit langen Zwischenpausen, für eine halbe bis eine Minute zählen und pro Woche nicht mehr als zwei Stunden insgesamt zuverlässig beobachten können. Der Nutzeffekt ist sehr gering; gute, brauchbare Zählungen gelingen nur bei 20 bis 40 Szintillationen pro Minute."

nique. An alternative technique based on ionization consisted in converting the sample into a gaseous form, for example, by oxidizing ^{14}C-labeled compounds to radioactive carbon dioxide and water, and subsequently using ionization chambers in order to monitor the decay events. This method of gas counting worked in principle, but one of the big disadvantages was the very tedious sample preparation procedure and the difficulties of quantifying the probes to be measured.

At the beginning of the 1940s scintillation counting was taken up again as a result of developments in another field: photoelectricity. Peter Galison, distinguishing an "image" tradition from a "logic" tradition in the history of registration methods in particle physics, describes this conjunction as follows:

> What transformed the scintillator's flash and Cerenkov's glow into basic building blocks of the logic tradition was the electronic revolution begun during the war. When attached to the new high-gain photomultiplier tubes and strung into the array of amplifiers, pulse-height analyzers, and scalers that emerged from the Rad Lab and Los Alamos, then and only then did the scintillator and Cerenkov radiation become part of the material culture of postwar physics (Galison 1997: 454).

The physicist and biophysicist Adolf Theodor Krebs, a staff member of the Kaiser Wilhelm Institute for Biophysics in Frankfurt since 1937, who in 1947 became director of the Division of Radiobiology of the U.S. Army Medical Research Laboratory at Fort Knox, was probably the first to develop an instrument in which the human eye was replaced by a highly sensitive, fast-responding photoelectric device for detecting and counting scintillations (Krebs 1941, see also 1955). Attempts at improving combined scintillation and photoelectric gadgets intensified toward the end of the war, mainly due to the construction of efficient and reliable photomultipliers. Radio Corporation of America in the United States and E.M.I. in Britain soon became leaders in this technology essential for weapons control and guidance systems as well as for civil mass communication. Scintillation counters are usually understood to consist of an appropriate scintillating crystal in conjunction with a photomultiplier (Curran and Baker 1944). Devices were constructed to measure α-particles (Coltman and Marshall 1947) as well as β-particles and γ-rays (Broser and Kallmann 1947a,b). Alternatively, the Geiger-like photon tube counting devices, essentially a combination of the classical scintillation arrangement with a photosensitive Geiger tube of special design, became popular. They could be used for either an α-β-γ survey, for the selective detection of α-particles in the presence of a β-particle and γ-ray background, for β-particle detection alone, or for γ-ray detection (Mandeville and Scherb 1950). The main problem with the instruments of the first type was to contain the dark current of the photomultiplier, that is, the spurious activity of the device; the instruments of the latter type had the disadvantage of a finite dead time of the Geiger tube between the discharges. In combination with new solid scintillators for appropriately counting α-particles, γ-rays, and β-particles (Hofstadter 1948, 1949), this development was recognized by contemporaries as "one of the most important advances in devices for the detection of nuclear radiations since the invention of the Geiger-Müller counter" (Morton and Mitchell 1949: 16) and as heralding a "new era" of nuclear development and research (Pringle 1950), both with respect to resolution and efficiency of the counting process, and to applications involving low specific activity. As early as 1949 a first conference on scintillation counting was held at Oak Ridge.

LIQUID SCINTILLATION COUNTING

A new and different direction in counting technology was charted when Hartmut Kallmann from the Physics Department of New York University,[9] in collaboration with Milton Furst, seriously began to work on his earlier observation that certain organic substances such as anthracene, in aromatic solvents such as toluene, worked as scintillators and when used in conjunction with an electron-multiplier phototube should be suitable for liquid scintillation counting (Kallmann and Furst 1950, 1951; Furst and Kallmann 1952). At Princeton University, George Reynolds and his colleagues worked on the new technology as well (Reynolds, Harrison, and Salvini 1950). As in the case of solid scintillation, the process basically involved the conversion of radioactive decay events into photons, and the photons into photoelectrons that could be amplified and counted. The energy of the decay events would be absorbed by the scintillator solvent, which would then transfer the energy to the scintillator solutes causing them to emit photons. These in turn could be collected in a photomultiplier tube and amplified. The energy transfer processes in the solvent system were only poorly understood at the beginning, and it took years to elaborate their physical details. The early work in this completely new field of liquid scintillation counting concentrated on the external counting of high energy radiation emitted from sources such as radium (Kallmann) or cobalt-60 (Reynolds).

In 1951, M.S. Raben from the New England Center Hospital and Tufts College Medical School in Boston and Nicolaas Bloembergen from the Nuclear Laboratory of Harvard University suggested "that a simple and geometrically ideal counting system might be obtained by dissolving the material to be counted directly in [the] liquid. This method would facilitate particularly the counting of soluble compounds labeled with a weak β-emitter, such as C^{14}" (Raben and Bloembergen 1951). First measurements showed that with internal counting, it might be possible to trace even Nanocurie amounts of carbon-14. That finding promised a gain in sensitivity, efficiency and accuracy of counting other weak β-emitters such as ^{35}S, and, for the first time, even ^{3}H. The reason was the homogeneous distribution of the radioactive sample and the virtually complete absorption of the emitted energy by the scintillator.

Some early internal sample liquid scintillation counters were basically an adaptation of the crystal scintillation spectrometers available at the end of the 1940s. They consisted of the sample in a glass bottle which was surrounded by a reflector and made contiguous to the photomultiplier tube by an optical coupling fluid such as silicon oil, glycerin, or Canada balsam. This arrangement was followed by a preamplifier, an amplifier, a pulse-height analyzer, and finally a scaler element. The expecta-

---

[9] Kallmann had done his thesis under Max Planck and had been a staff member of the Kaiser Wilhelm Institute for Physical Chemistry and Electrochemistry in Berlin-Dahlem since 1920. In 1933, he was dismissed after the Nazis had come to power. However, he was not allowed to leave the country and forced to work with I.G. Farben throughout the war (Oster 1966). Already in 1947 and meanwhile professor (1945–1948) at the Technical University of Berlin, he had announced his version of a scintillation counter (Broser and Kallmann 1947). A year later, he came to the United States, joining the U.S. Army Signal Corps Laboratories in Belmar, New Jersey, as a research fellow before being appointed as professor and director of the Radiation and Solid State Laboratory at New York University's Physics Department in 1949.

tion was that tritium could be measured by such an instrument up to an efficiency of 20%. However, the "dark current" (spontaneous thermionic emissions from the photomultiplier cathode) became prohibitively strong at the high voltages required to attain such efficiency (Hayes, Hiebert, and Schuch 1952). The noise could be somewhat reduced but not suppressed by the selection of appropriate multiplier tubes, by pulse-height discrimination, and by refrigeration. The last, in turn, put constraints on the scintillator solutes because of the temperature-dependence of their solubility. The single photomultiplier liquid scintillation spectrometer remained a transient adaptation of the previous solid scintillation counter.

A major advance was made by the group working at Los Alamos Scientific Laboratory, especially Newton Hayes and R. Hiebert, toward establishing internal sample liquid scintillation counting as the preferred method for measuring the activity of low energy β-emitters. They accomplished this both by exploring various scintillator solutions to meet typical applications in biology and medicine and by the development of improved and robust coincidence-type counting equipment (Hayes, Hiebert, and Schuch 1952; Hiebert and Watts 1953). The principle of coincidence counting went back to the days of Walther Bothe and Hans Geiger (Galison 1997: 438–54), and it had been adapted by the engineers of the Radio Corporation of America, R.C.A. Laboratories Division in Princeton, for use in a solid scintillation counter in 1949 (Figure 8.2, Morton and Robinson 1949). The method was then adapted by Kallmann as well as Reynolds to external liquid scintillation counting in 1950 and by Raben and Bloembergen to internal liquid scintillation counting in 1951 (Kallmann and Accardo 1950; Raben and Bloembergen 1951; Reynolds, Harrison, and Salvini 1950). In simplified terms, the noise generated by the electronic equipment was virtually eliminated by placing two photomultipliers opposite each other that simultaneously examined the same sample. After amplification only those pulses were counted that arrived 'in coincidence' at the pulse height analyzer and therefore could be assumed to arise from one and the same scintillation event caused by a decay electron rather than by noise of the system. Under these conditions, a single tube noise rate of tens of thousands of counts per minute (cpm) could be reduced to the acceptable order of tens of counts per minute.

Between 1952 and 1957, six internal LSC coincidence counters were built for use in Wright Langham's Biomedical Research Group at the Health Division of Los Alamos Scientific Laboratory (Hiebert and Hayes 1958; Langham 1958). Other people at Los Alamos thought bigger in terms of external LSC. Ernest Anderson built a machine for externally monitoring whole human bodies for such things as gross body composition with naturally occurring potassium-40 or for measuring the accumulation of radioactivity in the bodies of people exposed to radioactive fallout. In 1957, he duly noted the "sharp rise in public concern over the effects of low intensity radiation on man over the past few years." Exposure to radioactivity, its measurement, control, and prevention had become a vehemently debated issue with the spread of atomic power from weapon production and weapon testing to industrial plants and the biomedical sector. In a deliberately polemic and apologetic tone Anderson mocked that there might soon be "a legal prohibition of some of our most popular materials of construction, notably concrete and brick, on the basis of their high concentration of natural radioactivities such as radium and potassium" (Anderson 1958:

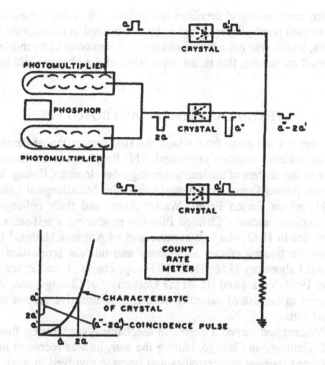

Figure 8.2. Coincidence circuit using crystal diodes (Morton and Robinson 1949).

211). A colleague of Anderson, Frederick Reines, was engaged in making giant liquid scintillation detectors for neutrino and for neutron detection (Reines 1958; Galison 1997: 460–3).

In the context of a discussion of instrumentation for research and the production of generic, multipurpose devices, this constellation is particularly interesting. For at the beginning the new technology of internal liquid scintillation counting served fairly special purposes in the whole context of radiation measurement, and it had quickly become a rather sophisticated assembly of different physical, organochemical, and technical parts into which biological, among other, material happened to be inserted. There appeared to be a long way to go, if there was such a way at all, to achieving the objective of producing an instrument for routine laboratory work that could be operated by inexperienced personnel. The potential "ease of preparing counting samples by simply dissolving the substance in solvent in a bottle" was hardly on the horizon (Davidson and Feigelson 1957: 3). The possibility of a generic use of the new device that could have been of appeal to a wide variety of laboratory workers concerned with isotope production, with monitoring, and with waste management in materials research as well as in medical diagnostics, in biology, chemistry, and pharmacology was remote. Additional technical feats included sample vial geometry, the elimination of luminescence, the appropriate choice of vial glass type, the optimization of photocathode sensitivity and the emission spectrum of the scintillator, and many more technical details of this caliber. The sample material in turn had also to meet certain prerequisites. It had to be solubilized in the organic liquid scintillation solvents, which was not a trivial matter; and it needed to be minimally colored in order to avoid quenching, that is, the depression of the photon-yield induced by the probe itself.

TESTING A COMMERCIAL PROTOTYPE

Los Alamos was not the place from which the first generation of commercial liquid scintillation coincidence counters originated. This happened at the University of Chicago, another of the centers of nuclear technology development during World War II. It was here that Arthur Compton had established the Metallurgical Laboratory (Met Lab) in 1941, where Enrico Fermi, Walter Zinn, and their colleagues built the world's first nuclear reactor – Chicago Pile-1 – producing a self-sustaining nuclear chain reaction late in 1942. Met Lab became part of Argonne National Laboratory in 1946, a center for nuclear reactor technology and nuclear propulsion engines (Argonne National Laboratory 1986; Holl 1997: esp. chapts. 1 and 2; see also Hewlett and Anderson 1962: Vols. I and II). At the University of Chicago, too, Willard Frank Libby pioneered in the use of naturally occurring radioactive carbon as a means of archaeological dating.

Lyle E. Packard had earned a degree in mechanical engineering from the Illinois Institute of Technology in Chicago. During the war, he was recruited into the Navy, where he obtained training in electronics and became involved in work on radio, radar, and sonar. After the war, in the spring of 1946, he was hired as an engineer for the Institute of Radiobiology and Biophysics at the University of Chicago by its director Raymond Zirkle. The institute was one of three new research bodies (the two others being the Institute for Nuclear Studies, now the Enrico Fermi Institute, and the Institute for the Study of Metals, now the James Franck Institute) that were established under the presidency of Robert Maynard Hutchins in the context of the University's peacetime program initiated immediately after the war (McNeill 1991: 123–4, 158). They replaced parts of the Manhattan Project that had been operating under the umbrella of the University of Chicago. Hutchins wanted the new Institutes "to advance knowledge and not primarily to develop the military or industrial applications of nuclear research." He pondered that "for the past six years the United States has abandoned both basic research and the training of a new generation of scientists. It is essential to our progress and our welfare that we overcome that deficiency" (The University of Chicago Publications Office 1991: 105).

In addition to the design and construction of temporary laboratory space for the first few years of the Institute's work, Packard supervised a small staff of engineers, technicians and machinists in conceiving and building special instrumentation and installing and maintaining equipment for the various research groups in the Institute. This involved a wide variety of very specialized items. For example, there was physiological equipment for studying axon cells in squids for a group headed by Kenneth Cole and George Marmont. Raymond Zirkle and William Bloom required custom systems for time-lapse photomicrography and various radiation equipment, including a Van de Graff generator for basic radiobiological studies. Along with several other groups, Leo Szilard had a number of special requirements.

Together with Aaron Novick, the physicist Szilard was then 'retooling' in biology and thought about building a "chemostat," an instrument that would keep a bacterial population growing over an indefinite period of time in order to be able to study their behavior subsequent to mutation (Feld and Weiss Szilard 1972[10]). After reluctantly leaving Columbia in 1942 and joining Compton's Met Lab in 1942, Szilard had taken an indefinite leave of absence without pay from the Manhattan Project in Chicago in the fall of 1945. He had been unsuccessful in trying to convince President Truman not to use the bomb against Japan. In October 1946, president Hutchins appointed him as professor of biophysics at the Institute of Radiobiology and Biophysics (Grandy 1996, esp. chapts. 5 and 6; Lanouette 1992). Packard recalls:

> When Szilard joined the University he had a number of patentable ideas, things that he wanted to preserve for himself and he excluded those from his contract. It was on one of those things in particular that I have worked personally with him, after-hours, weekends and so on. So I got to know him a little bit. Very, very interesting experience.

Szilard's laboratory had been designed by Packard and was located in the basement of a former synagogue of a Jewish orphanage. The building had been taken over by the University of Chicago and was ready by January 1948 (Feld and Weiss Szilard 1972). The first chemostats were placed in a thermostatically controlled room at 37° C. "We did that at Szilard's request in a very inexpensive way by controlling banks of commercial 1,500-watt heaters."[11]

In 1948, Packard also came into contact with a group of researchers at the Institute for Nuclear Studies who were using mica end-window Geiger counters for counting ^{14}C. The efficiency of these counters was very poor. Packard recounts that it was on the order of 10%. Nathan Sugarman at the Institute had devised an instrument without a window, and Packard engaged in efforts to construct a workable windowless counter of the sort where the sample could be pushed right into the counting chamber. It was to yield a higher counting efficiency, because it circumvented the absorption of the β-particles in the thin, but not thin enough mica plate of the window. The use of low energy radioisotopes whose electrons had only a very short range and thus were difficult to monitor was expanding at a rapid pace because of the Isotope Distribution Program activities. The reason was that they could be used to label a whole range of organic molecules and thus were of potentially unlimited application in metabolic studies in vivo and in vitro. The more these labeled compounds (among them nucleotides, amino acids, fatty acids, sugars, antibiotics) became disseminated, the more counting devices were needed. In addition, an urgent need for monitoring contamination arose. "At that point we had so many visitors coming to the University, visiting these new Institutes, and they would invariably come around and ask: 'Where could I get one of these?' 'Well, you can't get one, we make them here'."[12]

---

[10] See especially Part IV, Published Papers in Biology (1949-1964), with an introduction by Aaron Novick.

[11] Interview with Lyle E. Packard, Chicago, November 5, 1996.

[12] Packard, Interview 1996.

As a result of this demand, in 1949 Packard began to think about starting a company part-time. With the permission of the University administration, he set up a company, together with a partner, called "Research Equipment and Service." Their first product was a windowless counter that Herbert Anker had designed in the Biochemistry Department. It had a rather slow transfer mechanism with one single sample position to push in and to pull out. Packard sold these to other institutions as well as other departments at the University. A variation of the windowless counter was soon to follow, with a much more efficient circular sample device that had three positions: an internal counting position, a pre-flushing position, and a loading and unloading position exposed to the air. Another of the early products of the company was not directly related to the monitoring of radioactivity. It was something as unspectacular as a fraction collector.

At the Institute of Nuclear Studies, Packard had become friends with James Arnold, a former student and coworker of Willard Frank Libby who was soon (1950-1954) to serve as a member of the General Advisory Committee of the AEC and who, a decade later (1960), was to receive a Nobel Prize for having developed the concept and method of carbon-14 dating in archaeology (Libby 1952). Libby had gotten his first results with a solid sample Geiger-Müller counter, "a big thing about 4 inches in diameter, with all kinds of shielding around it."[13] James Arnold had heard of the potentials of liquid scintillation counting advocated by Kallmann and Furst from New York University and by Reynolds from Princeton. He had set out to explore the prospects for internal liquid scintillation counting in Libby's project of carbon-dating. His instrument had the features of a coincidence counter of the sort that Hayes was building for Langham's Biomedical Research Group at Los Alamos. Arnold entertained good contacts with Hayes and Anderson who made unpublished data and new scintillation materials accessible to him. Furthermore, he went on to try to push the internal sample idea to its very extreme by actually converting his very low activity samples into the solvent for the scintillator.[14] The Los Alamos work also caught the attention of George Leroy from the Argonne Cancer Research Hospital at the University of Chicago who consulted with Los Alamos on medical matters. While working at the University, Packard had followed Arnold's work with great interest. Leroy knew of this, and he knew that Packard was working in his own company by 1952. So he asked him to design and build a liquid scintillation system for him.[15] The Argonne Cancer Research Hospital had been founded in 1948 with money from the AEC and was operated by the University of Chicago. The Hospital was part of AEC's first efforts to fight "America's number-two killer disease" (Holl 1997: 75)

At that time, in 1952, Packard had left the University in order to work full-time in his company. His business, soon to be renamed "Packard Instrument Company," consisted only of himself.[16] Together with a newly hired coworker, Packard started to

---

[13] Packard, Interview 1996.

[14] "A method for converting samples to aliphatic hydrocarbon is being worked out, so that solutions of good efficiency can be prepared that are 80% sample" (Arnold 1954).

[15] Packard, Interview 1996; Packard to Rheinberger, August 26, 1998.

[16] His earlier partnership had been limited to the windowless Geiger counters.

Figure 8.3. First commercial Liquid Scintillation Counter made by Packard, sold to Argonne Cancer Research Hospital, circa 1953 (Collection of Edward F. Polic).

build a prototype liquid scintillation coincidence system in the front part of his apartment which he had transformed into a workshop. It took him about a year to get the first unit built, and it was delivered to the Argonne Cancer Research Hospital in 1953 (Figure 8.3). None of the basic components was completely new in this machine, but Packard knew how to engineer research equipment. And due to his previous experience at the University, he was aware of what biomedical users would require from such an instrument: versatility and easy operation. George Leroy and his coworkers were planning to evaluate double-label experiments with tritium and ^{14}C in this machine (Kabara, Okita, and LeRoy 1958). "That's when I came up with the idea of designing the production model especially for these two isotopes and naming it the Tri-Carb for ^3H and ^{14}C and gave it the model number 314." This requirement determined the unique design of the electronic circuitry for the production units that were to follow (Packard, Interview 1996).

The sale price of the prototype, including the refrigeration unit that basically consisted of an adapted commercial freezer, was $ 6,500, about five times the cost of a Geiger counter and a scaler at that time. Eugene Goldwasser, a biochemist who had become associated with Leroy in 1952 after two postdoctoral years with Herman Kalckar in Copenhagen, recalls:

> It was one of those peculiar times of history, at least from my perspective, when we had all the money we needed for research and George could go on and ask Lyle to build [an instrument] without worrying about where to get the money to pay for it. I came to Chicago from Copenhagen, and the first thing I had to do was sit in an unfinished room with stacks and stacks of catalogues and start a lab from nothing. And I figured roughly in 1952 I spent about a million dollars. [This] was all AEC money; there seemed to be no end to it. [That] was part of the original AEC charter from the Congress. They were to

promote the use of radioactivity in research and therapy, and to promote the development of instrumentation for study of radioactivity. So that my work which had little to do with atomic energy or cancer research [was] funded under their umbrella because I used isotopes.[17]

In Denmark he had experimented with radioactive adenine homemade by his mentor Kalckar (Goldwasser 1953). Now, he obtained his labeled compounds from Berkeley's Rad Lab (Goldwasser 1955).

It was indeed a very typical story on the one hand, and on the other hand, a particular epistemic and technical constellation. There was a mechanical engineer with electronics experience working in close contact with academic researchers on a piece of advanced research-technology that held the promise of very specified, local uses in archaeological dating and double-label experiments. When asked about the characteristics of such interaction, Packard replied:

> I don't think I can generalize my thoughts on the interactions of scientists and engineers. Based on nearly five decades of developing and manufacturing scientific instruments as well as five years at the University of Chicago functioning in somewhat varying roles, but all generally facilitating the requirements of scientists, I found extreme variations. To a certain extent it depends on the field of science. As might be expected, physicists typically are more concerned with technical specifications and details of the mechanics and electronics of what they want from engineers. Biological and medical researchers, I have found, usually interact with engineers on the basis of the function they wish to accomplish – how easily, how fast, how precisely, etc. – and typically are not interested in details of the equipment as long as it performs what it is supposed to do reliably. There are exceptions and, particularly in earlier times, I have seen extreme cases. For example, I've seen biological scientists who like to play at engineering spend months of their laboratory time improvising something like a homemade fraction collector (Packard to Rheinberger, letter of February 25, 1998).

Packard's company grew out of these variegated laboratory contacts, and his products initially were purchased principally with federal AEC and Public Health Service money to which there seemed virtually to be no limit in these first years of the Cold War.

Soon Packard relocated his business. He moved out to the suburbs of Chicago, to LaGrange, Illinois. Right after having built the prototype machine for Leroy, he had received orders for two more machines. One of them went to Jack Davidson of the Presbyterian Hospital at Columbia University. "That was really the first which I would call a production unit" (Packard, Interview 1996). Davidson used his machine to do a lot of optimization with respect to sample size and different mixes of solvents, primary and secondary solutes such as 2,5-diphenyloxazole (PPO) and 1,4-di(2-[5-phenyloxazolyl])benzene (POPOP) which came to be called 'cocktails' in laboratory jargon (Davidson and Feigelson 1957: 17). As a result, he rated LSC to be "a useful new technique," although "by no means the panacea for all counting problems" (Davidson and Feigelson 1957: 17). Its main advantages were the excellent sensitivity for very weak decay electrons, high precision, high absolute efficiency, and relative ease of sample preparation. Surveying the early literature one gains the

---

[17] Interview with Eugene Goldwasser, Chicago, November 5, 1996; Goldwasser, letter to Rheinberger, November 4, 1998.

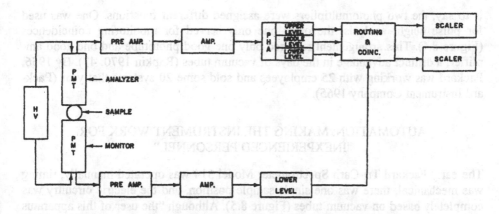

Figure 8.4. Packard Model 314 block diagram, 1954 (Rapkin 1970).

impression that these first commercial machines were themselves mainly part of exploring the method. They were components of a very small, but self-amplifying circuit. Basically, the application of the methodology consisted in its own optimization.

Clearly, in the early 1950s a liquid scintillation counter was not perceived as a potentially universalistic piece of equipment in the world of biomedicine. Nuclear-Chicago, one of the biggest and most experienced instrument builders, not only in the region, but in the whole field of radiation technology, had put its bets on solid sample and gas counting. According to the testimony of a user, its D-47 Micromil gas flow counter, with its ultra-thin Mylar window, came to approximately 40% efficiency for ^{14}C, had a low background due to its anti-coincidence circuitry, and its sample changer was "nearly failsafe."[18] But, although Nuclear-Chicago entered liquid scintillation counting only at the beginning of the 1960s (Rapkin 1970: 47), Packard's Tri-Carb system did not remain completely without commercial competition in those early years. Tracerlab in Waltham, MA, a company that also produced radioactive biochemicals, undertook some efforts in LSC construction, and so did Technical Measurement Corporation in New Haven, CT (for this instrument, see Utting 1958). Packard recalls an early and fierce competitive test between one of his and one of Technical Measurement Corporation's (TMC) units at the National Institutes of Health in Bethesda:

> Our biggest competition in the early days came at the NIH. It seemed to us, in a much smaller way of course, like something they sometimes had in military procurement, where two companies would be requested to provide special-purpose airplanes for a fly-off to see which one was better. In this case, NIH requested one of our Tri-Carb systems and one of the TMC units for side-by-side comparisons. After extensive testing, our Tri-Carb system was selected and purchased as the first of dozens that NIH would acquire during the next few years (Packard, Interview 1996).

Edward Rapkin, one of Packard's later coworkers, comments that it was one of the special features of the Packard unit that its system logic was "unsymmetrical." That

---

[18] Loftfield, letter to Rheinberger, September 24, 1998.

is to say, the two photomultipliers were assigned different functions. One was used for pulse height analysis, the other one only served for monitoring coincidences (Figure 8.4). This arrangement required only one good phototube and one good amplifier, a distinct advantage in the days of vacuum tubes (Rapkin 1970: 47). By 1956, Packard was working with 25 employees and sold some 20 systems that year (Packard Instrument Company 1965).

AUTOMATION: MAKING THE INSTRUMENT WORK FOR "INEXPERIENCED PERSONNEL"

The early Packard Tri-Carb Spectrometer Model 314 was operated manually, timing was mechanical, there was one single-sample position, and the electric circuitry was completely based on vacuum tubes (Figure 8.5). Although "the user of this apparatus [did] not need to understand all of the electronics involved," Davidson contended that "any intelligent use of the technique [required] familiarity with the general principles of the electronic equipment" (Davidson and Feigelson 1957: 3). The first major change of design was introduced in 1957. It consisted of incorporating an automatic 100 sample changer (Figure 8.6; Packard 1958). Prior to this, changing a sample had been a tedious and time-consuming exercise. Even the high voltage had to be switched off before removing one sample and inserting another one.

> Take it out, put the next sample in, close the light-tight chamber, close the lead shield, close the freezer door, turn the high voltage on. And then wait a little bit and then start your count. And your count was manual. So, you sat there and you watched the clock go around for a minute or two minutes or five minutes or whatever (Packard, Interview 1996).

The sample vial was a large fifty millimeter diameter glass vessel that was immersed in silicon oil to make a good optical connection to the phototubes. The first prerequisite for automation was reducing the size of the vials. This also made the silicon oil connection unnecessary. Packard and his associate Soderquist did the detailed design of a circular sample-changing device. The former sample shield had been a horizontal cylinder. It was changed to a vertical iron cylinder, and the turn-table arrangement was put on top of it. This model sold. In 1958 some 80 employees produced and delivered close to a hundred units. At this point, after TMC had discontinued its production of internal liquid scintillation counters, Packard had a market share of nearly 100%.

If there is one thing in particular which made the liquid scintillation counter really attractive, it is its transformation into an automatic machine. With the conventional counters in operation right after the war, "even when efficiently organized, each assay must have consumed an hour of professional time."[19] The possibility of doing serial counts involving hundreds of samples unattended and overnight opened the prospect for performing experiments of hitherto unheard-of dimensions that required frequent measurements and combined different types of assays. For instance, in kinetic experiments, such as binding studies, probes have to be taken at as many time

---

[19] Loftfield to Rheinberger, September 24, 1998.

Figure 8.5. First production Tri-Carb Liquid Scintillation Counter, detector with light-tight shutter – for 50ml sealed ampoule samples or 20ml vial samples, circa 1954 (Collection of Edward F. Polic).

intervals as possible. In addition, these assays usually have to be performed at different temperatures and, in order to be reliable, carried out in duplicate or triplicate thus easily resulting in hundreds of samples. The automatic machine also allowed testing long series of fractions derived from chromatographic purification or separation columns or from preparative ultracentrifugation runs; and it allowed laboratory workers to include in the design of their experiments as many controls as they deemed necessary for good and reliable work in order to obtain their differential signals. Packard comments:

> Putting this turntable on, and making it automatic, just opened up the possibility for what then were massive studies, I mean, you would take a rat and sacrifice it, and take all the parts of the rat. [It] made it possible for people to design different types of experiments than they ever could have designed before. Previously you never would have designed an experiment for a thousand samples. [Well] I don't know how many times, but certainly *many* times I have had people come up to me at trade shows, really prominent names in the field and say, 'you know, Lyle, all the work I have done in the last five years I could not have done if I hadn't had a Tri-Carb' (Packard, Interview 1996).

If at the beginning liquid scintillation counting had just been a potentially promising technology that would allow researchers to introduce tritium labels in their biomedical tracing arsenal and to exploit carbon-14 labels more efficiently, its automated variant enabled them to set up experimental systems of previously unthinkable dimensions and designs, and to scale up routine monitoring by orders of magnitude. On the epistemic side, a point was reached where the scope of the instrument broadened from locally enhancing sensitivity to gaining a generic impact on the way biomedical research could be conducted. On the routine side, a quantum leap in monitoring capacity was achieved. With that, automated liquid scintillation counting became one

Figure 8.6. First automated Liquid Scintillation Detector, steel shielding, dual elevators, 100 20ml vial samples in four circular rows, 1957 (Collection of Edward F. Polic).

of the instrumental bridges in joining the biochemical and biophysical world of preparation to the analytical world of molecular biology.

As Wright Langham from the Los Alamos Biomedical Research Group put it toward the end of the 1950s:

> Biological and medical investigations by nature call for counting systems with the greatest of versatility. Among the requirements are (a) analyses of large numbers of samples with a minimum of processing; (b) high sensitivity; (c) wide adaptability as to variations in sample size; (d) accommodation of wide variations in nature and chemical composition of the sample; [and] (f) dependable operation with a minimum of servicing.

Large sample numbers result from the necessity of doing multiple counts due to the inherent variability of biological probes, as well as from doing serial experiments of the kind just mentioned. A minimum of processing reduces the chances of accidental loss of activity. Every experimenter in the field knows that each processing step between the assay and the counts in principle is one step too much. High sensitivity allows for low doses of radioactivity often needed to avoid damage of the biological sample. Sample sizes often vary due to intrinsic necessities of the design of the experiment. And the same is true for variations in the nature of the sample. With the new automated machines, liquid scintillation counting promised to match all these needs. Langham went so far as to conclude: "Liquid scintillation counting is the most important recent development in the applications of radioisotopes to biology and medicine" (Langham 1958: 136–7).

Edward Rapkin, who had joined Packard in 1957, recalls: "It did also make the counter business very good. [Because] the thing that converted everything from an occasional sale of three or four a month to some months fifty was the automatic counter" (Rapkin in Packard, Interview 1996). In 1961, more than 700 Tri-Carb Spectrometers had been installed all over the world since the delivery of the first production unit in 1954.[20] Around the turn of the decade, direct sales to the United States government and its agencies accounted for about 15% of the total sales. The rest went to the universities, the hospitals, and the industry.

In 1959 Tracerlab made a second brief attempt to enter the market, this time with a 40-sample automatic counter. Tracerlab was the first company to use the new E.M.I. photomultiplier that soon became standard in the field, and it was the first to introduce transistorized preamplifiers which, however, in 1959 could not yet match the quiet performance of good vacuum tubes. These innovations notwithstanding, the Tracerlab machine lacked "important user requirements" such as high sample capacity and a light-tight detector (Rapkin 1970: 50). Electronic innovation alone could not outdo convenience in operation. In 1960, Packard introduced the 314A Tri-Carb which was entirely transistorized except for the preamplifiers. In 1961, the 314E Tri-Carb series came on the market. It was a completely transistorized liquid scintillation counter exhibiting an improvement of the system logic such that two different isotopes present in one sample could be counted simultaneously with separation efficiencies that previously required successive counts. Also in 1961, Nuclear-Chicago entered the liquid scintillation market. In 1962, it offered an instrument with a serpentine sample transport and a sample capacity of 150 vials. This machine was able to perform repeated counting cycles, and it allowed the grouping of samples so that the counting output could accommodate different users. The instrument provided three counting channels and thus allowed for channel ratio quench monitoring. A mechanical calculator could perform some data processing. Introduction of this Nuclear-Chicago counter was an effective challenge for Packard who introduced a new model 200 sample capacity Tri-Carb at the same time. Within five years, Nuclear-Chicago's market share rose to an estimated 20%, whereas Packard Instrument's went down from 85 to 63%.[21]

The market expanded rapidly. A glimpse on the publications in the *Journal of Biological Chemistry* is revealing in this respect. In 1959 only roughly one out of ten experiments performed with either ^{14}C or ^{3}H was evaluated by means of liquid scintillation counting. Conventional end window Geiger-Müller counters, vibrating reed electrometers, and gas flow counters dominated the scene, especially the Nuclear Chicago D-47 windowless gas flow counter. Only four years later, almost every third experiment involved a liquid scintillation counter, more than 80% of which were Packard Tri-Carbs, and by the end of the 1960s virtually every $^{14}C/^{3}H$-based experiment ended up being fed into a liquid scintillation counter. By that time, the application of the method had become so ubiquitous that it needed no longer to be mentioned in the "Methods" section of a paper.

Concomitantly, competition became fierce, and system-logic innovations and user-friendly improvements of operation as well as recording equipment became mandatory.[22] In 1962 a couple of people who had left Packard started a company called Vanguard that advertised a bench-top automatic LSC, but went out of business after delivering a few units. In March 1963, Rapkin left Packard and started a small

[20] Packard Instrument Company, Liquid Scintillation Counting Systems, Advertisement, September 1961.
[21] Estimates of Edward Rapkin. Packard to Rheinberger, August 26, 1998.
[22] Packard to Rheinberger, August 26, 1998.

company to produce an LSC he called ANSitron. It became the first unit in production to incorporate automatic external standardization. Packard had started developing ideas for automatic standardization to correct for quenching, and in June 1965 was issued a U.S. patent.[23] In 1966, Picker Nuclear acquired the ANSitron, but shortly went out of the business. Also in 1966, low cost models of LSC were introduced by Nuclear-Chicago, Packard, Picker, and Beckman. In 1967, both Packard and Beckman announced different and rather sophisticated systems that utilized their automatic standardization equipment to offset the effects of quenching before counting each sample, as opposed to just measuring how much the quench effect is.[24] Packard's company had grown from 50 employees in 1957 to over 500 in 1966, and the corresponding net sales had risen from half a million to $ 14 million, equalling about one thousand units per year. However, a still expanding market was able to accommodate all these competitors (Table 8.1). LSC counters began to spawn the laboratory spaces.

The combination of improved system logic, of facilitated operation for multiple users engaged in different experiments in large laboratory settings, of versatility in raw data processing, plus increasingly sophisticated scintillation cocktails for various kinds of probes, made liquid scintillation counting ubiquitous in molecular biological research, in biomedical diagnostics, in clinical settings, and in pharmaceutical firms. The production of increasingly more optimized liquid scintillation counters in the fifteen years between 1952 and 1967 coincided with the exponential growth of biochemical and biophysical research and the corresponding increase in the commercial production of radiochemicals in these years after the war, with its huge programs of fighting cancer (see Gaudillière 1998), and with the whole complex of nuclear medicine and related industries. Yet, we must see this as a two-way connection in which both fields contributed to co-generate each other. Without the massive advent of radioactive biomolecular tracers in general, among other technical feats to be sure, the molecularization of biology and medicine would have been different; and without the technology of liquid scintillation counting in particular, the extended in vitro assay designs and experimental systems that, for instance, made the code-deciphering work between 1961 and 1965 feasible, would not even have been thinkable. It is not that this technology opened up an entirely new field of investigation. It had a much more general impact on assay design and the range of metabolic analysis.

In this context, it is equally important to stress the ever more diversified synthesis of ^{3}H and ^{14}C labeled molecular compounds. At the beginning, they came mainly from the National Laboratory at Oak Ridge. Soon commercial producers took over, among them Tracerlab, New England Nuclear Corporation, Isotopes Inc., and the Radiochemical Center Amersham in England. Robert Loftfield recalls: "New England Nuclear set up a tritiation service: send us your compound, we will tritiate it

---

[23] The patent was on Method and Apparatus for Automatic Standardization in Liquid Scintillation Spectrometry. It covered both automatic internal and automatic external standardization, was licensed subsequently to the major companies in the field and became an essential requirement for every top LSC.

[24] Packard had its own Absolute Activity Analyzer (AAA) that actually printed out the disintegrations per minute (dpm) after counting each sample. And Beckman had its Automatic Quench Calibration (AQC) that adjusted system-gain to restore the counting efficiency of each quenched sample to that of a previously measured reference sample.

Table 8.1. Estimated % dollar market shares of main purchasers of Liquid Scintillation Counters (estimate from Edward Rapkin, 1998).

	% Dollar Market Shares (estimated)			
	1955	1960	1965	1970
Packard Instrument Company Inc.	95	85	63	45
Nuclear Chicago: Searle (first sale – 1960	–	10	20	15
Beck,man (first sale – 1965)	–	–	5	20
Wallac; LKB	ϒ only until approximately 1973			
Ansitron (first sale – 1964)	–	–	10	6 (Picker)
Intertechnique (first sale – 1960	–	–	–	15
Philips				
Picker Nuclear	–	–	2	acquired Ansitron 1966
Sharp Laboratories never sold LSCs (see Beckman)	–	–	–	–
TMC	5		discontinued	
Tracerlab		5	discontinued	
Vangard	sold <5 units total			
Nuclear Enterprises (Scotland)	very few units sold, none outside UK			
Berthold (Germany)	–	–	–	3 (Germany only)
Aloka (Japan)	–	–	–	
Russia	no commercial counters, only home built			
Belin (France)	sold a few units to French AEC 1958–62			

over a one week period and return it to you for purification or experimental use."[25] But the tritiation process was tricky because it frequently involved highly labeled by-products, and the tritium exchange with the solvent was difficult to control. This required rigorous checks that opposed routine application. The development of LSC and the synthesis, purification and application of tritium and carbon-14 labeled molecules proceeded parallel to one another and had to be mutually adapted over an extended period of time. Without mastering the 'software' problems of the compounds the hardware of the machines would not have been able to be put to much use in molecular biology and biomedical research. There was a constant mutual shaping and reshaping between the instrument, the molecular probes, and the epistemic agendas into which they became inserted.

---

[25] Loftfield, letter to Rheinberger, September 24, 1998.

BETWEEN INDUSTRY AND CUSTOMERS

In the previous sections I have shown that liquid scintillation counter prototypes arose from an interaction between researchers and engineers at universities and National Laboratory research sites. There was a direct correspondence between the users' needs and the technical solutions that the engineers could offer in terms of an assemblage of scintillation physics and chemistry, of photoelectronics, and of mechanical automation. Manufacturing and research virtually coincided both in time and in space. In fact, Packard's coworker Leo Slattery continually serviced Leroy's machine and gave advice on particular uses of the instrument in particular experiments (Okita, Kabara, Richardson, and LeRoy 1957[26]). The same relationship between engineer and scientific customer is documented for George Utting from the Technical Measurement Corporation (Blau 1957[27]).

As soon as the instruments began to be produced and sold as commercial items, this symbiotic and symmetric prototypical relationship was changed. Packard insists that from then on most of the instrument improvements came from "inside," which means from inside the company and from what other companies introduced into the market. But that does not mean that trying to find out what the scientists were doing with an instrument and getting a feeling for what they needed ceased to be significant. The feedback was taken over by a system of reporters who were both salespersons and service and repair personnel in one. Packard recalls: "We called them combination people." He claims that separating these two functions was counterproductive and would have been disastrous. "If we had ever tried to do this through reps, it just would not have worked. [I] think our own people could give the best installation, the instructions and the theory of operation, and all of that necessary support" (Packard, Interview 1996). Rapkin stresses that "the salesmen were all hired to be servicemen" and added: "One thing I think may have been the strength of the Packard Instrument Company in those days was its sales force. And they were good about reporting back new requirements and problems" (Rapkin in Packard, Interview 1996).

Gerhard Kremer summarized his experience in the field of LSC by emphasizing that the relations between research, technological refinement, customer, and marketing were decisive. He generalized his observations by stating that research and development engineers on the one hand and customers on the other tend to have different visions of perfecting an instrument. The necessary feedback between them has largely to be mediated and balanced by the salesmen who are competent servicemen at the same time. These mediators need to be scientifically and technically up to date and have a feeling for the customers' needs involving a whole 'psychology of competence.' Paradoxically, the more the technology becomes black-boxed for routine use, the more competence a salesperson must bring with him or her in order to be convincing. It is not uncommon therefore that engineers and even scientists with a PhD enter the sales business (Kremer, Interview 1996).

---

[26] Here, Slattery is acknowledged for "suggesting the discriminator-ratio method."
[27] Utting is acknowledged in this paper "for help and advice."

There was one part of the business, however, in which the know-how clearly travelled from the customers to the manufacturer. The refinements of sample preparation, including new recipes for scintillation cocktails, were largely due to tinkering out beyond in the diverse laboratories where people were struggling with their idiosyncratic experimental problems and trying to exploit the machines for their special purposes. Much of the early work of Newton Hayes at Los Alamos, James Arnold at the University of Chicago, and Jack Davidson at Columbia was devoted to this task. This is also the way plastic vials eventually came into use. Rapkin reports:

> A customer [Herbert Jacobsen, University of Chicago] told me that they were using plastic vials. And we tried it and it worked very well. [I] think by having a wall that was diffusing the light, there was a better chance the photomultipliers would get the light. So for tritium counting, the improvement was significant as a per cent of the total count (Rapkin in Packard, Interview 1996).

Although plastic has the disadvantage that it does not prevent the organic scintillation cocktail from diffusion and thus has to be disposed of quickly, the cheap polyethylene vials partially replaced the glassware with its additional inherent drawback of exhibiting at least some naturally occurring radioactivity (Rapkin and Packard 1961). Eugene Goldwasser gives another example, that of dual label counting in which he had been involved with George Leroy at the Argonne Cancer Research Hospital: "It's a two-directional kind of thing. Once it became known to experimenters that you could discriminate isotopes based on the magnitude of the pulse you get, then they would sort of talk to the people developing instrumentation saying: 'This is really what we would like to be able to do'" (Goldwasser, Interview 1996). Despite the early promises, dual label counting went through a decade of trouble-checking and deceptive experiences before it became a routine procedure.

The interaction between appliers and supplier was indeed vital. According to Robert Loftfield, Packard realized in 1958 or 1959 that many of his Tri-Carb machines, purchased by many inexperienced customers from generously distributed federal research money, were standing around in hospitals either unused or at best generating unimpressive data. Packard managed "to persuade the Atomic Energy Commission to set up an award sufficient to place Tri-Carbs in some 20 reputable laboratories where problems could be uncovererd and applications developed that would increase the usefulness of the Tri-Carb for other hesitant scientists."[28] One of these machines – "a beautiful machine: coincidence counting, automatic sample changing, cooled to about -10° C, automatic print-out, pre-selectable voltage gates, etc." – came to be located at the John Collins Warren Laboratories of the Huntington Memorial Hospital at the Massachusetts General Hospital in Boston, where Loftfield explored the pitfalls involved in the direct counting of paper chromatography strips (Loftfield and Eigner 1960).

Thus variable interfaces between the laboratory and industry are generated in such an epistemic-technical interplay. Highly developed research enabling technologies require special product management. This process is best carried out by people who operate and are at home at these interfaces, and it is typically materialized in objects and accessories that connect the core machinery to the experimental setup

---

[28] Loftfield, letter to Rheinberger, September 24, 1998.

and make it a generic device. On the other hand, without the cooperation of experienced researchers dwelling in their experimental systems, these connections inevitably collapse. Liquid scintillation counting is a good example for this mutual interaction. Ultracentrifugation with its different types and sizes of tubes and rotors would be another. The salesmen/reporters carry the new products into the laboratories, search for new applications in the laboratories, sense upcoming needs of the customers, and venture product modifications that feed back to the company's research and development program. A Common Shares Prospectus of Packard Instrument Company dating from 1961 notes:

> The Company [maintains] a laboratory to study product applications, to handle trial samples for prospective customers and to devise and test techniques for utilizing both its existing products and new products under development. The Engineering Department devotes its efforts to the development of new products and improvement of existing models. During the year 1960 the Company had approximately fifteen employees engaged in research and development and spent approximately $ 210,000 for this purpose, exclusive of quality control and normal product testing. Consultants are utilized where special skills and knowledge can be more effectively obtained than with full-time staff members.[29]

Thus the R&D share amounted to slightly less than 10% of the net sales ($ 2,964,161) and slightly more than 10% of the total number of employees (125).

AN INTERDISCIPLINARY AND INTERNATIONAL NETWORK

There is one more aspect to this story about research enabling technologies: It is networking. In addition to competition, there was also cooperation. On the technical part, liquid scintillation counting depended upon pure and reliably quantified sources of the various isotopes for testing instrument circuitry and solutions, for calibrating the counting procedure, and for suitable standard samples. On the biological part, suitably labeled compounds were necessary. Packard entertained close connections with Edward Shapiro and Seymour Rothchild of New England Nuclear Corporation, a company that had been founded by former Tracerlab employees and that was producing and purchasing labeled compounds. Tracerlab in turn had been one of the first private companies to be approved to receive isotope shipments from Oak Ridge as an agent for the purchaser and for synthesizing a variety of labeled molecules.[30]

Besides contacts to the isotope industry, Packard also sensed the need to be visible at the level of publications.

> When Edward Rapkin came to work with us, what we wanted was for him to become the leading liquid scintillation oracle and publish little newsletters for us, which he did. So from that time on, the person who knew everything that was being done, all the techniques, all those solvents, and all the cocktails, was Dr. Rapkin (Packard, Interview 1996).

Rapkin had been in the U.S. Army working with a mass spectrometer located in the Argonne National Laboratory when he first encountered the Packard Tri-Carb. After

---

[29] Packard Instrument Company, Inc., *Common Shares Offer* (1961), pp. 6-7.

[30] See, e.g., the advertisement in the October 1947 issue of *Nucleonics*, p. 85.

leaving the Army, but before joining Packard's Company in 1957, he had already experimented for Packard on the alkaline digestion of proteins with hyamine at Armour & Co., in an effort to make protein containing samples soluble in toluene based scintillation mixtures (Rapkin 1961). This was not a trivial effort since most of the biological samples contained variable amounts of proteins.

Publishing and disseminating a Technical Bulletin was, however, only one part of organizing publicity.[31] The more encompassing task consisted of setting up conferences to bring together scientists and engineers, research institutes and application laboratories, and thus make the company part of a circuit of communication that related instrument makers and instrument users, both academic and commercial. To this effect, starting in 1957, Packard, together with New England Nuclear's Seymour Rothchild and Atomic Associates, sponsored a long series of symposia on "Advances in Tracer Methodology" later edited by Rothchild in four volumes.[32] Packard himself gave papers at a large conference on Liquid Scintillation Counting held at Northwestern University in August 1957, sponsored by the National Science Foundation and the Technological Institute of Northwestern University, and at a conference with over two hundred participants on Organic Scintillation Detectors held at the University of New Mexico in August 1960 and sponsored by the University of New Mexico, the National Science Foundation, and the U.S. Atomic Energy Commission (Packard 1958; Rapkin and Packard 1961).

The 1957 Northwestern conference, in a nutshell, reflects the many facets of liquid scintillation counting as an explosively expanding research-technology towards the end of the 1950s. Participants came from research institutes of universities including Princeton University, Columbia University, New York University, the University of Chicago, and the University of California at San Francisco; from National Laboratories including Los Alamos, Argonne, and Brookhaven, the National Bureau of Standards, and the National Institutes of Health; from hospitals such as the Veterans Administration Research Hospital; from laboratories in France (Saclay, Gif-sur-Yvette) England (Atomic Energy Research Establishment, Harwell), and Israel (Weizmann Institute, Rehovoth); and from companies including Packard Instrument, Technical Measurement Corporation, Shell Oil Company, and Tracerlab. The constituency shows the wide range of attention for the new technology at that stage and the scope of its customers ranging from the epistemic core of physical, chemical and biomedical research to the wider realm of hospital diagnostics, precision measurement, standardization, and radiation control. Consequently, it brought together experimental physicists, chemists, radiologists, biochemists, archaeologists, medical researchers, electronics engineers, mechanical engineers, and instrument builders. They all either contributed to the physical, chemical, or engineering parts of the ma-

---

[31] Eighteen of these Technical Bulletins appeared between 1961 and 1969, one third of which were signed by Edward Rapkin.

[32] Between 1957 and 1966, a total of eleven conferences were sponsored in New York, Chicago, Washington, Los Angeles, San Francisco, Zurich, and Boston. See *Proceedings of the Symposium on Tritium in Tracer Applications*, sponsored by New England Nuclear, Atomic Associates, Packard Instrument, NY City, Nov. 22, 1957; *Proceedings of the Symposium on Tritium in Tracer Applications*, sponsored by New England Nuclear, Atomic Associates, Packard Instrument, NY City, Oct. 31, 1958; Rothchild 1963 (Vol. 1), 1965 (Vol. 2), 1966 (Vol. 3), and 1968 (Vol. 4).

chine, or they used it in their research, or participated in both realms. As Eric Schram and Robert Lombaert write, with a tone of understatement, in the introduction to their textbook: "The field of organic scintillation detectors may be said to extend to several sciences, physics, electronics, organic and biological chemistry" (Schram and Lombaert 1963: v).

The liquid scintillation counter was on its way to becoming the pivot point for a transdisciplinary, temporary and informal community of researchers, engineers, and industrialists. James Arnold, who was among the participants at the conference at Northwestern, remarked:

> As one of the early workers in the field, I am made rather complacent by the fact that it has ramified in so many unexpected directions. This is actually a rather good case history of the unexpected applications which result from 'pure' research. I do not think that either the people at Los Alamos or the others who were working in the field at that time would have been much more successful in predicting all the applications represented at this conference (Arnold 1958: 129).

After all, it should be added that it is a rather good case which shows that the distinction between pure research and its application does not help us much to frame its history. But this contemporary remark of one of the founders of the technology makes us once more aware of the fact that in the exploratory phase of the technology between 1950 and 1955, LSC did not amount to more than one option among other, much more established, procedures and counting devices. What made liquid scintillation counting really work and finally take over was the result of a techno-epistemic conjunction of heterogenous factors of different origin: a physico-chemical principle (liquid scintillation) of a potentially generic use in bioassays; photoelectronic industry driven by weapons production and mass communication; the pervasive use of weak β-emitters with all their intricacies throughout the biomedical complex in the aftermath of atomic fission; and mechanical automation effectively matching users' demands. Above all, an inherent disposition to 'wet' experimentation in biochemistry made a 'liquid' boundary between the measuring device and the probe a very adaptable and versatile assembly. Rapkin recalls:

> It used to be a very active field for discussions of technique. And so there were many, many conferences in the early days. Discussions about the best counting solutions, how do you measure steroids, all that kind of thing. It was a fairly active thing at one time (Rapkin in Packard, Interview 1996).

Journals such as *Nucleonics* (1947) were founded which aggressively promoted and advertised the spread of radiation technology and its industrialization. The journal's suggestive name was taken from a 38-page secret "Prospectus on Nucleonics," featuring postwar nuclear policies, that was submitted to Arthur H. Compton by a group of scientists chaired by Zay Jeffries and including Enrico Fermi in November 1944 (Luntz 1957: 79). The journal was announced as a "medium for the cross-fertilization of technical advances in all phases of nuclear technology, [a] meeting place for the exchange of ideas between engineers, physicists, chemists, life scientists and teachers."[33] The *International Journal of Applied Radiation and Isotopes* (1956) and the

---

[33] What is nucleonics? – The magazine. *Nucleonics* 1 (September 1947): 2.

Journal of Nuclear Materials (1959) followed. There were countless smaller meetings on scintillation counting over the years, such as the Symposia on Tritium in Tracer Applications already mentioned, and about a dozen big conferences besides the Northwestern Conference, including the Scintillation and Semiconductor Counter Symposia in Washington, the Annual Symposia on Advances in Tracer Methodology, the University of New Mexico Conference on Organic Scintillation Detectors in Albuquerque in 1960, the I.A.E.A. Symposium on the Detection and Use of Tritium in the Physical and Biological Sciences in 1961, the I.A.E.A. Conferences on Nuclear Electronics in 1958 and 1961, respectively, a conference on Liquid Scintillation Counting at the Massachusetts Institute of Technology in 1969,[34] and an international conference on Organic Scintillators and Liquid Scintillation Counting at the University of California, San Francisco, in 1970 (Horrocks and Peng 1971). Soon monographs came to complement the conference proceedings (Kobayashi and Maudsley 1974; Schram and Lombaert 1963).

One of the conferences jointly sponsored by Packard and New England Nuclear was organized in Switzerland and took place at the *Zürcher Kunsthaus*.

> The biggest [conference] I think we ever put on was in Zurich. God, they made a real deal out of that. [The] Bürgermeister came and talked, and we had one of the great big halls, right downtown in Zurich. It was a two or three day seminar, and then a big dinner. [People] from all over Europe came to that (Packard, Interview 1996).

Packard's company had expanded internationally. The first international sale was to France, at Saclay, Gif-sur-Yvette, the French atomic energy research site, in 1957. Sales all over Europe followed, including Yugoslavia and Hungary, even the Soviet Union. A chemical plant in Holland was added to the company to produce scintillation chemicals. The company even mounted a few whole body counters for monitoring radioactivity in living bodies, one of them at the University of Hamburg, Germany. Together with Niilo Kaartinen from the University of Turku, Finland, Packard constructed a sample oxidizer machine (Everett, Kaartinen, and Kreveld 1974; Kaartinen 1969). The Packard-Kaartinen combustion machine helped to convert otherwise insoluble material into a gaseous and then liquid form, and as a physical side effect of the combustion of organic materials to CO_2 and H_2O, it provided an elegant and very efficient means of separating tritium and carbon-14 in critical dual label experiments with, for example, low ^3H and high ^{14}C content. The machine was widely used, although its mechanical operation remained somewhat troublesome due to frequent soot deposits and the delicacy of the mechanics of closing and opening the combustion chamber.

To promote international business, Packard set up Packard Instrument Sales Corporation in 1957, and in 1959 a wholly owned foreign subsidiary called Packard Instruments International S.A. Under the direction of James Kriner, and with Zurich as headquarters, a widespread network of international offices began to be established. By the end of the 1960s, Gerhard Kremer, who had started with the German office, became Director of International Operations in Zurich.[35]

---

[34] To this meeting, Rapkin contributed a valuable historical paper. See Rapkin 1970.

[35] Years later Kremer became President of the parent company and operated out of the corporate headquarters near Chicago, while still retaining his permanent residence in Zurich, where he returned to

> We had the most advanced sales and service arrangement in our field, by having a corporation in Sweden, a corporation in Italy, a corporation in Belgium, a corporation in Germany, in France, in England, in Israel. And each of these corporations had a local person who was the top man. Whether he was the president or the director general or whatever other title, it was always a local person, with local staff, and everybody was bilingual, at least bilingual: their own language and English. And, all the business was done in the local currency with local bank accounts and it worked out just really beautifully (Packard, Interview 1996).

The essential point was that it was general business practice in those times to use distributors. Operating directly was unusual in the scientific instrumentation business, and the high value of the dollar helped to keep costs low. In 1969, Packard's worldwide sales and service network included, in addition to those just mentioned, representatives and sales engineers in Australia, Canada, Denmark, Japan, Norway, and in the Republic of South Africa.

Packard sold his company in 1967. When I asked him why he did this, he answered:

> Basically, the reason for selling the company was that we needed additional financial support to continue to expand, compete and maintain our dominant market position. An undesirable alternative would have been to retrench both technical and marketing development and concede the leading market share to the bigger companies, Searle and Beckman, that had entered our market with acceptable products by acquiring our smaller competitors, Nuclear-Chicago and Sharp Laboratories. [The] merger (with American Bosch Arma Corporation) did provide us with the necessary financial support to compete with the larger companies we were facing in 1967. [It] is very tough to compete when big companies buy their way into a business and are willing to lose money until they win or drop out (Packard, letter to Rheinberger, February 25, 1998).

INSTEAD OF A CONCLUSION

Lyle Packard stayed with the business for three and a half more years and then left at the end of 1970. He then did extensive sailing, bought a couple of islands in the Caribbean, and enjoyed life. Later, he became engaged in several smaller companies and is currently the Chairman and Chief Executive Officer of Advanced Instrument Development, Inc. (AID), a firm that produces special X-ray equipment for medical diagnostics. Packard Instrument Company, today part of Packard Bioscience Company (formerly Canberra Industries, Inc.), remained with LSC and has kept its leading position in the technology of liquid scintillation counters, sample preparation solutions, and other accessories up to this day.[36] But the company now increasingly focuses on developing a new generation of microplate fluorescence and luminescence technologies.

The overwhelming dominance of radioactive tracing in molecular biology and medicine is beginning to cease toward the end of the millennium. The days of almost unlimited growth of nuclear research-technology are past. Alternative tracing meth-

---

work as a corporate director and senior executive for a few years before finally retiring from the Company in 1997.

[36] Dr. Kremer proudly insists that the actual Model 2700 Tri-Carb Liquid Scintillation Spectrometer is probably the best LSC which has ever been built in 45 years.

ods based on fluorescence, or luminescence, and other tracing tools that circumvent the use of radioactivity altogether are the order of the day. An advertisement of Packard Instrument Company from June 1998 reads: "Make the move to non-isotopic assays!" Although there is a range of fundamental applications where radioactive tracing may remain indispensable for a long time to come, visualization through labeling has become a field for competing alternatives. Research-technologies, even comparatively long-lived ones, have their historical conjunctures and eclipses. Scientific objects and the ways they are manipulated, likewise, come and go with the technologies. And companies only stay if they keep abreast of these epistemic moves in the changing technoculture of research.

Max-Planck-Institut für Wissenschaftsgeschichte, Berlin, Germany

ACKNOWLEDGMENTS

I thank Mr. Rainer Stangl, a Viennese Product Manager of Packard, for his help in contacting former representatives of Packard Instruments Company. My special thanks go to Professor Eugene Goldwasser, Dr. Gerhard Kremer, Lyle E. Packard, Edward Polic, and Dr. Edward Rapkin for the interviews they kindly granted me, and their kind responses to further requests. Jean-Paul Gaudillière, Myles Jackson, Lily Kay, Robert B. Loftfield, Patricia Nevers, Xavier Roqué, Sahotra Sarkar, Otto Sibum, and Norton Wise are acknowledged for valuable comments on earlier drafts of the manuscript.

REFERENCES

AEC (Atomic Energy Commission) (1947, September). Radioisotopes. *Nucleonics* 1: 64–9.

Alvarez, Luis W. and Cornog, Robert (1939). Helium and hydrogen of mass 3. *The Physical Review* 56: 613.

Anderson, Ernest C. (1958). The Los Alamos human counter, in C.G. Bell and F.N. Hayes (eds.), *Liquid scintillation counting* (pp. 211–19). New York: Pergamon Press.

Argonne National Laboratory (1986). *Argonne News* 30(5): 3–15.

Arnold, James R. (1954). Scintillation counting of natural radiocarbon: I. The counting method. *Science* 119: 155–7.

Arnold, James R. (1958). Archaeology and chemistry, in C.G. Bell and F.N. Hayes (eds.), *Liquid scintillation counting* (pp. 129–34). New York: Pergamon Press.

Balogh, Brian (1991). *Chain reaction. Expert debate and public participation in American commercial nuclear power, 1945–1975.* Cambridge: Cambridge University Press.

Blau, Monte (1957). Separated channels improve liquid scintillation counting. *Nucleonics* 15(4, April): 90–1.

Broda, Engelbert (1960). *Radioactive isotopes in biochemistry.* Amsterdam: Elsevier.

Broser, Immanuel and Kallmann, Hartmut (1947a). Über die Anregung von Leuchtstoffen durch schnelle Korpuskularteilchen I. *Zeitschrift für Naturforschung* 2a: 439–40.

Broser, Immanuel and Kallmann, Hartmut (1947b). Über den Elementarprozess der Lichtanregung in Leuchtstoffen durch alpha-Teilchen, schnelle Elektronen und gamma-Quanten II. *Zeitschrift für Naturforschung* 2a: 642–50.

Cohn, Waldo E. (1968). Introductory remarks by chairman, in S. Rothchild (ed.), *Advances in tracer methodology*, Vol. 4 (pp. 1–10). New York: Plenum Press.

Coltman, J.W. and Marshall, Fitz-Hugh (1947, November). Photomultiplier radiation detector. *Nucleonics* 1(3): 58–64.

Crookes, William (1903). The emanation of radium. *Proceedings of the Royal Society of London*, Vol. 71: 405–8.

Curran, Samuel C. and Baker, W.R. (1947). A photoelectric alpha particle detector. *U.S. Atomic Energy Commission Rpt. MDDC 1296*, 17 Nov. 1944, declassified 23 September 1947.

Davidson, Jack D. and Feigelson, Philip (1957). Practical aspects of internal-sample liquid-scintillation counting. *International Journal of Applied Radiation and Isotopes* 2: 1–18.

Elster, Julius und Geitel, Hans (1903). Über die durch radioaktive Emanation erregte scintillierende Phosphoreszenz der Sidot-Blende. *Physikalische Zeitschrift* 4: 439–40.

Elzen, Boelie (1986). Two ultracentrifuges: A comparative study of the social construction of artefacts. *Social Studies of Science* 16: 621–62.

Everett, L.J., Kaartinen, Niilo, and Kreveld, P. (1974). An advanced automatic sample oxidizer – new horizons in liquid scintillation sample preparation, in P.E. Stanley and B.A. Scoggins (eds.), *Liquid scintillation counting* (pp. 139–52). New York and London: Academic Press.

Feld, Bernard T. and Weiss Szilard, Gertrud (eds.) (1972). *The collected works of Leo Szilard. Scientific papers, Volume I*. Cambridge, MA: The MIT Press.

Furst, Milton and Kallmann, Hartmut (1952). Fluorescence of solutions bombarded with high energy radiation (Energy transport in liquids), Part III. *The Physical Review* 85: 816–25.

Galison, Peter (1997). *Image and logic. A material culture of microphysics*. Chicago: The University of Chicago Press.

Gaudillière, Jean-Paul (1998). The molecularization of cancer etiology in the postwar United States: Instruments, politics, and management, in S. de Chadarevian and H. Kamminga (eds.), *Molecularizing biology and medicine: New practices and alliances 1910s – 1970s* (pp. 139–70). Amsterdam: Harwood Academic Publishers.

Goldwasser, Eugene (1953). The incorporation of adenine into ribonucleic acid in vitro. *Journal of Biological Chemistry* 202: 751–5.

Goldwasser, Eugene (1955). Incorporation of adenosine-5'-phosphate into ribonucleic acid. *Journal of the American Chemical Society* 77: 6083.

Grandy, David A. (1996). *Leo Szilard. Science as a mode of being*. Lanham, MD: University Press of America.

Hayes, F. Newton, Hiebert, R.D., and Schuch, R.L. (1952). Low energy counting with a new liquid scintillation solute. *Science* 116: 140.

Heilbron, John L. and Seidel, Robert W. (1989). *Lawrence and his laboratory: A history of the Lawrence Berkeley Laboratory*, Vol. I. Berkeley, CA: University of California Press.

Hevesy, Georg von (1948). Historical sketch of the biological application of tracer elements. *Cold Spring Harbor Symposia on Quantitative Biology* 13: 129–50.

Hewlett, Richard G. and Holl, Jack M. (1989). *Atoms for peace and war, 1953-1961*. Berkeley, CA: University of California Press.

Hewlett, Richard G. and Anderson, Oscar E. Jr., (1962). *A history of the United States atomic energy commission*. Volume I, *The New World 1939/1946* and Volume II, *Atomic shield 1947/1952*. University Park, PA: The Pennsylvania State University Press.

Hiebert, R.D. and Watts, R.J. (1953, December). Fast-coincidence circuit for H^3 and C^{14} measurements. *Nucleonics* 11(12): 38–41.

Hiebert, R.D. and Hayes, F. Newton (1958). Instrumentation for liquid scintillation counting at Los Alamos, in C.G. Bell and F.N. Hayes (eds.), *Liquid scintillation counting* (pp. 41–9). New York: Pergamon Press.

Hofstadter, Robert (1948). Alkali halide scintillation counters. *The Physical Review* 74: 100–1.

Hofstadter, Robert (1949). The detection of gamma-rays with thallium-activated sodium iodide crystals. *The Physical Review* 75: 796–810.

Holl, Jack M. (1997). *Argonne National Laboratory, 1946-96*. Urbana and Chicago: University of Illinois Press.

Horrocks, Donald L. and Peng, Chin-Tzu (eds.) (1971). *Organic scintillators and liquid scintillation counting*. New York and London: Academic Press.

Hughes, J.A. (1993). *The radioactivists: Community, controversy, and the rise of nuclear physics*. PhD Thesis, University of Cambridge.

Kaartinen, Niilo (1969). *Packard Technical Bulletin*, No. 18. Downers Grove, IL: Packard Instrument Co., Inc.

Kabara, Jon J., Okita, George T., and LeRoy, George V. (1958). Simultaneous use of H^3 and C^{14} compounds to study cholesterol metabolism, in C.G. Bell and F.N. Hayes (eds.), *Liquid scintillation counting* (pp. 191–7). New York: Pergamon Press.

Kallmann, Hartmut and Accardo, Carl A. (1950). Coincidence experiments for noise reduction in scintillation counting. *Review of Scientific Instruments* 21: 48-51.

Kallmann, Hartmut and Furst, Milton (1950). Fluorescence of solutions bombarded with high energy radiation (Energy transport in liquids), Part I. *The Physical Review* 79: 857–70.

Kallmann, Hartmut and Furst, Milton (1951). Fluorescence of solutions bombarded with high energy radiation (Energy transport in liquids), Part II. *The Physical Review* 81: 853–64.

Kamen, Martin D. (1963). Early history of carbon-14. *Science* 140: 584–90. Appeared also in *The Journal of Chemical Education* (1963), and as an introduction to S. Rothchild (ed.) (1965), *Advances in tracer methodology* Vol. 3, New York: Plenum Press.

Kamen, Martin D. and Ruben, Samuel (1940). Production and properties of carbon 14. *The Physical Review* 58: 194.

Kay, Lily (1988). The Tiselius electrophoresis apparatus and the life sciences, 1930-1945. *History and Philosophy of the Life Sciences* 10: 51–72.

Kobayashi, Yutaka and Maudsley, David V. (1974). *Biological applications of liquid scintillation counting*. New York: Academic Press.

Krebs, Adolf (1941). Ein Demonstrationsversuch zur Emanationsdiffusion. *Annalen der Physik* 39(5): 330–2.

Krebs, Adolf (1953). Szintillationszähler. *Ergebnisse der exakten Naturwissenschaften* 27: 361–409.

Krebs, Adolf T. (1955). Early history of the scintillation counter. *Science* 122: 17–18.

Langham, Wright H. (1958). Application of liquid scintillation counting to biology and medicine, in C.G. Bell and F.N. Hayes (eds.), *Liquid scintillation counting* (pp. 135–49). New York: Pergamon Press.

Lanouette, William (1992). *Genius in the shadows: A biography of Leo Szilard; The man behind the bomb.* New York: Scribner's.

Lenoir, Timothy (1997). *Instituting science. The cultural production of scientific disciplines.* Stanford, CA: Stanford University Press.

Lenoir, Thimothy and Lécuyer, Christopher (1997). Instrument makers and discipline builders: The case of nuclear magnetic resonance, in T. Lenoir (ed.), *Instituting science. The cultural production of scientific disciplines* (chapter 9, pp. 239–92). Stanford, CA: Stanford University Press.

Lenoir, Timothy and Hays, Marguerite (2000). The Manhattan Project for biomedicine, in P.R. Sloan (ed.), *Controlling our destinies: Historical, philosophical, social and ethical perspectives on the human genome project* (pp. 29–62). South Bend, IN: University of Notre Dame Press.

Loftfield, Robert B. and Eigner, Elizabeth A. (1960). Scintillation counting of paper chromatograms. *Biochemical and Biophysical Research Communications* 2: 72–5.

Luntz, Jerome D. (1957, September). The story of a magazine and an industry. *Nucleonics* 15(9): 78–83.

Libby, Willard F. (1952). *Radiocarbon Dating.* Chicago: University of Chicago Press.

Mandeville, C.E. and Scherb, M.V. (1950, November). Photosensitive Geiger counters: Their applications. *Nucleonics* 7(5): 34–8.

McNeill, William H. (1991). *Hutchins' University. A memoir of the University of Chicago 1929-1950.* Chicago: The University of Chicago Press.

Morton, G.A. and Mitchell, J.A. (1949, January). Performance of 931-A type multiplier as a scintillation counter. *Nucleonics* 4(1) 16–23.

Morton, G.A. and Robinson, K.W. (1949, February). A coincidence scintillation counter. *Nucleonics* 4(2): 25–9.

Okita, George T., Kabara, Jon J., Richardson, Florence, and LeRoy, George V. (1957, June). Assaying compounds containing H^3 and C^{14}. *Nucleonics* 15(6): 111–14.

Oster, Gerald (1966). A young physicist at seventy: Hartmut Kallmann. *Physics Today* 19: 51–4.

Packard Instrument Company (1961). *Common shares offer.*

Packard Instrument Company (1965). *Annual Report 1965, Ten year financial highlights.*

Packard, Lyle E. (1958). Instrumentation for internal sample liquid scintillation counting, in C.G. Bell and F.N. Hayes (eds.), *Liquid scintillation counting* (pp. 50–66). New York: Pergamon Press.

Pringle, Robert W. (1950). The scintillation counter. *Nature* 166: 11–14.

Raben, M.S. and Bloembergen, Nicolaas (1951). Determination of radioactivity by solution in a liquid scintillator. *Science* 114: 363–4.

Rabinow, Paul (1996). *Making PCR. A story of biotechnology*. Chicago: The University of Chicago Press.

Rapkin, Edward (1961). Hydroxide of hyamine 10-X. *Technical Bulletin Number 3*, Revised June, 1961.

Rapkin, Edward (1970). Development of the modern liquid scintillation counter, in E.D. Bransome (ed.), *The current status of liquid scintillation counting* (pp. 45–68). New York: Grune & Stratton.

Rapkin, Edward and Packard, Lyle E. (1961). New accessories for liquid scintillation counting, in G.H. Daub, F.N. Hayes, and E. Sullivan (eds.), *Proceedings of the University of New Mexico Conference on Organic Scintillation Detectors, August 15-17, 1960* (pp. 216–31). Washington, DC: U.S. Government Printing Office.

Rasmussen, Nicolas (1997). *Picture control. The electron microscope and the transformation of biology in America, 1940-1960*. Stanford, CA: Stanford University Press.

Regener, Erich (1908). Über Zählung der α-Teilchen durch die Szintillation und die Größe des elektrischen Elementarquantums. *Verhandlungen der deutschen physikalischen Gesellschaft* 10: 78–83.

Reines, Frederick (1958). Giant liquid scintillation detectors and their applications, in C.G. Bell and F.N. Hayes (eds.), *Liquid scintillation counting* (pp. 246-57). New York: Pergamon Press.

Reynolds, G.T., Harrison, F.B., and Salvini, G. (1950). Liquid scintillation counters. *The Physical Review* 78: 488.

Rheinberger, Hans-Jörg (1997). *Toward a history of epistemic things. Synthesizing proteins in the test tube*. Stanford, CA: Stanford University Press.

Rheingans, Friedrich G. (1988). *Hans Geiger und die elektrischen Zählmethoden, 1908-1928*. Berlin: D.A.V.I.D. Verlagsgesellschaft.

Rothchild, Seymour (ed.) (1963). *Advances in tracer methodology*. Vol. 1. New York: Plenum Press.

Rothchild, Seymour (ed.) (1965). *Advances in tracer methodology*. Vol. 2. New York: Plenum Press.

Rothchild, Seymour (ed.) (1966). *Advances in tracer methodology*. Vol. 4. New York: Plenum Press.

Rothchild, Seymour (ed.) (1966). *Advances in tracer methodology*. Vol. 3. New York: Plenum Press.

Ruben, Samuel and Kamen, Martin D. (1940). Radioactive carbon of long half-life. *The Physical Review* 57: 549.

Schram, E. and Lombaert, R. (1963). *Organic scintillation detectors. counting of low-energy beta emitters*. Amsterdam: Elsevier.

The University of Chicago Publications Office (1991). *One in Spirit. A Retrospective View of the University of Chicago on the Occasion of its Centennial*. Chicago: The University of Chicago Publications Office.

Trenn, Thaddeus J. (1976). Die Erfindung des Geiger-Müller-Zählrohres. *Deutsches Museum, Abhandlungen und Berichte* 44: 54–64.

Trenn, Thaddeus J. (1986). The Geiger-Müller counter of 1928. *Annals of Science* 43: 111–35.

Utting, G.R. (1958). The design of a commercial liquid scintillation coincidence counter, in C.G. Bell and F.N. Hayes (eds.), *Liquid Scintillation Counting* (pp. 67–73). New York: Pergamon Press.

Weart, Spencer R. (1988). *Nuclear fear. A history of images*. Cambridge, MA: Harvard University Press.

Whitehouse, W.J. and Putman, J.L. (1953). *Radioactive isotopes. An introduction to their preparation, measurement and use*. Oxford: Clarendon Press.

CHAPTER 9

JEAN-PAUL GAUDILLIÈRE

MAKING MICE AND OTHER DEVICES: THE DYNAMICS OF INSTRUMENTATION IN AMERICAN BIOMEDICAL RESEARCH (1930–1960)

In 1937, in a manner typical of the New Deal approach to governmental duties, Washington policy makers considered the creation of a federal agency that would support medical research on the cancer problem. During the spring of that year, as a contribution to this public debate, the major news magazines depicted on their covers the fight against the dread disease. In March, *Life* displayed a picture of hundreds of inbred mice produced by the Jackson Memorial Laboratory at Bar Harbor. Mice were presented as the "ideal laboratory tool for the propagation and study of cancer" under the general headline "Mice replace Men on the Cancer Battlefield."

A few months later, during the congressional hearings which eventually led to the passage of an act establishing the National Cancer Institute, Clarence Cook Little – then director of the Jackson Memorial Laboratory – advised congressmen that:

> So, the first thing you should do, I should say, in spending your money would
> be to insure a constant and adequate supply of controlled, known animal mate-
> rial on which investigations could be carried out . . . if that work has to be done
> with them [the animals] then there has got to be knowledge of how to produce
> animals which are nearly as uniform as it is possible for any living higher ani-
> mal to be. In other words, we can produce as nearly a chemically pure animal
> and as nearly alike fellows as it is possible to produce . . . during this past year
> the little laboratory where I work has sent out over 65,000 such animals all
> over the United States and to Europe for research in cancer and in other ex-
> perimental medicine. There is a tremendous demand for them. So the first
> thing to do in the spending of this money would be to have under governmen-
> tal control, or at least see that there is an insurance of, a definite certainty of
> the availability of animal material of a controlled nature.

Thirty years later, Life's prophecy and Little's demand had become reality since medical research observers estimated that the U.S. laboratories were consuming roughly 35 million mice a year, while the Jackson Memorial Laboratory was both the most important supplier of laboratory mice and the center maintaining genetically homogenous strains of ascertained purity to be employed worldwide as reference animals.

175

B. Joerges and T. Shinn (eds.), Instrumentation: Between Science, State and Industry, 175–196.
© 2001 *Kluwer Academic Publishers.*

Most scientists who work with mice explain this success by citing the biological properties of the organism: mice are small animals, easy to handle, cheap to feed; they breed readily and often, they are mammals as we are and they are highly susceptible to many of the diseases that afflict human beings. In spite of all these natural advantages, mice have for a long time not been considered as a suitable material for medical research precisely because they are not humans but laboratory animals. So how is it that medical research has increasingly taken mice for men and women?

Medical scientists view such models as irreplaceable because animal organisms can be manipulated, have a high degree of plasticity, and can more easily be standardized than humans. The history of laboratory mice is, therefore, to a large extent a history of production and technological developments.

This paper argues that the increasing production and use of mice has to be analyzed within a broader context: the development of a large biomedical complex that changed the nature of biomedical instruments. In the United States, this change started in the 1930s but was greatly advanced by World War II and its accompanying scientific mobilization. To address this transformation, the paper focuses on instrument-making and its putative isolation from research tasks. Following the American mouse-makers from the 1930s to the 1960s and comparing their fate with the work of the engineers and scientists associated with the development of other research tools like the ultracentrifuge and the electron microscope, I will suggest that the visibility of biomedical instrumentation after World War II was a matter of changing scale, industrial patterns of production, and instrument-centered research. In other words, new ways of experimenting originated in forms of research that seem to fit the notion of an emerging 'research-technology' pattern or role within the biological sciences.

PATRONAGE AND INSTRUMENTS IN THE 1930s: THE ROCKEFELLER FOUNDATION, ULTRACENTRIFUGES AND MICE

Historians of the twentieth century biological sciences have often commented upon the role that the Rockefeller Foundation and its Natural Sciences Division played in the creation of a 'new biology' operating at the molecular level (Kay 1993; Kohler 1991; Abir-Am 1982). Warren Weaver's "molecular biology" program, which was launched in the early 1930s, has accordingly been seen as a key element in the emergence of a form of biological research centered on the development and uses of physico-chemical instruments. In a nutshell, Weaver, trained as a mathematical physicist, approached the life sciences by fostering a biology "without biologists." While in charge of the Natural Sciences Division of the Rockefeller Foundation he funded the efforts of physicists and chemists interested in biological systems or in biological applications of their know-how. Instrumental innovation rapidly became a key aspect of this venture, as illustrated by the major investments of the late 1930s: UV spectroscopy at Chicago, X-ray crystallography at London University and Caltech, isotopes and cyclotrons at Berkeley and Paris, ultracentrifuge at Uppsala and at the Rockefeller Institute, electrophoresis in Stockholm (Kohler 1991: chapt. 13). Although retrospective analysis suggests that the "molecular biology" program was a highly intentional and articulated renovating enterprise, each project followed its own logic. This pattern is well illustrated by the history of the ultracentrifuge.

Making Ultracentrifuges

Early stages in the history of the ultracentrifuge can be traced back to the work of Theodor Svedberg in Sweden (Elzen 1986, 1988). Svedberg was a colloid chemist at the University of Uppsala, who adopted the cream separator as a means of analyzing heterogeneous colloid particles. He complemented the industrial device with an optical system to watch and photograph materials as they sedimented. To Svedberg, achieving faster and faster ultracentrifuges was essential to analyze the fine distribution of colloid size and to refine the description of molecular structures.

The Rockefeller Foundation actually endorsed Svedberg's apparatus in the late 1920s, as it appeared that the use of the ultracentrifuge would significantly alter existing views about proteins, a special class of colloids of great biological importance. The Rockefeller Foundation not only favored Svedberg's costly machine but also his vision of the ultracentrifuge as the most important apparatus for revealing the structure of big molecules. Up to the late 1930s, the foundation reinforced Svedberg's hegemony by encouraging chemists and biologists to send material to his laboratory and by agreeing to fund only the construction of two machines of Svedberg's oil-turbine type (one at the University of Wisconsin in 1937, and one at the Lister Institute in London in 1936) (Creager 2001; Elzen 1988: chapt. 2).

As carefully analyzed by B. Elzen, however, by the mid-1930s, a competing instrument had been built by physicists at the University of Virginia – Jesse Beams and Edward Pickles – who adopted material and social technologies different from Svedberg's.

The Beams' machine scheme included an optical device similar to Sevdberg's, but the principal objective was separating rather than visualizing sedimenting material. This air-driven ultracentrifuge was a relatively small instrument that, in Beams words, any good university machine-shop could build (Elzen 1988: chapt. 3). Rather than targeting particular groups of users and specific physical results, Beams stressed the flexibility of the apparatus, which might make it useful to people working on atom-light interactions, colloids, organic synthesis, hormones, viruses, etc. Beams and his colleagues thus spent most of their time in the 1930s building a wide-range of ultracentrifuges, designing specific improvements, constructing new prototypes or pieces of machinery, and working out specific applications. For instance, Pickles' thesis focused on a new design for the air-driven ultracentrifuge in order to make it amenable to fragile biological specimens. Rather than resting on a stator and being directly driven by air jets, the rotor containing the sedimenting material was hung in a water-cooled vacuum chamber and rotated by a piano wire connected to an air-driven turbine positioned above the chamber.

In addition, rather than participating in the production of the ultracentrifuge, Beams and Pickles circulated blueprints and promoted their machine by engaging in long-term research collaborations. While Svedberg critically changed the definition of colloids and proteins by having chemical and biological samples sent to his physico-chemical laboratory for ultracentrifuge analysis, Beams and Pickles made ultracentrifuges cross the threshold of biological laboratories. This is best exemplified by Pickles' career at the Rockefeller Foundation.[1]

In 1935, after obtaining his Ph.D., Pickles was invited to join the New York laboratory of the International Health Division of the Rockefeller Foundation, a department distinct from the Natural Sciences Division.[2] Johannes Bauer, a specialist of yellow fever, wanted Pickles to work on the purification and characterization of viruses. Bauer's interest was then triggered by the highly visible work of another Rockefeller Institute scientist, Wendel Stanley, who had just announced that he had crystallized a plant virus – the tobacco mosaic virus – and that this virus was nothing but one single huge protein (Creager 2001: chapt. 1; Kay 1986). Pickles and his machine could become assets for the biochemist and bacteriologists who adopted a similar understanding of viruses as simple macromolecular entities that could be purified like chemicals.

By the late 1930s, Pickles had the assistance of a draftsman, a mechanic, a junior scientist, and a technician. The other scientists at the International Health Division supplied him with material, while he did all the ultracentrifuge work, computing molecular weight and size. In proposing a new design for the machine, Pickles thus found himself driven into sophisticated discussions about biology-related research problems, i.e. the hypothetical aggregation of virus particles, the fuzziness of sedimentation boundaries, the assumptions employed to compute molecular data, and the biological testing of the integrity of ultracentrifuge-prepared viruses. In other words, for the first part of his scientific life, Pickles was a "biophysicist" at the Rockefeller Institute.[3] Professionally, he had become an expert at running and developing an important piece of equipment within a biomedical institution in order to produce both technical and biological results. These research products were usually published in biological or bacteriological journals like the *Journal of Experimental Medicine*. However, in the eyes of the Rockefeller officers – including Bauer, the head of the International Health Division – Pickles remained a resource-man, a competent and irreplaceable instrument maker. For instance, on the eve of World War II, he was enrolled into an IHD project aiming at the large-scale production of influenza vaccines, for which Pickles was to design new freezing and drying apparatuses.[4]

The Jackson Laboratory and the Origins of Inbred Mice

Similar patterns of action may be found when analyzing the development of inbred mice as tools for biomedical research. This was achieved by biologists who viewed themselves as geneticists selecting new experimental systems and using them.

One important and early feature in the history of the laboratory mouse is that mice were not studied for themselves but as models of human diseases, cancer in the first place. The American geneticist Clarence Cook Little stabilized the first inbred strains of mice in the 1910s by repeated brother-sister mating. Little started to use his

---

[1] On Pickles' career see Creager (2001); Terry Shinn in this volume.

[2] Mr. Pickles Appointment, 19 Sep 1936, Rockefeller Archives Center (RAC)(a).

[3] On the role of physicists and instrumentation in the making of 'bio-physics' see Rasmussen 1997b.

[4] On the IHD influenza project see Gaudillière 1998, May.

"genetically homogenous" animals to make tumor transplantation less variable (Gaudillière 1999). Karen Rader has shown in her work on the development of mice as research material how Little was driven into the creation of a private research center: the Jackson Memorial Laboratory (Rader 1995). Before the World War II, Little's laboratory gradually became a mouse production site. A few aspects of this early evolution are worth considering in relation to the research-technology perspective.

The Jackson Memorial Laboratory was created as a research site where scientists would explore "man's knowledge of himself, of his development, growth and reproduction . . . and of his in-born ailments through research with genetically controlled experimental animals" (cited in Holstein 1979). In Little's eyes, as well as in those of his wealthy patrons, research on the etiology of cancer was critical to the development of the Jackson laboratory. The center was to a large extent a setting for biology-based experimental medicine. Gathering new tools was viewed as a means to this target. Typical was, for instance, the recruiting of Leonell Strong, a student of T.H. Morgan, who had developed a particular inbred strain in which 90% of the females were affected by mammary tumors. These animals embodied a genetic cause for cancer. Locally, they were massively employed to mimic human breast cancer and render evidence for the putative relationship between cancer genes, hormones, and environmental factors (Gaudillière 1999).

Production tasks escalated as scientists at the Jackson Memorial Laboratory, then known simply as "Jax", started to distribute their mice to outside workers (Rader 1995: chapt. 3). A few figures highlight the process. In 1932, three years after the construction of the first building at Bar Harbor, a dozen scientists and technicians were breeding 20,000 mice a year for local use only. In 1933, the U.S. Public Health Service Cancer Laboratory at Harvard Medical School was the premier outside destination for Jax mice, yet the PHS imported barely a few thousands animals. In 1937, however, as debates on the creation of the National Cancer Institute raged, Jax workers were already selling 40,000 mice a year.[5] Major consumers were not cancer specialists but experimental pathologists using homogeneous "normal" white mice. In contrast to the Jackson workers, they did not work on cancer but on the dynamics and treatment of infectious diseases like yellow fever, typhoid, tuberculosis, etc. For instance, the scientists at the International Health Division of the Rockefeller Division were then buying 20,000 Jackson mice a year.[6]

Up to the late 1930s, sales covered roughly 10-20 % of the Jackson budget. This means that the classical system of patronage, which dominated American biological research, remained the norm. The growth of the Jackson Memorial Laboratory was paid for by individual philanthropy and by the Rockefeller Foundation.[7] The pressure of the Rockefeller Institute scientists as well as Little's ability to assemble a signifi cant group of mammalian geneticists praising the value of his mice explain why Weaver and the Natural Sciences Division kept funding a laboratory that they (in

---

[5] Clarence C. Little to Warren W. Weaver, 4 December 1937, RAC(b).
[6] Johannes Bauer to Warren W. Weaver, 24 March 1937, RAC(c).
[7] On this, see Warren W. Weaver's correspondence, RAC(d).

contrast to the local participants) did not view as a research center but as a resource center making mice.[8]

Although selling the mice did not solve the Jackson lab's financial problems, scaling-up the production changed the face of the laboratory. Initial work had been organized around a handful of "research-maintenance" units, i.e. one scientist and one or two technicians using and breeding a few strains, which they would eventually pass to other researchers. By the late 1930s, following the signing of the first big supply contract with the Rockefeller Foundation, two settings emerged (1954). On the one hand, there were laboratories linked to maintaining "research stocks." On the other hand, there was a production site where the most-demanded 'generic' mice were multiplied by caretakers. Genetically controlled breeding pairs regularly moved from the former to the latter.

Although C.C. Little viewed the production activities as something that has "a good effect on morale and keeps everyone in close contact with the animal stock," he did not think of himself and of his colleagues as instrument makers.[9] This did not mean that the research agenda was not affected by the production practices, but connections remained loose. For instance, specific interests in breeding and husbandry actually produced objects such as John Bittner's agent causing mammary tumors in the C3H strain (on this research, see Gaudillière 1999). This agent was transmitted by the milk of nursing mice. The Jackson workers nonetheless remained scientists working on a wide range of topics including growth, development, and the etiology of human diseases. They collaborated with half a dozen medical institutions, but the bulk of exchanges were loosely related to the circulation of mice.

To conclude this section, one may suggest that in the United States before World War II, ultracentrifuge-makers and mouse-makers were scientists making instruments rather than instrument makers. Although Pickles and Little spent considerable amount of time developing devices and methods – and for the latter, even engaged in sales – they employed instruments and techniques as inventions linking two areas of scientific practices, and they contributed to the construction of a new range of biological entities: proteins, viruses, or genetic pathologies. Rather than participating in a self-referential group of research-technologists, they viewed themselves as physicists or biologists operating at the boundary between subspecialties and investigating new scientific problems. During and after World War II, the changing scale of the biomedical enterprise was, however, to critically affect the careers of these scientists making instruments.

SCIENTIFIC MOBILIZATION DURING THE WAR: ELECTRON MICROSCOPY AND BIG BIOMEDICINE

One may say of "big biomedicine" what David Hounshell once said of "big science": "I can't describe it, but I know it when I see it" (Hounshell 1992). It is now commonplace to claim that scientific mobilization in the war realigned government, uni-

---

[8] On this, see the contrasting fate of research applications and of requests for support to the housing and maintaining of stocks; RAC(e).

[9] Clarence C. Little to Warren W. Weaver, DATE RAC(f).

versities, philanthropic foundations, and industries in their support of research. It thus provided the conditions for the emergence of postwar projects entailing massive funding for large-scale equipment operated by teams of scientists and engineers.[10] Although the restructuring of biomedical sciences has been less investigated than the development of the physical sciences, the existing historiography suggests that big biomedicine shares, in the United States at least, many traits with big physics, beginning with the rapidly changing scale of funding and the increasing commitment of the federal government to the development of research (Patterson 1987; Starr 1982).

A good example of the impact of this reorganization on the nature of biological instrumentation is the history of the electron microscope. Nicolas Rasmussen has followed in detail the early developments of biological electron microscopy, particularly the relationship between the Radio Corporation of America (RCA), which constructed the first prototypes available in the United States, and the handful of biologists who began using these prototypes during the war (Rasmussen 1997b).

Early electron microscopes were actually built for the study of metals. The basic technology was worked out by physicists. Industry, however, quickly entered the game. In 1939, the German firm Siemens offered the first commercial model. At the same time, on the other side of the Atlantic, RCA started to consider electron microscopy as an element in a wide variety of electron technologies. RCA successively recruited two physicists who had some experience in building prototype electron microscopes. The firm's special attention to biological applications was a way to enlarge the putative market for the machine.[11] In 1941, RCA's chief of research decided to offer biologists a machine to work with and to develop biological uses. Anderson, a physicist with three years experience in colloid biochemistry at Caltech, was hired to operate the machine at RCA's research headquarters. Then, the collaboration of prominent biologists was secured by suggesting that the National Research Council establish a special committee to supervise the uses of the RCA prototype.

The first difference from our prewar examples is that, beyond his comments on the performance of the machine, Thomas Anderson had little say on the construction of the instrument. His role was in assisting biologists and visualizing new objects of research. Anderson nonetheless became a biologist while "normalizing" electron microscopic pictures, i.e. defining the right techniques for preparing or observing specimens and selecting the good images of previously invisible entities.[12] Since this was done under the close supervision of the NRC committee and in collaboration with the biologists bringing the specimens to be examined, it is not surprising that images were generally calibrated against existing knowledge.

A second element to be noted is that RCA developed the electron microscope during the war. This context resulted in several peculiarities. First, there is little doubt that the unusual pattern of collaboration between the firm and the NRC was

---

[10] On the postwar transformation of the physical sciences, see Galison and Hebbly 1992; Kevles 1987; Kleinman 1995.

[11] Rasmussen points to another motive, namely the need to advertise expertise in electronic imaging in the context of a fierce battle on television standards (Rasmussen 1997: chapt. 1).

[12] On the linkage between the making of new entities and the development of electron microscopic techniques, see Rasmussen 1997; Gaudillière 1998.

favored by the war's mobilization, with its multiple projects linking academic settings, governmental agencies, and industries. This in turn reinforced the collective regulation of practices: For a few years, almost every American biologist publishing electron micrographs collaborated with Anderson and RCA. Finally, the war contracts helped disseminate the RCA machines: by 1946, RCA model B had been installed in half a dozen laboratories in the United States.[13]

A third element that contrasts with the early history of the ultracentrifuge is that the electron microscope became a tool of multiple uses by being integrated in contrasted arrangements of instruments rather than by changing the design of the machine. This diversity was already visible at the RCA research department, since one aim of the company was to have visitors reveal a wide range of (biological) uses. When the machine entered biological laboratories, the electron microscope was adapted to local panoplies of instruments by inventing new techniques and new modes of evidence-making. The biochemist Wendell Stanley, for instance, developed a "molecular arrangement" by having the electron microscope complement ultracentrifuges, electrophoresis apparatus, and chemical analysis in order to visualize viruses (Creager 2001; Rasmussen 1997b: chapt. 5). In contrast, his colleague at the Rockefeller Institute, Albert Claude, abandoned biochemistry for cell biology by focusing on electronic imaging and by supplementing the RCA machine with microtomes (for ultrathin sectioning of tissues) and cell cultures (Rasmussen 1997b: chapt. 3).[14]

The last element to consider is the appearance of electron microscopy as a research specialty with two branches, one biological and one physical. Although the number of machines rapidly increased in the 1940s, the electron microscope, in contrast to the ultracentrifuge, was not transformed into a simple bench-top apparatus that any biologist could use following brief training. For almost two decades, electron microscopy became the province of highly skilled technicians, engineers, or biologists. The autonomy of this group is illustrated by the rise of the Electron Microscopy Society of America, an outgrowth of the RCA network created in 1949. Similarly, in the 1950s, electron microscopy congresses gathered biologists discussing both techniques and results.[15]

To complement this overview of the rise of industry-related and instrument-based research activities, one should briefly comment on the changing fate of the ultracentrifuge. Though the instrument existed before the war, the scientific mobilization altered the meaning of ultracentrifugation too. First, the machine was increasingly used to isolate and prepare macromolecules.[16] In many discussions, the purity of such entities of medical interest became synonymous with homogeneous centrifugation images. Using an ultracentrifuge for mere preparative purposes required much less a physicist's skill than the analytical use of the machine. This in turn favored the dis-

---

[13] For examples of these relations between war research and electron microscopy at MIT and at the University of Michigan, see Rasmussen (1997b: chapts. 4 and 5). On the role of influenza vaccine research and electron microscopy, see Gaudillière (1998).

[14] On 'experimental arrangements' and instrumentation, see Creager and Gaudillière (2000, in press).

[15] On the EMSA, see Rasmussen (1997: chapt. 1). For the latter, see Gaudillière (2000, in press).

[16] Two examples of such developments are Edwin Cohn's production of purified plasma proteins and the Rockefeller-based research on influenza vaccines.

semination of the ultracentrifuge in biochemical and bacteriological laboratories with little or no experience in "biophysics." Second, the growth of research funding during and after the war created an enlarged demand that was filled by for-profit instrument makers. As a consequence, a few physicists who concentrated on the use of analytical ultracentrifuges as a means to study the structure of macromolecules reinforced their commitment to technological and commercial developments. For instance, Pickles, who had been mobilized by the Navy to advance biomedical ultracentrifugation, after the war turned down an offer for an academic position in biophysics at the University of Michigan to establish Spinco, a California-based company that specialized in the invention and production of new ultracentrifuges for research purposes.[17] Similarly, Stanley's chief ultracentrifuge specialist, Howard Schachman, collaborated as long-term consultant of the same company.

Examining these examples, one may characterize the new dynamics of biomedical instrumentation by several features: 1) the development of multipurpose – generic – machines; 2) a new form of instrument-making increasingly associated with standardization and commercial production; 3) the birth of instrument-centered specialties.

These features all point to reinforced connections between biomedical researchers and the industry. In this respect, one key element of the postwar dynamics of biomedical research was the rising importance of the pharmaceutical industry. The search for "magic bullets," for molecules that would render man immune to most scourges, emerged out of the war research on penicillin and other antibiotics. Large scale chemotherapeutic ventures had little effect on the status of instruments such as the electron microscope, but they made a critical impact on other areas of biological research and instrument development.

POSTWAR RESEARCH-TECHNOLOGY: MASS PRODUCTION, INSTRMENT-CENTERED RESEARCH AND FLEXIBLE USES OF MICE

Instrument

During the war, the Jackson Memorial Laboratory became the main supplier of mice for the contractors of the U.S. Office for Scientific Research and Development, which organized war-related research. Most consumers did not order the highly-controlled inbred strains but the randomly bred common white mice of the Swiss type. The genetic purity of these animals was untested, but researchers working on the control of infectious diseases viewed them as cheap material, normal and homogeneous enough. By 1944, the shipments of the Jackson Laboratory reached 500,000 Swiss mice while sales peaked at $60,000 – more than the entire prewar budget of the laboratory.[18]

Little was thrilled by this booming market for multipurpose animals. He reinforced the service function of the laboratory and endorsed its public image as mouse-

---

[17] Edouard Pickles to Thomas Francis, 8 March 1945, RAC(g). There is no history of Spinco; for a preliminary analysis, see Creager 2001.

[18] Clarence C. Little to Warren W. Weaver, 3 May 1946, RG(h).

producer. In 1941, Little approached the Rockefeller Foundation with plans for the establishment of new production lines, for example, for selecting inbred strains of guinea pigs, rabbits, and dogs, which would fulfill the diverse needs of biologists and pathologists.[19] Discussion stalled for a couple of years, but a plan for the breeding of small mammals was finally endorsed by Weaver and the Natural Sciences Division.[20] It is worth noting that the Medical Sciences Division actually funded the research counterpart of this plan in the form of a large research program in eugenics, focusing on the analysis of the genetic control of behavior by means of inbred lines of dogs.[21] Little's second move toward enlarged production was taken right after the war as it became obvious that some research needs created by the war would endure. In 1946, he again approached the Rockefeller Foundation with a plan for three new breeding buildings.[22] With more than 30 strains of mice in production, Little then thought that his center was in a position to supply biomedical researchers with any tool of choice.

It is highly probable that the Jackson Laboratory would soon have become the first mass producer of inbred strains of mice if it had not been for two setbacks: a) the fire which destroyed most of the lab in 1947; b) the research policy of the Rockefeller Foundation officers, who rated biomedical research as now being the province of the Public Health Service and giant voluntary health organizations such as the American Cancer Society. They decided to leave American biological research to its embarrassment of riches. Little's patronage then shifted to the National Cancer Institute and the cancer society. In the late 1940s, these supports were, however, mobilized to repair the damage of the 1947 fire. The shift toward mass production did not emerge out of a gradual postwar scaling-up of the Jackson activities but a decade later as a consequence of the reorganization of NCI-based cancer research.

Screening Drugs and Producing Mice

In the 1930s, the search for anti-cancer drugs was seen as a marginal, slightly disreputable subject doomed to failure (Bud 1978). Surgery and radiation were the treatments of choice. By contrast, after World War II, professionals as well as the lay public shared the feeling – partly based on the recent success of antibiotics – that the control of malignant diseases by drugs was imminent and that it would be achieved by the establishment of large-scale cooperative programs modeled on the collaboration between state research agencies, hospitals, and industries that characterized the development of penicillin during the war (Gelhorn 1953).

In the early 1950s, this feeling was transformed into a political issue. The continuous pressure of Congress, together with growing demands of non-NCI cancer specialists and of the chemical industry, led in 1955 to the development of the Cancer Chemotherapy National Service Center (CCNSC) (Endicott 1957; Zubord 1984; Zubord, Schepartz, and Carter 1977;). The explicit aim of the new structure was, in the words of one of its main organizers, "to set up all the functions of a pharmaceuti-

---

[19] Clarence C. Little to L. Webster, 25 Jan 1941, RAC(i).
[20] Warren W. Weaver to Clarence C. Little, April 1943. RAC(k).
[21] Medical Sciences Division. RAC(l).
[22] Warren W. Weaver to Clarence C. Little, 21 March 1946, RAC(m).

cal house": the synthesis of candidate molecules, the pre-clinical testing of the anti-cancer activity of these chemicals, and the clinical trials of the most promising drugs. The CCNSC was formally part of the NCI, but decision-making power was delegated to panels that included numerous extramural scientists and physicians: a chemistry panel, a clinical-studies panel, a pharmacology panel, an endocrinology panel and a screening panel. Congress rapidly allocated large funding to this organization: $5.6 million in 1956, $20 million in 1957, and $28 million in 1958.

Preclinical screening of drugs, unlike clinical research, can be relatively easily adapted to standardized, industry-like patterns of production.[23] The achievement of uniformity among experimental mice was one of the important elements in such adaptation. The CCNSC decided that three transplantable tumors would be employed in all the laboratory tests of drugs. The basic scheme was that molecules inhibiting the growth of these tumors in mice would be good candidates for clinical trials. The decision to employ tumors transplanted in inbred strains of mice immediately created a need to enlarge the production of such animals. From 1956 on, the CCNSC collaborated with the Jackson Memorial Laboratory to develop minimal standards for laboratory animals and develop a mouse production infrastructure.

Mass production was first conducted in Bar Harbor and was later extended to commercial laboratories (Zubord et al. 1966). All the animals used in screening tests had to be supplied by producers accredited by the CCNSC, and all the demands for the supply of mice were processed through CCNSC's Mammalian Genetics and Animal Production Section. The control of genetic purity was achieved by the Jackson workers, who supplied certified breeding pairs of mice to other commercial breeders. The CCNSC thus opened a large specialized market, which made possible the large scale production of "more uniformly healthy, well fed mice, with known genetic background and variability" (The National Program of Cancer Chemotherapy Research 1959).

A few years later, in 1959, Gelhorn, one of the leading U.S. specialists in the domain of drug therapy for cancer, strongly criticized CCNSC methods of screening for anti-tumor drugs. In his testimony before the meeting of the National Advisory Cancer Program Gelhorn contended that the expanded screening program did not lead to the discovery of new, clinically important classes of agents. The method employed by the CCNSC, he added, was inefficient: "[T]he mass and mechanized type of screening now employed is less likely to be productive than the observations of the individual investigators" (Gelhorn 1959). Scientists associated with the CCNSC program observed in the meantime poor correlation between screening results obtained in the three-screen system and the results of their first clinical trials. Therefore, they proposed enlarging the assay system in order to include other tumors and other laboratory animals.

These changes in the organization of CCNSC services did not, however, end the controversy between scientists and doctors who advocated an industrial-type selection of cancer-inhibiting compounds and the supporters of a more traditional style of investigation, based on individual expertise (Zubord et al. 1966: 375–6). Facing the criticism that too large an amount of public money was spent on a program that was

---

[23] On unification and standardization of clinical trials by the CCNSC, see Löwy 1995.

only efficient at eliminating drugs and failed to uncover new anti-cancer drugs, CCNSC directors argued that besides concrete achievements in the organization of efficient testing for anti-cancer drugs, their program brought important benefits to the scientific community as a whole. One of the most important results of the chemotherapy screening program, they explained, was

> the development of enough high-quality animal resources to meet the needs of the program and the entire scientific community. The program was a key factor in anticipating and providing such resources for the major expansion of biomedical research in the past decade (The Cancer Chemotherapy Program 1965).

The impact of this industrial research operation on the community of 'mousers' was actually tremendous. The CCNSC-based mass production did make some inbred strains widely available to biomedical scientists but it also significantly altered the nature of the Jackson operations and products. In 1955, the Jackson Laboratory sold 200,000 mice. Five years later, the production of inbred animals or first-generation hybrids for cancer research reached one million.

Scaling-up required more production rooms, more technicians and caretakers, and cheaper and simpler means for bedding, feeding, and housing the mice. Growth was not only a problem of quantity but a problem of quality as well. In lieu of guaranteeing the uniformity of dozens of thousands of animals, the laboratory was to operate on a plant size and standardize many hundreds of thousands of mice. Raising the production of mice by two orders of magnitude meant increased risks of contamination, diseases, and unnoticed variations in the living material. Scaling-up required increased control of the living conditions, increased control of the personnel, and a well-organized and routinized control of quality. In other words, mass production changed the definition of tasks, the organization of the laboratory, and the division of labor. The production of standard testing-mice for cancer research thus established the animal, technical and human infrastructure that was indispensable for the formation of a new community of mouse geneticists.

From Quantity to Quality: Making Mouse Mutants at the Jackson Laboratory

A reformed production scheme operated from 1956-57 onward, after the signing of a major contract with the CCNSC (Chemotherapy Review Board 1959). This scheme relied on the coordination of three departments: a) the "foundation stocks" where scientists and technicians maintained pedigreed and systematically controlled animals from every inbred strain; b) the "research expansion stocks" and "production expansion stocks" where technicians produced hundreds of breeding pairs a year; c) the "production stocks" where caretakers and supervisors put these pairs to their reproductive work.

In the newly built mouse-production rooms, not only new mice but new careers could be made. Caretaking was not 'taylorized' in the sense of time-motion studies, but task distribution became much more specific. Local recommendations completed the general rules set up through the CCSNC. Critical to the success of the organization was keeping records of the mouse circulation, as when moving pairs from the

Table 9.1. Mouse mutants described after 1945.

	Jackson	Other Centers	Pathological mutants	Neurological mutants	Skeleton mutants	Total
1946-1950	0	12	6	0	2	12
1951-1955	11	37	24	9	9	48
1956-1960	15	49	34	12	10	64
1961-1965	48	68	52	19	11	116

(Source: *Biology of the Laboratory Mouse.* E. Green (ed.), Jackson Memorial Laboratory, 1966.)

production-expansion stocks to the production rooms.[24] Breeding cards showing the complete record of an animal's life were kept on permanent file, coded and transferred onto IBM cards for statistical analysis. Increased control of the mice both generated and required a form of knowledge closer to that of agricultural engineers involved in livestock production. A few engineers were employed to conduct studies of breeding behavior, longevity, nutrition, causes of deaths, and growth rates. They soon joined colleagues in the other production centers to develop a laboratory-animal care society.

In contrast to what might have been expected from the concerns with homogenous, large-scale production, the development of the research-expansion stocks actually opened a new space between the laboratory and the plant. In this space, a few researchers were responsible for systematic quality control based on grafting assays and for the selection of scientifically interesting stocks. Following the mid-1950s reorganization, the Jackson workers gave special attention to variants and entered a new world inhabited by dozens of mutant strains of mice. Odds of observing morphological or physiological mouse variants remained low until the 1950s. With the changing scale of production and increased attention to abnormalities in large stocks, a threshold was passed: The 'gray zone' between the laboratory and the plant started to produce numerous good mutants for chromosome mapping and for biomedical research. This pattern may be illustrated with the origins of the mouse mutants described after 1945.

An important aspect of the mutational explosion of the late 1950s/early 1960s was that some of the Jackson researchers could redefine their roles as scientists on the basis of a strong coupling between the making of new tools (in this case the se-

---

24 As a booklet for Jax workers said: "Each complete litter not used to make new breeding pairs within the PES should be put into a pen marked with small cards noting strain, pedigree number of parents (female number is always written first), birthdate of offspring, number of males and females .... Arrange these animals in trios and in brother-sister pairs for the production stocks supply assigning to each mating an identifying number, which is recorded in the strain ledger in a completely different serial numbering system, related to the PES individual animal pedigree number only through the parents data recorded in this ledger. These identifying numbers follow these animals throughout their existence in the production colony."

lection of new mutants) and an instrument-centered research program (in this case the mapping of mouse chromosomes).

Mouse genetics previously contrasted the expansion of the mainstream genetics typified by T. H. Morgan's work on the fruit fly (on Drosophila and Morgan see Kohler 1994; on the early days of mouse genetiscs: Rader 1995: chapt. 4). In the 1920s and 1930s, activities in Morgan's group revolved around three aspects: a) the description of fruit-fly morphological mutants, b) the management of multiple crossings aiming at the localization of these genes, c) the design of chromosomal maps from the statistical evaluation of these crossings. Lacking a significant number of mutants, mammalian genetics followed a different path, with incidental studies of genetic linkages and chromosome location. In the early 1940s, when the Jackson mousers started to circulate a *Mouse Newsletter* modeled on the *Drosophila Newsletter* originating in Morgan's laboratory, information did not focus on mapping tools and results but on inbred lines. Discussions about mutants and mapping results slowly expanded in the 1950s as the number of laboratories in the field increased, but there was no qualitative change until the emergence of more numerous mutants in the early 1960s, or, more precisely, until the multiplication of mutants at the Jackson Laboratory and the establishment by the National Science Foundation of a "mouse mutant stocks center" associated with the Jax research-expansion stocks.

As described by its head, Marjorie Green, the mutant center was to maintain all "good mutants, i.e. easily recognizable single genes" including special "stocks" with multiple mutations that were intentionally "constructed or under construction for efficiently testing a new mutation for linkage with markers of the known linkage groups of the mouse."[25] In addition, the center was to maintain in significant number some mutants of special interest to the biologists for "physiological, embryological, or pathological studies." Lastly, it was to contribute to the organization of the mouse-genetics community by making all these stocks available and by publishing an expanded newsletter with classified bibliographies.

In 1960, when the center was established, 73 mutants out of the 190 existing variants composed the core stocks. In 1970, the center maintained 125 "useful" mutants out of 400 classified variations. The centrality of the mutant center is also illustrated by publication data. In 1950-1954, 32 papers on linkages between genes and chromosomal localization were published, four of them authored by Jackson scientists.[26] In the years 1955–1959, the respective figures were 34 and 6. By contrast, in the years 1960–1964, the number of linkage reports doubled (61) with one third (22 papers) coming from the Jackson mutant center. The role of the Jackson as "specialty-organizing center," "reference center" and "supply center" was further advanced by the establishment of a system of Jackson fellowships supporting the training of young biologists at Bar Harbor.[27]

The mutant center did not only provide reference mutants for outside geneticists but also supplied the Jackson production sector with marketable mutants. All these

---

[25] Jackson Memorial Laboratory, 'Research Proposal to the National Science Foundation,' November 1960. Jackson Laboratory Archives.

[26] These figures are derived from the volume edited by the staff of the Jackson Laboratory, 1966: chapt. 8.

[27] Jackson Laboratory, Annual Reports 1963–64, Jackson Laboratory Archives.

mutants were models of human pathologies, such as the "obese" mice, the "muscular dystrophic" mice or the "hairless" mice. The formation of scientific gatherings characterized by common use of one (or a few) among the Jax products parallels the development of these mutants.

Research-Technology, Standard Mice and Flexible Uses

This description of the Jackson Laboratory in the 1960s reveals a biotechnology enterprise showing many features of a 'research-technology' venture (Shinn, this volume). The Jackson workers mediated between a plant and laboratories. The setting combined research, development, and production tasks. The main products were mouse breeding-pairs constructed and/or selected for particular research purposes. Consequently, the Jackson workers were perceived by outside biologists as suppliers of instrumental resources. In other words, the Jackson Laboratory was an "interstitial," "mediating" community of "instrument-developers."

Looking at the hybridization of roles taking place within the laboratory, however, suggests another – complementary – picture. The setting was not only linked with biological research centers in academia by means of contracts, supply orders, counseling and other services only. The Jackson was also an academic research setting. Local scientists applied for and obtained grants from governmental research agencies (NSF, NIH and AEC). They participated in all sorts of seminars and academic conferences; they published most of their articles in biological and medical journals. Integration of research and technology thus remained the norm, and one may say that the Jackson researchers (as well as their academic clients) remained biologists working on a medical frontier.

Does this mean that besides increasing production tasks, there was little change in the Jackson laboratory spaces? Our description of the mutant explosion suggests that a new feature of this medical frontier was the importance of instrumentation. As emphasized in the case of electron microscopy after 1945, instrument-based specialties surfaced in various domains with the changing scale of the biomedical enterprise. The development of mouse genetics is another example of the same trend. It had important consequences for the meanings of mice: The new infrastructure contributed to the appearance of many specialized tools employed to model the causes of a wide range of diseases. The flexible but highly specific use of these models contrasted both with the openness of generic tools like the Jackson white mice and the highly specialized use of the mapping mutants circulated by the Jackson mutant center. It is worth noting that these new models of human pathologies were currently developed and used within the same settings although some specialization may be perceived. This tension will be illustrated with the history of "obese" mice – inside and outside of the Jackson Laboratory.

The definition of obesity is highly contested, as it stands at the boundary between the normal and the pathological, between nature and nurture, between life choices and medicine. At its simplest, medical obesity is a pathology caused by overeating and leading to overweight. The obese strain was an old mutant which arose in 1949. Because of their abnormal growth patterns, fatness, and eating behavior, the new animals were classified "obese" mice. Classical breeding experiments led the Jackson inventors to attribute the animal disorder to a single recessive gene (Ingalls, Dickie, and Snell 1949). Early studies were conducted at Bar Harbor by Margaret Dickie, a junior assistant who was making her way from quality control into research by following "her" mutants. Building on the current hormonal theory of obesity, the Jackson researchers first conducted a series of physiological experiments ranging from the removal of adrenals, pituitary glands and/or thyroid to the injection of purified and synthetic hormones while checking changes in body weight and fat deposits. The general conclusion was that endocrine factors were certainly involved in the disease but that no specific causal relationship could be established.

As their project took them to the realm of physiology, where they lacked both skills and legitimacy, Jackson scientists searched for outside collaboration. André Mayer, then Harvard professor of nutrition, was a good candidate. By 1951, Mayer and the Jackson researchers had worked out a convenient division of labor. Mayer's laboratory, supplied with the obese mice, conducted most of physiological and biochemical studies comparing obese mice with other animal models of obesity (Mayer 1963a). The Jackson workers concentrated on genetics and reproduction.

At the Jackson, research and production were never very far apart. For instance, to look for hormonal causes of sterility, the Jax workers invented a delicate surgical technique for transferring eggs from a usually sterile obese female into a non-obese foster mother. Beginning in 1954, this technique was employed to increase production yield from 1/4 obtained when crossing heterozygotes to 1/2 obtained when ovaries from an obese mouse were implanted in normal females that were later fertilized by heterozygous males (Lane 1954; Jackson Laboratory Annual Reports 1956/7 and 1957/8). For ten years, however, obese mice were restricted to these two settings: Sales of the strain to other researchers did not begin until the reorganization of the late 1950s and the establishment of a "mutant production unit" in 1959.

What was the impact of obese mice on the science of the human disease? "Obese" as an inbred mutant strain was both an opportunity and a challenge. Like other Jax products it was reproducible, homogeneous, seemingly easy to manipulate. But it was far from being the sole model of obesity available. One key specificity was the fact that the Jax model embodied a causal relationship between one recessive gene and overweight. Such a simple genetic causation did not fit the medical culture and the broad social meaning of obesity. Consequently, obese mutants did not win the status of good experimental models by replacing their competitors but by displacing them through technical and instrumental developments.

In Mayer's laboratory, diet studies that mimicked clinical practice came first: The mice were fed with various weight-reducing preparations including some regimen already in use with humans. A typical series of experiments was conducted by having the mice face what was called a "cafeteria situation" (Mayer 1956). Obese animal were isolated in a specially built cage with three small levers by means of which the animals could get different diets from three bottles. Weekly measurements of the amounts of food consumed were organized. The experiment led to the conclusion that the obese animals did not eat more than the non-obese mice but that they ate differently. Obese mice selected high-fat diets. To a nutritionist this was not by chance: The mice were behaving like diabetic patients selecting nutrients with a low carbohydrate content. This led to a difficult attempt to collect enough blood to measure blood sugar at different ages. The diabetic character of obese mice was then confirmed: In old age, the concentration of blood sugar was abnormally high and it was little affected by the inoculation of insulin. Thus the diabetes-like disease manifested in old obese mice was taken to be a consequence of their obesity. By 1953, Mayer's obese mice were showing a hereditary obese-hyperglycemic syndrome, which supported the notion that new forms of "metabolic" obesity existed alongside the "regulatory" forms commonly exemplified by laboratory rats. Obese mice "proved" that obesity could be an authentic genetic disorder and suggested that the late-onset form of human diabetes may often be a secondary symptom of obesity.

Another nice illustration of the use of the Jackson tools may be found in Mayer's investigation of the role of exercise in the triggering of obesity. Like the 'cafeteria' series of experiments, this sequence highlights both the instrumental logic of research and the local development of tools to be associated with the mice. Building a special motor-driven treadmill (whose plans later circulated within nutritionist circles), Mayer and his assistants exercised obese mice (and rats) between three and ten hours daily and compared them with animals left unexercised. Systematic use of this platform revealed that low-exercising mice overfed themselves: eating more than normally active mice and growing fat. Building on this analogy, Mayer claimed that the most important mechanism involved in human obesity is the central brain regulation that balances appetite, body composition, and energy expenditure. Since 90 % of the cases of human obesity involve no or undiagnosed metabolic abnormality, their disease was taken to reflect an unbalanced relationship between appetite and caloric consumption, whose origins might be found in urban Americans' failure to exercise (Mayer 1955).

One final comment on the uses of "obese" may be that the division of labor between Mayer's and the Jackson laboratories was based on perceived local scientific skills. It was strongly constrained by the status of production tasks within the two settings. In contrast to the Jackson, Mayer's laboratory was a shop: Instrumental innovation was almost exclusively for local use. The laboratory performed some service functions, but these concentrated on the nutritionist's expertise in food regimens.

This difference proved to be of a critical importance in the early 1960s, when a changing political and cultural environment first socialized and later gendered obesity.[28] In the late 1950s, building on his exercise experiments with mice and rats, Mayer performed surveys of high school students aimed at detecting the relationship between activity and overweight. Obese students were described as consisting of students who ate little less than normal students, but exercised much less and enjoyed "sitting activities." Moreover, motion pictures of obese and non-obese children "swimming or playing volley ball or tennis revealed that the obese were in motion only a fraction of the time during which the non-obese moved." Mayer then began collaborating with psychologists in giving a new social meaning to the autocatalytic loop of low exercise, overeating and neurological deregulation. The key elements in the vicious circle driving young girls into obesity from constitutional or environmental predisposition were undermined confidence, insecure personality, family

---

[28] For an overview of changing perceptions of obesity: Bordo 1993.

blame, and advertising that featured a slender body (Mayer 1963b). As a result of this displacement, obese mice that were not used in behavioral or psychological studies quietly left the local experimental scene.

At the same time as Mayer redefined his research agenda, the monopoly on obese mice vanished. As a consequence of administrative technical developments that resulted in the building of a mutant production center, Jackson Laboratory started to sell obese animals. The mutant strain then crossed the threshold of many biochemical laboratories. There, it was used in ways which mimicked the practice of the Jackson biochemists: obese mice were combined with a newly discovered "diabetic" mutant in order to conduct metabolic studies. Obese mice were employed to study the action of insulin, the relationship between fat cells and pancreatic cells, the relationship between fat biosynthesis and the production of insulin.[29] Enlarged circulation within biochemical circles and supplementary instrumentation – all features originating in the Jackson production practices – thus turned the meaning of obese upside down: instead of having diabetic-like symptoms caused by obesity, obese mice were affected with a form of insulin resistance leading to obesity and late-onset diabetes.

CONCLUSION

In order to evaluate the changing nature of biomedical instrumentation after World War II, one should distinguish different dimensions in the notion of "instumentation": a) the human and material resources invested by research groups in the building of instrument prototypes or in the design of methods compared with the resources invested in collecting and analyzing data; b) the extent to which the study of a given range of phenomena depended on a specific and irreplaceable group of instruments; c) the balance between the making of generic devices circulating in many laboratories and the making of tailor-made instruments for local use; d) the scientific division of labor as reflected in the institutional location and collaborative patterns of researchers making instruments; e) the existence of serial and standardized production; f) the patterns of collaboration with commercial firms.

From this perspective, research-technology may be broadly characterized as an activity centered around the development of multiple tools and techniques within the context of collaborative projects that gather scientists of different domains and result in the production of generic instruments (often by commercial and/or industrial firms). Our comparison of the ultracentrifuge, the electron microscope, and inbred mice then suggests that there was no or little research-technology before the 1930s in the biological sciences and that related activities expanded after World War II, as the roles and meanings of instrumentation evolved. One may accordingly propose a "big picture" based on the American case.

During the first quarter of the twentieth century, most instruments for biological research were devised by the scientists themselves: Animals were bred in laboratory animal rooms, while recording instruments were built in local machine shops. Important exceptions to this pattern were a) chemicals and reagents like dyes, which were

---

[29] See, for instance, the papers presented at the New York Academy of Science meeting on 'Adipose Tissue Metabolism and Obesity,' in *Annals of the New York Academy of Sciences* 131 (1965).

purchased from the chemical industry; b) microscopes and electrical devices which were usually supplied by small instrument-making companies; c) machines for medical usage, such as the X-rays apparatuses sold by electrical firms.

New research niches for investigators who spent considerable amounts of time developing devices and finding biomedical uses for them emerged in the 1930s. For instance, a new generation of larger "biophysical" instruments – i.e. spectroscopes, ultracentrifuges, isotopes – were developed by physicists with strong interests in biological applications. These scientists operated in hybrid spaces at the juncture between physics, biology, and medicine, usually supported by private patronage with the Rockefeller Foundation in the leading role. They cultivated neither disciplinary autonomy nor the notion of a specific technological community. They perceived instrumental development as a means to investigate new "boundary" objects or boundary phenomena like viruses or genetic disorders. Rather than being associated with serial production and commercialization, their activities remained within the academic world.

The nature of this nexus changed after 1945 as medical research escalated into a form of "big biomedicine" associated with an exponential growth of (both private and governmental) funding and with the establishment of large-scale research programs aligning university laboratories, hospitals, and industrial settings. From the research-technology perspective this expansion resulted in two changes of significant consequences.

First, the enlarged scale of the research enterprise created niches for scientists operating at the boundary between instrument-producing plants and biological laboratories. As illustrated by the creation of Spinco or the growth of the Jackson Laboratory, a rapidly expanding research market opened venues for the serial construction and sale of apparatuses and tools. This in turn favored the development of generic devices, such as the preparative ultracentrifuge or the genetically standardized mice. Research-technology activities thus became more important and visible. This process, however, did not imply that independent communities of "research-technologists" emerged. It is true that the changing scale of the biomedical research market occasionally resulted in gatherings of instrument developers. For instance, conferences and journals targeted the engineers involved in the breeding and standardization of laboratory animals. The example of inbred mice nonetheless suggests that in most instances the scientists responsible for instrumental innovation remained deeply involved in the study of biomedical objects.

Second – and more important – was the emergence of what should be called "instrument-centered specialties" (such as electron microscopy or mouse genetics). Such fields reveal that: (a) A whole range of biological objects and phenomena were invented simultaneously with the development of new instrumental setups; (b) biologists could then make careers focusing on the use and development of such instruments; (c) in contrast to instrument makers, these scientists produced highly specific packages of tools and methods rather than generic devices; (d) collective resource centers (rather than production sites) played a critical role in disseminating these technologies and in regulating their uses.

Labeling this form of instrument-centered biomedicine as research-technology as opposed to the umbrella notion of "technoscience" is useful in stressing the specific research dynamics originating in the building, handling and circulating of these instruments. When speaking of research-technology, however, on should remember that a new identifiable research-technology community did not form and did not isolate itself from the biology world. In contrast a whole range of biological specialties were affected by the new roles of instrumentation and by the "molecularization" process that accompanied the uses of biophysical machines that characterized the postwar era.

INSERM, Paris, France

REFERENCES

Abir-Am, Pnina (1982). The discourse of physical power and biological knowledge in the 1930s: A reappraisal of the Rockefeller Foundation's policy in molecular biology. *Social Studies of Science* 12: 341–82.

Annals of the New York Academy of Sciences 131 (1965). Adipose Tissue Metabolism and Obesity.

Bordo, Susan (1993). *Unbearable weight: Feminism, Western culture and the body*. Berkeley, CA: University of California Press.

Bud, Robert F. (1978). Strategy in American cancer research after World War II: A case study. *Social Studies of Science* 8: 425–59.

Creager, Angela (2001, in press). *The life of a virus: Wendell Stanley, TMV, and material models in biomedical research*. Chicago: The University of Chicago Press.

Creager, Angela and Gaudillière, Jean-Paul (2000, in press). Experimental arrangements and technologies of visualization, in J.-P. Gaudillière and I. Löwy (eds.), *Transmission: Human diseases between heredity and infection. Historical approaches*. Amsterdam: Harwood Academic Publishers.

Elzen, Bolie (1986). Two ultracentrifuges: A comparative study of the social construction of two artefacts. *Social Studies of Science* 16: 621–2.

Elzen, Bolie (1988). *Scientists and rotors*. PhD thesis, University of Twente, Twente, NL.

Endicott, Kenneth M. (1957). The chemotherapy program. *Journal of the National Cancer Institute* 19: 275–93.

Galison, Peter and Hebbly, Bruce (eds.) (1992). *Big science, the growth of large-scale research*. Stanford, CA: Stanford University.

Gaudillière, Jean-Paul (1998). The molecularization of cancer etiology in postwar United States, in S. de Chadarevian and H. Kamminga (eds.), *Molecularizing biology and medicine* (pp. 165–193). Amsterdam: Harwood Academic Publishers.

Gaudillière, Jean-Paul (1998, May). *Rockefeller strategies during the Second World War: Molecular machines, viruses, and machines*. Paper presented at the conference on 'Medicine as a Social Instrument: The Rockefeller Foundation, Medical Research and Public Health.' Paris.

Gaudillière, Jean-Paul (1999). Circulating mice and viruses: The Jackson Memorial Laboratory, the National Cancer Institute, and the genetics of breast cancer, 1930-1965, in M. Fortun and E. Mendelsohn (eds.), *The practices of human genetics, sociology of science yearbook 1999* (pp. 89–124). Dordrecht, The Netherlands: Kluwer Academic Publishers.

Gaudillière, Jean-Paul (2000, in press). *L'invention de la biomédecine: la reconstruction des sciences du vivant en France après la guerre*. Paris: Editions des Archives Contemporaines.

Gelhorn, Alfred (1953). A critical evaluation of the current status of clinical cancer chemotherapy. *Cancer Research* 13: 202-15.

Gelhorn, Alfred (1959). Invited remarks on the current status of research in clinical cancer chemotherapy. *Cancer Chemotherapy Reports* 5: 1-12.

Holstein, Jean (1979). *The first fifty years at the Jackson Laboratory*. Bar Harbor: The Jackson Laboratory.

Hounshell, David (1992). Du Pont and the management of large-scale research and development, in P. Galison and B. Hebbly (eds.), *Big science, the growth of large-scale research*. Stanford, CA: Stanford University Press.

Ingalls, Ann M., Dickie, Margaret, and Snell, G. (1949). Obese, a new mutation in the house mouse. *The Journal of Heredity* 41: 317-18.

Jackson Laboratory (1954). *Genes, mice and men: A quarter-century of progress at the R.B. Jackson Memorial Laboratory*. Bar Harbor: Jackson Memorial Laboratory.

Jackson Laboratory, *Annual Reports* 1956–57 and 1957–58.

Jackson Laboratory, Roscoe B. Jackson Memorial Laboratory (ed.) (1966). *The biology of the laboratory mouse*. Bar Harbor: The Jackson Laboratory.

Jackson Laboratory (1986). Two ultracentrifuges: A comparative study of the social construction of two artefacts. *Social Studies of Science* 16: 621–2.

Jackson Memorial Laboratory (1960, November). *Research proposal to the National Science Foundation*. Jackson Laboratory Archives.

Kay, Lily (1986). W. M. Stanley's Crystillization of the Tobacco Mosaic Virus, 1930-1940. *Isis* 77: 450–72.

Kay, Lily (1993). *The molecular vision of life*. Oxford: Oxford University Press.

Kevles, Daniel (1987). *The physicists: The history of a scientific community in Modern America*. Cambridge, CA: Harvard University Press.

Kleinman, Daniel L. (1995). *Politics on the endless frontier*. Durham: Duke University Press.

Kohler, Robert (1991). *Partners in science: Foundations and natural scientists, 1900-1945*. Chicago: University of Chicago Press.

Kohler, Robert (1994). *The Lords of the Fly: Drosophila genetics and the experimental life*. Chicago: University of Chicago Press,

Lane, Priscilla W. (1954). Fertile, obese, male mice. *Journal of Heredity* 45: 56–8.

Löwy, Ilana (1995). *Between bench and bedside*. Cambridge, CA: Harvard University Press.

Mayer, Jean (1955). An experimentalist's approach to the problem of obesity. *Journal of the American Dietetic Association* 31: 230-35.

Mayer, Jean (1956). Appetite and obesity. *Scientific American* (November 1956), 21–7.

Mayer, Jean (1963a). Obesity. *Annual Review of Medicine* 14: 111–32.

Mayer, Jean (1963b). Obese adolescent girls, an unrecognized minority group. *American Journal of Clinical Nutrition* 13: 35–9.

Minutes of the *CCNSC Chemotherapy Review Board*, List of mouse supply contracts (11959). NCI Archives, AR 002397

The National Program of Cancer Chemotherapy Research: Information statement (1959). *Cancer Chemotherapy Reports* 1: 99–104.

Patterson, James T. (1987). *The dread disease, cancer and modern American culture*. Cambridge, MA: Harvard University Press.

Rader, Karen (1995). *Making mice: C.C. Little, The Jackson Laboratory and the standardization of mus musculus for research*. PhD thesis. Bloomington, IN: Indiana University.

Rasmussen, Nicolas (1997a). The midcentury biophysics bubble: Hiroshima and the biological revolution in America, revisited. *History of Science* 35: 245–99.

Rasmussen, Nicolas (1997b). *Picture control: The emergence of electron microscopy and the transformation of American biology, 1940–1960*. Stanford, CA: Stanford University Press.

Rasmussen, Nicolas (1998). Instruments, scientists, industrialists and the specificity of influence: The case of RCA and biological electron microscopy, in J.-P.Gaudillière and I. Löwy (eds.), *the invisible industrialist: Manufactures and the construction of scientific knowledge* (pp. 173–208). London: Macmillan.

Rockefeller Archive Center (= RAC):

RAC(a), RU 5, Mr. Pickles' Appointment, 19 Sep 1936, Series 4, Box 22, Folder 260.

RAC(b), RG 1.1, Clarence C. Little to Warren W. Weaver, 4 December 1937, Series 200, Box 143, Folder 1774.

RAC(c), RG 1.2, Johannes Dauer to Warren W. Weaver, 24 March 1937, 100 IHD Laboratory

RAC(d), RG 1.1, Warren W. Weaver's correspondence, Series 200, Box 143, Folder 1774.

RAC(e) RG 1.1, Series 200, Box 143, Folder 1774.

RAC(f), DATE RAC, RG 1.1 Series 200, Box 143, Folder 1774.

RAC(g). RG 5, Edouard Pickles to Thomas Francis, 8 March 1945, Series 4, Box 22, Folder 260.

RG(h) RG 1.1, Clarence C. Little to Warren W. Weaver, 3 May 1946, Series 200, Box 144, Folder 1777.

RAC(i) RG 1.1, Clarence C. Little to L. Webster, 25 Jan 1941, Series 200, Box 143, Folder 1774.

RAC(k). RG 1.1, Warren W. Weaver to Clarence C. Little, April 1943, Series 200, Box 143, Folder 1774.

RAC(l). RG 1.2, Medical Sciences Division. Series 200, Box 113, Folder 1189.

RAC(m), RG 1.1, Warren W. Weaver to Clarence C. Little, 21 March 1946, Series 200, Box 144, Folder 1777.

Starr, Paul (1982). *The social transformation of American medicine*. New York: Basic Books.

The Cancer Chemotherapy Program (1966). *Cancer Chemotherapy Reports* 50: 397–401.

Zubord, C. Gordon (1984). Origins and development of chemotherapy at the National Cancer Institute. *Cancer Treatment Reports* 68: 9–19.

Zubord, C. Gordon, Schepartz, Saul A., and Carter, Stephen C. (1977). Historical background of the National Cancer Institute's drug development trust. *National Cancer Institute Monographs* 45: 7–11.

Zubord, C. Gordon, Schepartz, Saul. A., Leiter, John, Endicott, Kenneth M., Carrese, Martin L., and Baker, Christopher G. (1966). History of the cancer chemotherapy program. *Cancer Chemotherapy Reports* 50: 349–81.

PART IV

STANDARDIZED LANGUAGES

CHAPTER 10

ANN JOHNSON

FROM DYNAMOMETERS TO SIMULATIONS: TRANSFORMING BRAKE TESTING TECHNOLOGY INTO ANTILOCK BRAKING SYSTEMS

In the 1990s, the public's perception of antilock braking systems has been formed by the automobile industry's advertising strategies. Automobile manufacturers market these braking systems to automobile buyers in all price ranges, either as standard or optional equipment. In 1998, over half of all vehicles sold in North America included antilock braking systems. Since these systems do add to the cost of the automobile, they are heavily promoted in the U.S., Europe, and Japan. The automobile industry presents antilock braking systems, or ABS, as a triumph of engineering over dangerous road conditions and unpredictable human drivers.[1] ABS is often shown in advertisements as a great *idea* whose time has finally arrived – completely effacing the difficult process of inventing these devices that prevent the wheels of an automobile from locking when the brakes are applied. Simply put, ABS was a difficult product to bring to the market precisely because ABS is a *device* and not an *idea*. ABS's history as an idea is far older than its existence as a product. Moving from the idea of a machine that would detect and prevent an automobile's wheels from locking to a prototype of a system which would actually do this is a story centered around the development of technologies to measure the performance of a vehicle's braking system. Consequently, research-technology is the critical link between ABS as promising idea and ABS as device. The process of creating antiskid devices constitutes a case study that shows the important role research-technologists have played in inventing new products. Perhaps ABS should be considered a triumph of research-technologists over the difficult task of measuring and reacting to the moving wheel.

The problem of skidding is not a new development; automobiles have skidded for over a century now, for as long as there have been automobiles. As a result, research

---

[1] For the sake of clarity and brevity throughout this paper, I will refer to antilock braking systems generically as ABS, even though this term is legally the trademark for the system produced by Robert Bosch GmbH. The generic use of the term 'ABS' is common practice within the automotive industry. Furthermore, it is the practice of engineering journals, especially those in Europe, to refer to authors of articles using only their first initials and surnames, instead of their complete names. I have followed this practice and herein refer to authors as they actually appeared in the articles cited.

B. Joerges and T. Shinn (eds.), Instrumentation: Between Science, State and Industry, 199–218.

into the problem of skidding, why it occurred, and what could be done to prevent it flourished even before the Second World War (Bradley and Wood 1930). Both the British and the German governments in the 1930s approved patents for skid preventing devices.[2] However, the devices protected by these patents were never commercially produced, and their designs did not provide a foundation for subsequent devices to prevent skidding. While these patents do furnish examples of the history of the *idea* of antilock braking systems, they are not the mechanical ancestors of the devices available today. Antiskid devices first became commercially available on passenger cars in the 1960s – albeit to a very limited market of consumers able to afford a Rolls Royce or Jensen. Moreover, it is not coincidental that the two first automobile manufacturers to provide antiskid devices were British automobile manufacturers, as modern, post-World War II ABS development began in Britain in the 1950s. However, the development of new research-technology in Britain in the 1950s does have an important history, in that it constituted a corrective response to the types of testing instruments in common use during and immediately following the war. The British pushed to invent better devices to measure the performance of a vehicle on a test track, in large part because the earlier methods of measurement were confined to the laboratory.

FALSE HOPES: THE INERTIA DYNAMOMETER

In 1951 the Society of Automotive Engineers (hereafter SAE) created Brake Sub-Committee 3 to produce a report detailing the "Construction and Operation of Brake Testing Dynamometers" at the 1953 annual meeting of the SAE. Dynamometers had become a common instrument in the development and maintenance of automobiles in the postwar world. Engineers involved in the work of designing dynamometers for automotive use even named their work "dynamometry," by which they meant "the art of applying a dynamometer to the problems of automotive service work" (Godbey 1948: 1). The 1953 SAE report described the state of the art in the early 1950s. It detailed the operation and protocols for using the inertia dynamometer as well as the type of information the inertia dynamometers could be expected to provide. These engineers championed bench testing as a more simple and elegant option to the expensive, messy, inaccurate, and often dangerous work of road testing. For the engineers who produced this report, road testing meant losing control of the conditions of testing, e.g., temperature, speed, and load. Controlling conditions carefully meant experiments were more easily replicable and the results more useful for the sake of comparison. As the main purpose of the inertia dynamometer was to compare the performance of different systems, creating similar conditions in which to test systems proved to be of utmost importance. The authors wrote,

> The inertia dynamometer is invaluable in permitting observation and recording – in
> many stages – of the pertinent characteristics of a vehicle brake which indicate its prob-

[2] "An Improved Safety Device for Preventing Jamming of the Wheels of Automobiles when Stopped," U.K. patent 382,241 to Werner Mom, 20 October, 1932. 'Vorrichtung zum Verhüten des Festbremsens der Räder eines Kraftfahrzeuges,' Reichspatentamt Patentschrift Nr. 671,925 to Robert Bosch G.m.b.H., 15 September, 1936.

able performance on the vehicle. In view of the fact that the inertia dynamometer cannot duplicate all road conditions or complications resulting from the installation of the brakes on a vehicle, it is limited to the role of screening or providing data for comparisons of brake operating characteristics measurable within its scope. The inertia machine cannot eliminate the test work required to evaluate brakes on the vehicle, but it can supplement and shorten the amount of road testing necessary (SAE 1953: 329–30).

In short, when research-technologists defined their projects in ways that made replicable experiments and similar testing conditions critical, they constructed an argument for bench testing. However, when duplicating the unpredictable nature of the street became their goal, bench testing appeared to be an undesirable, cost-cutting alternative. To the engineers, using existing equipment showed a lack of commitment to improving testing machinery on the part of their employers. This did not mean the dynamometers were themselves inexpensive. The costs of these inertia dynamometers ranged up to $150,000 and the largest had flywheels over five feet, weighing 15,000 pounds. Clearly, these machines were designed for large, well-endowed corporations or research agencies. Buying a $15,000 dynamometer showed a clear corporate or agency commitment to bench tests. Managers did not make this type of investment as a step to inventing better testing methods.

Other groups interested in improving ways to test braking systems balked at the inertia dynamometers. For research in skidding, the inertia dynamometer had clear drawbacks. The inertia dynamometer measured torque, a variable that had proven difficult to determine in road tests. Brake torque was the primary comparison used in the design of braking systems in the 1950s. However, researchers working on skidding did not find torque to be nearly as useful as brake designers did. Moreover, inertia dynamometers tested one brake at a time, so braking system balance and distribution could not be determined.[3] Furthermore, the sub-committee admitted, "considerable judgment and experience are required to predict from dynamometer test results the actual vehicle performance involving two or more brakes" (SAE 1953: 341). For the most part, engineers investigating skidding were looking at unpredictable behavior, so the need to extrapolate predictions from the inertia dynamometer proved to be an insurmountable constraint. The inertia dynamometer proved useful for research into the frictional surface of brake drums and shoes, but engineers were never able to use it to provide information which predicted the behavior of an actual automobile. The control over experimental conditions that the dynamometer supporters championed meant that the dynamometer could never provide the information skidding research needed. Investigation into skidding required real-time, real-space testing of actual vehicles, not bench testing. The claims of Brake Sub-Committee 3 aside, progress in engineers' understanding of skidding relied on improvements in road and track testing of real autos. These road tests were more closely related to the work of civil engineers involved in the comparison of road surfaces than it was to the work of braking system designers.

---

[3] This drawback was eliminated in the later 1950s when first two-wheel, then four-wheel inertia dynamometers were invented and marketed. Still, the community of engineers working on skidding viewed all dynamometers as obsolete behemoths that failed to simulate or accurately predict the behavior of vehicles on the street. See Gulick (1963: 253).

The dynamometer, which had looked like one of the keys to designing better braking systems in 1950, had proven incapable of quantifying skidding behavior by 1955. Nevertheless, bench testing had been an important domain for research-technologists in the development of improved braking systems. Learning the shortcomings of bench systems formed an important step in developing better machines to measure performance outside the laboratory. As the next stage, field testing proved to be far more complex and ultimately more important to the development of antiskid devices. Consequently, a different group of research-technologists asked a different set of questions in the process·of designing instruments that could better measure the behavior of a skidding car in the unpredictable conditions of road tests. The effort to replace expensive and complicated road testing with bench tests distinguishes the use of research-technology in the early 1950s from what superceded it. By the late 1950s, the effort had switched to making instruments that facilitated less expensive but more accurate and accessible road testing.

SKIDDING AND THE RESEARCH PROGRAMS OF THE ROAD RESEARCH LABORATORY

Immediately following World War II, the British Road Research Laboratory, a branch of the National Physical Laboratory, turned a great deal of its attention to the problem of automotive accidents with an eye toward preventing them. Engineers oversaw projects on a number of different issues related to auto safety, including safety belts, seating, tire and road design, and the performance of braking and handling systems. In 1953 engineers at the Road Research Laboratory (hereafter, RRL) in Slough, England, undertook two unrelated projects both aimed at measuring the braking performance of vehicles in use. One set of engineers examined police records of all accidents recorded in Britain since the war (Giles and Sabey 1959. See also *Research on Road Safety* 1963). Their goals were to identify where accidents occurred, determine why, and propose solutions. The RRL also entrusted them with presenting an overview of the safety of British roads. The project of the second group was to measure the actual braking performance of a sampling of automobiles. They tested automobiles owned by government agencies, as well as randomly chosen, privately owned vehicles (Lister and Starks 1955). By 1958, both groups had published results of their respective studies and their results were in agreement: skidding was a considerable problem on British roads and the performance of braking systems in use was appallingly poor. The director of the RRL, W.H. Glanville, formulated a cause and effect argument between these two reports: skidding, a dangerous, but common, phenomenon on British roads was caused, in large part, by the poor design and condition of braking systems (Shelburne 1959. See also *Research on Road Safety* 1963: 337). Skidding would occur less frequently, and therefore pose less of a threat if braking systems were both better designed and better maintained.

Despite the similar goals and findings of these two groups, the projects were in fact completely dissimilar from the points-of-view of the participants. The former project involved largely statisticians and civil engineers. In examining accident records, they sought to locate especially dangerous conditions on the roads in question. Their project involved a great deal of interpreting these records and they often called in psychologists to discuss the behavior of drivers. The latter project, testing in-use vehicles, was a job for engineers and mechanics. These technicians had to design an apparatus for measuring the performance of an auto's braking system. Since this testing was to be representative of cars found on Britain's roads, it had to be random. Measuring instruments had to be portable, attach directly to a vehicle in service, and provide consistent measurements. The system had to be quickly and easily installed and could not modify the braking system or handling characteristics of the vehicle under examination. It could not bolt on or leave any permanent marks on the car. This research program presented an array of problems that took years to solve satisfactorily. Nevertheless, the instruments invented to measure the performance of a randomly-chosen braking system proved critical to the development of ABS technology in the decade that followed.

One must look even earlier to find the origins of skidding research in British governmental agencies. On behalf of the Ministry of Transport, the National Physical Laboratory created a research program in 1911 to investigate the problem of skidding in automobiles. After halting the program during World War I, the program was restarted by the Department of Scientific and Industrial Research (DSIR), and given a facility at Harmondsworth in 1933 (Pyatt 1983: 119). The DSIR ran research programs for various government agencies, and the skidding investigations were undertaken by the DSIR for the Ministry of Transport. The laboratory at Harmondsworth was called the "Road Research Programme" and used scale models to investigate the relationship between the motion of a vehicle and which of its wheels were locked (Bradley and Wood 1930: 46–50). Models were used because the facility at Harmondsworth did not have grounds sufficiently large enough to skid full-size cars at realistic speeds. Using models also kept costs down. By building a small, rigid chassis with four wheels and four brakes, of the two-shoe, internal-expanding variety, the engineers were able to experiment with the dynamics of the vehicle by locking each wheel independently. This go-cart was towed, then let go to travel under its own momentum. The brakes were applied and the cart came to a stop; the engineers then measured the distance from the initial application of the brakes to the resting position of the cart. The engineers only discovered what had already been known. The shortest braking distance with the least deviation occurred when the cart stopped with all four wheels locked. However, a driver would possess no directional control, although the car would continue in a straight line and not spin. The engineers at Harmondsworth also wondered whether this model be satisfactory for full-scale vehicles? What were the scale effects in translating the behavior of the cart of a full-size car? More importantly, what happened when the chassis was not rigid, when the vehicle had a suspension and a steering system? What other variables did the cart ignore and what role did they play in determining the behavior of an actual vehicle? Unfortunately for these researchers, the Second World War prevented them from answering their own questions. In 1940, the National Physical Laboratory halted the Road Research Programme at Harmondsworth in an effort to divert resources into research more central to the war effort.

In 1946 when DSIR programs returned to normal function, the Road Research Programme was upgraded to the Road Research Laboratory and given its own facility in Slough in Berkshire. Through the 1950s, the RRL was the best-endowed and most extensive road and automotive safety facility in the world.[4] In the decade immediately following the war, under W. H. Glanville's tutelage, the RRL assembled a considerable staff of engineers able to carry out projects befitting the facility the DSIR supported. For the research on braking, R. D. Lister proved to be an important hire in 1946. Lister had been trained as an engineer, receiving his B.Sc. before the war. He held associate memberships in both the Institution of Mechanical Engineers and the Institution of Civil Engineers. Lister led the project surveying the condition of autos currently on the road (Lister 1959; Lister and Starks 1955; Lister and Stevenson 1952; Starks 1954–55). Working with several colleagues, he carried out an extensive survey of cars in use. This study provided many challenges in developing methods of testing, which would play a large role in his later work on antiskid devices. The first and perhaps most obvious problem was how to measure the stopping ability of a vehicle, without reference to torque. Engineers could measure torque easily with the inertia dynamometer, but road testing provided no direct way to measure torque. RRL researchers investigating skidding were more interested in braking distance as an expression of stopping power.

Over about a decade, Lister and others at the RRL came up with seven different methods for measuring braking distance and calculating deceleration. They gathered enough data on each method to calculate deviation figures for each method. The most accurate method became the hallmark of the RRL skidding prevention program, the Fifth Wheel Apparatus. The previous standard of accuracy in measuring stopping distance was the direct measurement of stopping distance. In this method, a switch was fitted to the brake pedal that completed a circuit when depressed. The circuit then fired a charge that left a chalk mark on the pavement within four inches of the spot where the brake pedal had been engaged. Engineers measured the distance between that mark and the final, stopped position of the vehicle. Direct measurement set the standard for the accurate determination of a vehicle's stopping distance against which all other methods were compared. However, it required that the braking system be adapted to accommodate the switch, so it was not practical for in-use cars. The Fifth Wheel method, on the other hand, used a trailer that could be attached to any vehicle in a couple of minutes. The vehicle was driven down a track where the fifth wheel, usually a bicycle wheel, could count the number of revolutions between the beginning of braking at a fixed point and at stopping point. A magnet system in the track triggered the wheel to start counting when braking began. This method yielded less than 1.1% deviation from the direct measurement method, and was a practical solution for testing vehicles off the street. Lister also developed procedures for measuring the braking force at each wheel.

Lister's findings in this program told researchers a great deal about the quality of braking systems on the road and directed further research into improving braking systems. In general, the RRL found that in a significant proportion of vehicles tested, the condition of braking systems was horrible. In a random check in 1950, the average stopping distance from 30 M. P. H. was 70 feet. The shortest stopping distance,

---

[4] Report of Sub-Committee B: Accidents and the human element in skidding (1959).

achieved with all four wheels locked, was 43 feet.[5] Most disturbingly, 10% of cars could not stop within 100 feet and 1% could not stop within 200 feet. Maximum deceleration achievable was $0.57g$ on average. 28% of cars had at least one wheel with no braking force at all (*Research on Road Safety* 1963: 380–2). These results were especially upsetting considering that well over half of the vehicles tested had been manufactured since the war. Lister took these results to the human engineering group. Responsibility for the failures in braking performance lay between manufacturing quality control and upkeep of the vehicles. Driver education could address the latter. But these poor results distressed Lister, and this experience can be seen in his subsequent search for devices that promised to reduce the risk of skidding without being maintained or activated by the driver.

ENTER DUNLOP RUBBER COMPANY

In developing the Fifth Wheel Apparatus, Lister had gained considerable experience in devising a new apparatus by modifying an older design and using readily available parts. This attitude of modification and modularization stayed with Lister into his next project. Because of his experience with the Fifth Wheel Apparatus, Lister worked on many different models of autos and braking systems. He gained first-hand knowledge of the way drivers used the braking systems on their vehicles and how this usage affected the condition of the brakes. This knowledge would become an important element in the next project in which Lister became involved. Lister's position in a government agency also provided an ideal platform from which to reach commercial braking system producers. Lister obtained far more cooperation with corporations than an engineer employed by a potentially competing firm would have.

In 1953 Lister obtained an aircraft antiskid unit from the Dunlop Rubber Company called the Maxaret system. J. W. Kinchin, an engineer at Dunlop in England, headed the team that developed the Maxaret system, and it appeared on the aviation market in 1952. Hydro-Aire, Lockheed, Westinghouse, and Goodyear also introduced similar devices between 1949 and 1952. The aircraft antiskid systems were designed to prevent wheel lockup for both economic and safety considerations. On a landing plane a locked wheel could cause a tire to develop a flat spot after only 1 second of skidding and could blow out within 3 seconds. Blowouts were extremely dangerous as they often sent the plane careening off the runway. In addition the weight of the planes and the pilot's position made skidding impossible to sense, eliminating the option of training pilots to modify their actions in a skid (*The Engineer* 1952). In addition the tires on aircraft were (and still are) very expensive, and airline companies usually planned on retreading them a dozen times or so. A blowout

---

[5] In RRL tests, locked wheel braking consistently gave better stopping distances, despite the fact that locking the wheels actually decreases friction and therefore should lengthen braking distance. The reason for this is practical. If a car could maintain its brakes at a point of impending skid, that is just before the wheels were about to lock, its braking distance would be shorter. However, to prevent skidding the car cannot actually maintain impending skid; one wheel will lock up because of irregularities in the brake lining, the road surface or imbalances in the brake pressure. One of the keys to antiskid devices is that they keep braking efficiency at the impending skid level in order to maintain stability as well as shorten braking distance.

ended the life of the tire and frequent blowouts became very costly, very quickly. In addition, other devices were invented that rotated the wheels of a jet before touch-down. The higher coefficient of rolling friction prevented the friction between the brake and wheel from exceeding that between the wheel and runway surface, making the locking of the wheels less likely.

Dunlop's Maxaret unit consisted of a small rubber-tired wheel positioned against the inside rim of the vehicle's wheel. Inside the wheel were a drum, flywheel, and drive spring. The spring held the drum in contact with the flywheel. When decelera-tion exceeded a certain preset point the spring retracted and the drum stopped rotat-ing. The flywheel, which had 60° of free rotation, continued to rotate relative to the drum. A valve was triggered by this action to close the pressure supply line to the brake-actuating cylinder. Pressure was then exhausted from the actuating cylinder. The brake was released until the wheel regained its angular velocity, at which point the spring reset itself and the cycle started over. This system could achieve 6-1/2 to 7 cycles per second (*Automobile Engineer* 1958; *The Engineer* 1958; *Research on Road Safety* 1963: 414–19; Lister and Kemp 1958; Kinchin 1962).

Lister installed the Maxaret on a 1950 Morris 6 equipped with hydraulically-actuated, self-energizing, drum brakes on front and rear.[6] This involved significantly adapting the device, because the Maxaret was designed to operate with disc brakes, which were standard in the aviation industry. Each wheel required its own unit, and this eliminated braking distribution problems. The system simply detected decelera-tion exceeding a set threshold and modulated pressure at that wheel. Both treaded and smooth tires were tested. The hydraulic-assisted system had to be modified to permit flow of the fluid during the modulation process. The threshold was set at .65g, a much higher level than the unit used on aircraft. The tests were very successful in retaining directional stability, but not in shortening braking distance. On all but a few surfaces, the Maxaret lengthened braking distances, because it could not keep the braking efficiency high enough. However, Lister and Kinchin agreed it was promis-ing that braking distances were shortened when the vehicle was stopped from its highest speeds. Nevertheless, longer braking distances, even at low speeds, would not be acceptable. Lister's experiment had pinpointed the areas for further work.

There were even more problems. In general, the system did not respond quickly enough or to fine enough gradations of wheel slip. These difficulties were directly traceable to the aviation genesis of the device. On an aircraft, the device did not need to respond to mild wheel slip. In addition, the unit made the steering of the vehicle sluggish. These were problems of the device being out of scale. If a device of the Maxaret's type were to succeed at preventing wheel slip in autos, it would have to be designed specifically for the automotive market, which opened a debate about what kind of automotive braking system the device should be designed to work with.

Ever since an internal combustion engine was placed on a chassis in the last quar-ter of the nineteenth century, and the need for an automobile braking system arose, there have been a great variety of automobile braking systems. Consequently, any antiskid device would have to accommodate multiple types of systems, and this fac-

---

[6] Kinchin was already working with Jaguar and used a XK120 with disc brakes in similar tests to those Lister was performing. See Kinchin (1962: 204). See also Starling 1959a,b and Sherlock 1958.

tor, in itself, posed problems. Many British engineers were convinced that the disc brake would answer the questionable brake performance seen on in-use vehicles, albeit at a considerable cost to the consumer. The disc brake was common on expensive luxury and sports cars, while economy cars continued to use front and rear drum brakes until the 1980s, and rear drum brakes continue to be common today. If antiskid devices worked only with disc brakes, then only very expensive models, like Jaguars and Rolls-Royces would have them. Should only the most expensive cars have antiskid devices? While most engineers knew that the units would initially appear on the most expensive cars, the hope of eliminating skidding as a major factor in injury accidents required a larger market saturation. However, these devices could not be designed to be "one size fits all," and they had to be designed in conjunction with a specific braking system. Dunlop's preference was to design an automotive Maxaret system to work with the disc brake. The decision reflected the general British preference for the disc brake, but also the disc brake operated at higher hydraulic pressures that an antiskid device could modulate more easily. In addition, Jaguar supported the disc brake, and Jaguar was the automobile manufacturer most interested in installing an antiskid system in the early 1950s (Kinchin 1962: 204; Randle 1990). Despite engineers' goals of universal use, the interests of the auto manufacturers involved in the early application of antiskid units lay in the performance market, which favored disc brakes of reasons independent from antiskid devices. Consequently, antiskid units also prioritized disc brakes. Building a functional, but specialized, prototype became the primary goal of the engineers and their employers deemed universal applicability less important at the outset.

By 1958, Dunlop engineers had designed a Maxaret unit especially for automobiles. Lister, who remained in the employ of the RRL, not Dunlop, tested this unit on a Jaguar Mark VII with disc brakes on both front and rear (Lister 1963). A pneumatic wheel no longer drove the automotive Maxaret unit. Instead, a geared ring inside the disc brake was in contact with the wheel rim. The exhaust valves could be opened without shutting off the master cylinder supply. A modulator valve allowed the brake to be pre-pressurized to allow the shoes to nearly contact the disc before actuation, meaning that the brake achieved full application almost immediately. Consequently more cycles of modulation occurred. Although this addressed one of the concerns that had grown over the initial Maxaret device, it actually caused a major problem. The brakes shuddered when applied, causing a potential loss of steering control, as the driver lurched forward. This had not occurred during the first trials, because the brakes did not engage quickly enough, although the delayed action was a problem in its own right. On the plus side steering sluggishness was no longer a problem, as the unit permitted finer tuning and could distinguish between cornering and braking more effectively. Lister tested the braking distance of the vehicle on five different surfaces and at two different speeds on each surface. In all but one test, the Maxaret shortened the vehicle's stopping distance versus locked wheels, while retaining directional stability. Indicating his continued concern with the average quality of vehicles on the road, Lister also saw an advantage in that the device could correct for severe brake imbalance, whether it was due to misuse, poor design, or lax maintenance of the vehicle.

Since the work of developing the Maxaret was so closely tied to R. D. Lister and he remained in public service, the results of the tests on the Maxaret were widely published and presented at no less than a dozen conferences. Reaction to the device was split between unbounded enthusiasm and curmudgeonly dismissal. Most of the detractors simply did not believe that an antiskid device was needed; proper driving techniques were sufficient. Lister's experience in testing vehicles in use and his knowledge of accidents from the early studies of the RRL on skidding were enough to tell him this was not true. Therefore, he continued to pursue collaborative antiskid device development with Dunlop and Jaguar into the 1960s.[7] The Maxaret's problems proved to be even more difficult than they originally appeared, however. Vibration problems, from the sharp pre-pressurized actuation and subsequent cycles, became the most common difficulty in all attempts to transfer the aircraft-style brake to automobiles. In the U.S., both Ford Motor Company and Chrysler ran into similar obstacles. The Maxaret and other early devices in development at other companies had tremendous unit-cost problems, as well. When the Maxaret was offered on the 1966 Rolls Royce, it increased the cost of the car by 10%. The Maxaret system alone cost 200% of the price of a Volkswagen Beetle! The question of whether four antiskid units were necessary or whether the application of antiskid to only the rear wheels would be effective arose because of the concern over the high cost of the Maxaret system.

By the 1960s, companies were developing systems designed to work on either four wheels or, to cut costs, two wheels. The RRL's experiments with the Maxaret brought optimism about the possibility of an antiskid device, and this was a most important first step. Given the importance of Lister's involvement with Dunlop and the Maxaret system, subsequent research and development on antilock braking systems followed a similar path. Commercial applications depended upon the work of research-technologists, many of whom worked in the public sector. The transfer of both instruments and expert knowledge between these two groups and two sectors of the economy formed a critical link. Research on skidding moved remarkably seamlessly between public and private sector organizations, especially in the period before ABS was commercially marketed.

FORMATION OF A RESEARCH-TECHNOLOGY COMMUNITY

Lister's work in the 1950s changed the way engineers involved with measuring the performance of a braking system thought about the qualities they were trying to quantify. Discussions of skidding often became discussions of intangible qualities such as "feel for the road," "road grip," "effective handling." How should these qualities be measured? William Thomson's famous quote that "when you can measure what you are speaking about and express it in numbers you know something about it; but when you cannot measure it . . . you have scarcely, in your thoughts advanced to the stage of science" seems particularly appropriate here (Thomson, as quoted in Wise 1995: 5). Until these qualities were quantified, engineers had no way

---

[7] Lister also worked with Lockheed Brake Company Limited on a rear-wheel only system. See Lister (1963: 48).

to be sure they were discussing the same qualities or the same standards. The Fifth Wheel Apparatus was a machine for measuring braking distance, but designing devices to prevent skidding required a way to define whether a driver was in control of a vehicle or not. Lister's Fifth Wheel Apparatus constituted a necessary first step in this direction, but did not itself address this concern.

For the development of further instruments that lent themselves to providing numerical references to these qualities, automotive engineers looked to the aircraft industries, as Lister had in the initial experiments with the Maxaret. The design of planes had long been plagued by the problem of translating intangible qualities into numerical equivalents. In addition, airplane designers often subscribed to the maxim that control and stability were inversely proportionate. That is, the more control a pilot had, the less inherently stable a plane's design was; and the more stable an airplane was in the air, the less control a pilot had. Because of this peculiar quality, as well the on-going struggle between pilots and engineers over ownership of these design decisions, the handling qualities of planes had been scrutinized with the goal of quantifying this relationship since before WWII. Automobile designers looked to aeronautical engineers for guidance in both the construction of theories of vehicle dynamics, often modified from theories of aerodynamics, and the design of instruments which would allow them to establish numerical correspondents to the intangible qualities of handling. Both Edwin Layton and Edward W. Constant have presented other cases where instruments for testing formed a middle ground between scientific, theoretical knowledge and engineering design (see Layton 1971; also Constant 1983). Research-technologists devoted themselves to creating these machines for measurement that allowed the translation of theories of vehicle dynamics into the basis for designing antiskid devices.

The Cornell Aeronautical Laboratory in Buffalo, New York, emerged as the critical location for the modification of theories of airplane dynamics to vehicle dynamics. William Milliken and David Whitcomb led the investigation interested in transforming mathematical models of airplane stability into mathematical models of a vehicle's handling characteristics (see Milliken Jr. and Whitcomb 1957). However, the construction of these theories depended on a significant amount of new quantitative information about the dynamic behavior of an automobile. Engineers at Cornell Aeronautical Lab required new instruments for providing this data. The engineers producing these instruments were not imports from aerospace engineering; rather, they were designing braking systems. The network of technical people working on the problem of skidding was becoming larger and representing a broader array of disciplines and more industries.

One engineer who played a particularly important role in creating machines to measure the intangible qualities of automobile handling was Jean Odier of S. A. Française Ferodo in Paris. Ferodo, with both British and French divisions, constituted the world's largest commercial researcher into friction in the 1950s and 1960s. In an article highlighting their new English research facility that opened in 1959, Ferodo's research areas were described as follows: "fundamental research on friction and raw materials; the development of improved types of brake linings; the testing of such new linings in the laboratory and on the road; and experimental production" (*Automobile Engineer* 1959). Odier, a graduate of the École des Mines, became the director of research in the physical and chemical division by 1962. Odier's early work was highly theoretical, but this experience led him to yearn for better correlation between theory and measurable, track results (Odier 1961, also 1955). Consequently, he turned his attention to building machines to make these measurements in the 1960s when he became a director of research at Ferodo. Odier's primary contributions to the development of antiskid devices consisted of his efforts in two areas: building systems to simulate road conditions in order to facilitate more accurate measurements; and his use of the results of these tests to construct mathematical models of vehicle dynamics. Odier wanted to create machines for reproducing the conditions of the road inside the laboratory. Unlike the engineers who worked in dynamometry in the 1940s and early 1950s, Odier and his colleagues were not trying to eliminate the messy, unpredictable aspects of road testing, but rather to replicate this kind of uncertainty in the laboratory. Doing this work in the laboratory, rather than on a testing track, had several advantages. Not only did lab testing cost less and allow engineers to test a greater number of systems more conveniently, bench tests reduced the need for skilled test drivers to risk life and limb skidding cars on the track.

Odier's contribution to testing rigs consisted of a dynamometer which could simulate high speed, high deceleration, and high centrifugal force. Recorders at each wheel measured and graphically recorded several variables. Displacements, speed, acceleration, deceleration, and torque were detected at every wheel, while vibration, reaction of the ground, pivoting torque, braking ratio, driver corrections, and air resistance could be determined for the vehicle as a unit (Odier, Molinier, and de Briel 1968, 1972, 1976). In addition, Odier's testing rig allowed engineers to safely run tests that would have endangered a human driver. Ferodo's dynamometer, which was available for rent to other brake-producing companies, provided a more solid theoretical basis for designing braking systems. Engineers constructing theories of vehicle dynamics, such as those at the Cornell Aeronautical Laboratory, used the results of Odier's work to set their parameters and provide test variables. Odier was also one of the primary contributors to the translation of qualitative effects into measurable quantities useful in the construction of vehicle dynamics. Still, one could argue that his oversight of the construction of the simulation dynamometer had a far greater effect than did his or anyone else's theoretical work. Once again, the research-technology mattered the most.

Odier's work forms one piece of the development of a core group of engineers working around the world on the development of better machines used to understand exactly what was happening to a vehicle's handling under different conditions. The state of the art in vehicle dynamics in the 1960s can be found by looking at the proceedings of two conferences held in the late 1960s, one by the Automotive Division of the Institution of Mechanical Engineers in London in 1966 and one held by Wayne State University in Detroit in 1964-5. Engineers working on the problem of measuring handling qualities have stated the problem they faced:

> There are two distinct sides of the objective measurement of vehicle handling: the first is the problem of providing the equipment needed to make the measurements; and the second is the problem of which measurements to make. The second of these involves not only which variables to measure but also the decision on what tests or manoeuvers to

perform so that the measured movement of the vehicle during these tests can be used to describe its handling (Hales, Barter, and Olivier 1967).

Engineers at the Highway Safety Research Institute at the University of Michigan worried about these issues of measurement. The Highway Safety Research Institute was another location for the development of theories of vehicle dynamics and the machines needed to provide the measurements on which these theories depended. In the early 1960s, Leonard Segel came to work at the Highway Safety Research Institute, where he joined Ray Murphy. Segel had worked with Milliken and Whitcomb at the Cornell Aeronautical Laboratory in the 1950s, and thus created a human link between these two institutions (Segel 1956–7). In the early 1960s, he joined the Highway Safety Research Institute, where he worked with Ray Murphy. In a career trajectory similar to Odier's, Segel and Murphy turned from building theories to building machines to provide the measurements their theories required. The career paths of Odier and Segel further reinforce Edward Constant's claim that, "testing hardware and procedures embody substantial, technologically relevant, esoteric scientific information and therefore constitute a major mechanism for science-technology interaction" (Constant 1983: 196). This interaction is embodied not only in the machines; it can be seen equally clearly in the research-technologists who invent the hardware.

By the mid-1960s, Segel and Murphy had modified an existing car to have a variable braking system (Segel and Murphy 1968). This system allowed the driver to experience many different types of braking systems in the case vehicle. The system also possessed various recorders for correlating the test driver's qualitative assessments with a graphical or numerical reading provided from the vehicle. Segel and Murphy envisioned the variable braking vehicle as a "research tool which will supplement analytical studies of the braking process. The vehicle will serve to validate theoretical predictions of braking performance for a large variety of braking system configurations" (Segel and Murphy 1968). Segel and Murphy oriented their investigations in a theoretical direction, largely because this was the mission of the Highway Safety Research Institute. However, the testing machines they invented and publicized were widely used to establish quantitative tests for the qualities braking system designers were seeking in the 1960s. And unlike Odier's attempt to replicate road behavior in the laboratory, their machine was for use on the road. The variable-braking vehicle was a simulator of a very different kind – one that simulated a specific type of braking system not a road condition.

Perhaps the most important design aspect of the variable braking vehicle was the use of signal processors to simulate the behavior of different braking systems. This signal processor was an analog computer, sized to fit into the folded-down rear seat of a station wagon. This black box took an electrical signal generated by the driver's foot on the brake pedal and delivered a proportional amount or pattern of braking torque to each wheel. The variable braking vehicle was used by several different automobile, tire, and brake manufacturers in the development of new braking systems, and tire tread and composition designs. Still, the more important role played by the variable braking vehicle was the precedent it set for inserting a computer between the action of the driver and the response of the braking system. Segel and Murphy worked this problem out in real time and real space, creating a system which could modify its response based on pre-programmed algorithms which for Segel and Murphy defined the salient characteristics of particular braking systems. This system became a model for the designers of antiskid devices. The variable braking vehicle was a machine for making measurements, but its role became to show engineers that a computer could be inserted into the control of a braking system. Research-technology was lighting the path to the creation of an antilock braking system with electronic control.

CONSTRAINTS AND PROBLEMS IN BRAKE TESTING

Engineers working on the improvement of braking systems often faced constraints on their ability to design better systems. However, between the 1950s and 1960s the nature of these constraints changed. The primary obstacle to designing better brakes in the 1950s was a limited understanding of the way vehicles responded to different braking conditions. Overcoming this shortfall of information required improvements in research-technology, which led to theories that are more sophisticated and better mathematical models, which again drove the search for more accurate instruments and improvements in research-technology. The engineers involved in this process of research and development recognized the importance of this feedback cycle between theory and experiment.

Developments in research-technology changed the nature of designing braking systems in general, not only in the area of antiskid devices. First, research-technologists also overcame an industry bias against road testing. Lister's work spearheaded the effort to improve the accuracy and convenience of the road test with this Fifth Wheel Apparatus. Lister's connections to Jaguar, Dunlop, and Lockheed brought his bold approach to road testing into corporate research and development by the early 1960s. As a result of new measuring technologies, new designs were produced more quickly and in far greater number in the 1960s. Instead of facing a dearth of information, as they had in the previous decade, engineers in the 1960s had too much information. Their problems lay in separating the useful information from the noise. Engineers needed instruments that provided only the variables their design methods required. Too much interpolation took too much time. Consequently, instrument designers had to be increasingly sophisticated about producing devices in which the variables being measured were changeable. Devices that were successful gave engineers the information they needed in the package that fit their design needs. The testing devices of the 1960s, best represented by Odier's dynamometer and Segel and Murphy's variable braking vehicle, were much more generic than the devices Lister had developed. They provided precise measurements of a wide variety of phenomena. The researchers using Odier's or Segel and Murphy's machines had to decide what conditions to simulate, as they were able to simulate a nearly infinite number of braking conditions in Odier's case, or braking systems in Segel and Murphy's. In a review of testing technology presented at the Second International Skid Prevention Conference in 1977, Hanns Zoeppritz, an engineer working for the Federal Republic of Germany, made a clear statement of the genericity of testing instruments by the 1970s:

> The most important questions always for the choice of measuring equipment or a particular method is the question: What shall we measure? ... As a consequence of the different goals indicated by the basic question – what shall we measure – different measuring methods are used in road construction and in the automobile and tire industries. This does not exclude using the same equipment (Zoeppritz 1977).

Zoeppritz's report also shows the sprawl of the testing machines into several different industries, from automobile manufacturers and braking component firms to tire producers and Zoeppritz's own work in designing road surfaces. Furthermore, the instruments detailed here show the slide that constantly took place between public agencies and private firms. In looking at the technologies developed between the mid-1950s and when Zoeppritz made his report in 1977, the development of flexible machines which lent themselves to measuring multiple variables for different types of projects stands out. The almost constant publication of these machines in a wide range of journals indicates a wide audience for this research-technology.

BRIDGING THE GAP FROM RESEARCH–TECHNOLOGY TO ANTILOCK BRAKING SYSTEMS

While much historical work on research-technology creates a distinction between the production of instruments and the production of new knowledge using these instruments, the case of brake testing technology and ABS effaces this distinction. In the development of ABS, many engineers working on antiskid devices recognized that the measurements their instruments were taking were exactly the measurements ABS needed to take. At its heart, ABS is a system for measuring, comparing, and responding. The development of ABS required precise measurements of angular velocity and ABS itself depended on precise measurements of wheel deceleration. Both testing instruments and ABS had to record change over time and react quickly to actuate mechanical components. Engineers involved in ABS development recognized that their greatest resource was the technicians and engineers at the testing facilities. ABS evolved in an environment of testing, measurement and evaluation.

ABS had to function without constant maintenance and set-up, and this comprised the chief difference between ABS and laboratory testing instruments. ABS had to be as reliable, accurate, and precise as the research-technology from which it developed, but also had to be more rugged, maintenance-free, and absolutely failsafe. An instrument failing in the laboratory created far different consequences than an instrument failing on the highway. In developing instruments which measured and compared angular deceleration and torque on four wheels and computers which could nearly instantaneously compare these values, engineers had solved many of the metrology problems inherent in ABS. But packaging these systems of measurement in a small black box in a passenger car was no small challenge. ABS faced very difficult environmental conditions, too. It had to function under wet conditions, in high and low temperatures, and with significant vibration. Depending on its location on the vehicle, ABS also faced being hit by gravel or stones, or even smashed. These problems provided plenty of challenges in the development of ABS, even once research-technologists had perfected the measurement systems.

Outside of the problem of producing ABS for a reasonable cost, which proved the greatest challenge in bringing ABS to the market, engineers struggled most often with problems of how to assess the performance of ABS, since ABS itself constituted a testing system. Testing had to be performed by external instruments – not by the ABS itself – and focused on verifying the responses of the ABS system. Had the system reacted to the conditions in the best and most predictable way possible? This question evolved in three different directions. First, engineers defined performance standards, i.e., what was the best reaction for the conditions? Then they focused on two different, but complementary, ways to test the system versus these standards: road testing and computer simulation. The technologies used for road tests and for generating quantities for the variables used in computer models were quite similar and were both similar to the ABS itself. This convergence brought greater coherence to the community of engineers working on brake testing. Testing machinery proliferated to provide double and triple checks on the responses and accuracy of measurement of the ABS system's performance.

The work on ABS at Teldix GmbH in Heidelberg, West Germany provides one case study of a company that brought an ABS to the market. Teldix, working with Robert Bosch GmbH and Daimler-Benz AG, introduced the first commercially available electronic antilock braking system in 1978. This system, to which the actual trademark *ABS* belongs, originally debuted on a Mercedes-Benz sedan. However, Daimler-Benz's contract was not exclusive and the Teldix design, subsequently produced by Robert Bosch GmbH., has been used by dozens of automobile manufacturers. The three leaders of the team that developed Teldix's ABS represented different disciplines and different approaches to the project. They also benefited greatly from the work done and published before 1964, when research and development began at Teldix. Heinz Leiber was a mechanical engineer, some of whose previous work had been in developing high performance hydraulic valves for navigational systems at Teldix, which identified itself largely as an avionics (that is, aviation electronics) firm until the ABS project. Joining Lieber in 1966 was Wolf-Dieter Jonner, a self-identified research-technologist. Jonner would lead research in road and laboratory testing of ABS and this experience also provided the knowledge he needed to create algorithms for Teldix's ABS. The electrical engineer on the project was Hans Jürgen Gerstenmeier, whose experience was largely in gyroscope and navigational systems design, when he joined the ABS project in 1968. The development of an electronic control system characterized his work on ABS. At Teldix, ABS development was an interstitial activity, lying between the disciplines of precision mechanical and electrical engineering, on the one hand, and between the firm's departments of laboratory testing, electronics, and hydraulics, on the other hand. ABS never fit into an existing organizational scheme, even when the project was transferred to Bosch in the 1970s.

Because work on the testing of brakes had moved into the electronic era in the 1960s, Teldix's intention from the beginning of the ABS project was to produce an electronic ABS (Binder 1973; *Bosch Technische Berichte* 1971). Teldix was a firm uniquely qualified to move ABS designs into an electronic arena since it was an independent subsidiary of Robert Bosch GmbH, one of the leading automotive electronics firms in the world. However, first Teldix had to produce a system which reduced stopping distance in comparison with completely locked wheels. In the 1950s and 1960s locked wheels produced the shorted stopping distances, despite having the considerable disadvantage that the driver lost directional control of the vehicle. The ABS team produced such a prototype in 1966 when they developed a new type of wheel sensor that measured angular deceleration through the rotation of the wheel at the disc brake. This sensor sent an electrical signal to a mechanical control unit, which could modulate the pressure of the brakes using new valve technology Leiber had developed at Teldix (Leiber and Limpert 1969; *The Elmer A. Sperry Award 1993* 1994: 9).

Once Leiber, Jonner and Gerstenmeier had a promising mechanical system, they focused their attention on creating an electronic system to compare the values coming in from the sensors and control the valves. It was this control system which made Teldix's work newsworthy and which made Daimler-Benz take an interest in the project from an early date. Although the system that Teldix eventually introduced to the market in 1978 possessed several generations of modifications from their successful prototype in, Teldix earned most of its credit for constructing a functional electronic control system. Eventually Teldix, working with Bosch, collaborated with Siemens and American Microsystems to develop purpose-built circuitry for their ABS. For their achievement of producing the first marketed electronic ABS, Jonner, Leiber, and Gerstenmeier won the 1993 Elmer Sperry Award for Advancing the Art of Transportation. In many ways, this was an award for their transformation of research-technology into a mass-market product.

CONCLUSION

In this article, I have attempted to show that research-technology and the engineers who created instruments and machines for measurement built an important foundation for the development of antilock braking systems. The most common popular presentations of ABS rightly emphasize its contribution to automotive safety, and this creates a tendency to see its development as part of a larger emphasis in the postwar world to create safer automobiles, including the development of seat belts, airbags, collapsible steering columns, and steel reinforced autoframes. While intellectually this may have been the motive for developing ABS, none of these other developments in auto safety is related to ABS from an engineering perspective. No other so-called safety feature is so fundamentally an instrument for measurement and comparison. From the beginning, preventing a vehicle from skidding required a nearly instantaneous measurement of the relative decelerations of the wheels. The knowledge of how to determine wheel speed came from the devices used to test automobile braking systems, tire designs, and road surfaces. In the twenty years between the First International Skid Prevention Conference in 1958 and the introduction of Teldix's electronic ABS in 1978, the development of ABS remained a project with close ties to research-technology. In the place where one might anticipate that corporate desires might determine the course of research, one finds instead a small, interstitial community of research-technologists both asking and answering the questions that defined the trajectory of ABS's development.

While one might be tempted to look at the development of ABS as the commercialization of a type of research-technology, instead it represents the transformation of research-technology into a marketable product. None of the measuring systems documented in this paper was transformed unproblematically into ABS, despite the fact that they may have measured the same variables, such as angular velocity, torque, or deceleration, in similar ways. ABS had to function under much harsher conditions. Consumers expect their automobiles to last for 10 years or 200,000 kilometers with minimal maintenance, as opposed to the constant assessment and tinkering that a research and development team's instrument might receive. In addition, engineers involved in designing ABS had to create systems that functioned reliably and predictably without the knowledge of the driver. ABS constituted a black box in the most literal sense of that term, as well as the sense of being a non-interactive, input-output device.

The research-technology produced by government agencies and universities, such as the RRL and the University of Michigan, as well as private companies such as Ferodo, was a giant step closer to ABS than an idea in an executive's mind, but still required a complete transformation. Furthermore, even once a functional product was developed, moving from the prototype phase to the production line and affordability was a difficult set of steps. The commercialization of ABS consisted of the development from prototype to production line. Interestingly, an entirely different community of practitioners took over the process of making ABS commercially viable. Once a prototype was constructed, the work of the research-technologists was complete; they then moved on to inventing a second generation of ABS, not to mass-producing the first generation.

Fordham University, Bronx, NY, USA

ACKNOWLEDGMENTS

I am grateful for the financial support of the National Science Foundation for a dissertation improvement award that supported the research in this paper. I would also like to thank *the Centre National de la Recherche Scientifique* for their financial support in bringing me to the 1997 Sociology of the Sciences Workshop.

REFERENCES

Automobile Engineer (1958, July). Anti-skid device. 48: 248–54.
Automobile Engineer (1959, June). Ferodo Research Centre. 49: 204.
Automobiltechnische Zeitung (1971). Das Bremsregler-System ABS von Daimler-Benz/Teldix. 73(3): 104.
Binder, Kurt (1973). Die Elektrotechnik im Kraftfahrzeug – ein Beitrag zur Verkehrssicherheit. *Bosch Technische Berichte* 4(3): 120–2.
Bradley, J. and Wood, S.A. (1930). Some experiments on the factors affecting the motion of a four-wheeled vehicle when some of its wheels are locked. *Proceedings of the Institution of Automobile Engineers*, Vol. XXV: 46–62.
Constant, Edward W. (1983). Scientific theory and technological testability: Science, dynamometers, and water turbines in the 19th century. *Technology and Culture* 24: 183–98.
Giles, C.G. and Sabey, B. (1959). Skidding as a factor in accidents on the roads of Great Britain, in *Proceedings, First International Skid Prevention Conference* (27–40). Charlottesville, VA: Virginia Council of Highway Investigation and Research.

Godbey, Roy S. (1948). The electric dynamometer. *SAE Paper 480152.* New York: Society of Automotive Engineers.

Hales, F.D., Barter, N.F., and Oliver, R.J. (1967). Assessment of vehicle ride and handling, in *Instrumentation and Test Techniques for Motor Vehicles* Vol. 182 Part B (p. 19). London: Institution of Mechanical Engineers.

Kinchin, J.W. (1962). Disc brake development and antiskid braking devices, in *SAE Paper 620525* (pp. 203–11). Warrendale, PA: Society of Automotive Engineers.

Layton, Edwin T. Jr. (1971). Mirror-image twins: The communities of science and technology in nineteenth century America. *Technology and Culture* 12: 562–80.

Leiber, Heinz and Limpert, Wolf D. (1969). Der Elektronische Bremsregler. *Automobiltechnische Zeitung* 71(6): 182.

Lister, R.D. and Starks, H.J.H. (1955). Experimental investigations on the braking performance of motor vehicles, in *Proceedings of the Institution of Mechanical Engineers, Automobile Division 1954-55* (19–30). London: Institution of Mechanical Engineers.

Lister, R.D. (1959). Brake performance measurement, in *Automobile Engineer* 49(7): 7.

Lister, R.D. (1963). Some problems of emergency braking in road vehicles, in *Symposium on Control of Vehicles during Braking and Cornering* (pp. 47–52). London: Institution of Mechanical Engineers.

Lister, R.D. and Kemp, R.N. (1958). Skid prevention. *Automobile Engineer* 48 (October 1958): 382–91:

Lister, R.D. and Stevenson, R.G. (1952). Fifth wheel for measuring speed and braking distance. *Motor Industries Research Association Bulletin* 4: 1.

Milliken, William F. Jr. and Whitcomb, David A. (1957). General introduction to a programme of dynamic research, in *Proceedings, Institution of Mechanical Engineers 1956-57* (pp. 287–8). London: Institution of Mechanical Engineers.

Odier, Jean (1955). Contribution à l'étude des freins et à l'utilisation correspondante rationnelle des matériaux de frottement. *Bulletin de la Société Française des Mécaniciens* 16, 2e trimestre.

Odier, Jean (1962). Contribution to the study of vehicles during braking, presented at the 1962 FISITA conference. Published in *Proceedings of FISITA IX* (p. 69). London: Institution of Mechanical Engineers.

Odier, Jean, Molinier, P., and de Briel, J.T. (1968). Conception et étude d'une nouvelle machine d'essai automobile simulant la tenue sur route. *Journal de la Société des Ingénieurs de l'Automobile* 10 (October): 531.

Odier, Jean, Molinier, P., and de Briel, J.T. (1972). A dynamometer on which the dynamic behavior of a passenger car can be simulated, in *Proceedings of the Institution of Mechanical Engineers,* Vol. 186 (pp. 7–8).

Odier, Jean, Molinier, P., and de Briel, J.T. (1976). Recent progress in braking tests by use of a car dynamometer, in *Institution of Mechanical Engineers Conference on Road Vehicles* (p. 161). London: Institution of Mechanical Engineers.

Proceedings of the First International Skid Prevention Conference (1959). *Report of sub-committee B: Accidents and the human element in skidding* (p. xxix). Charlottesville, VA: Virginia Council of Highway Investigation and Research.

Pyatt, Edward (1983). *The national physical laboratory: A history.* Bristol: Adam Hilger, 1983.

Research on Road Safety (1963). London: HMSO.

Segel, Leonard (1956). Theoretical prediction and experimental substantiation of the response of the automobile to steering control, in *Proceedings, Institution of Mechanical Engineers* (p. 310).

Segel, Leonard and Murphy, Ray W. (1968). The variable braking vehicle: Concept and design, in *Proceedings of the First International Conference on Vehicle Mechanics* (p. 276). Amsterdam: Swets and Zeitlinger.

Shelburne, Tilton E. (1983). Introduction,' in *Proceedings, First International Skid Prevention Conference* (p. vii). Charlottesville, VA: Virginia Council of Highway Investigation and Research.

Sherlock, S.E. (1958). The disc brake, in *SAE Paper 580114* (pp. 1–2). New York: Society of Automotive Engineers.

Sinclair, David and Gulick, W.F. (1963). The dual brake inertia dynamometer – A new tool for brake testing, in *SAE Paper 630464.* New York: Society of Automotive Engineers.

Society of Automotive Engineers (1994). *The Elmer A. Sperry Award 1993* (program text) (p. 9). Warrendale, PA.

Starks, H.J.H. (1954). Experimental investigation of the performance of motor vehicles, in *Proceedings, Institution of Mechanical Engineers, Automotive Division 1954-55 No. 1* (p. 31).

Starling, J.O. (1959a). Disc brakes, Part I. *Automobile Engineer* 49 (January 1959): 10–14.

Starling, J.O. (1959b). Disc brakes, Part II. *Automobile Engineer* 49 (February 1959): 53–9.

The Engineer (1952, 3 October). Automatic landing wheel braking. 194: 460.

The Engineer (1958, June 20). Wheel lock inhibited braking. 205: 940.

Thomson, William (1995). Electrical units of measurement, as quoted in M.N. Wise (1995), Introduction, in *The Values of Precision* (p. 5). Princeton: Princeton University Press.

Transactions of the SAE (1953). Construction and operation of brake dynamometers: Report of brake subcommittee No. 3, in *SAE Paper 530130* 61 (pp. 329–30). New York: Society of Automotive Engineers.

Zoeppritz, Hanns (1977). An overview of European measuring methods and techniques, in *Transportation Research Record*, No. 621 (75–82). Washington DC: National Academy of the Sciences.

CHAPTER 11

ALEXANDRE MALLARD

FROM THE LABORATORY TO THE MARKET: THE METROLOGICAL ARENAS OF RESEARCH-TECHNOLOGY

The construction of metrological systems is an important topic in the study of research-technology. Research-technologists have to elaborate materials and procedures that help certify the reliability of the instruments they develop. Building an instrument's metrology is a task involving much more than strictly technical work: diplomacy, financial risk-taking, social know-how and certain negotiation skills are among the ingredients intervening in the metrological elaboration of research-technology. Particularly, this work requires the anticipation of the way in which an instrument's reliability and precision will be evaluated by different audiences, in order for the device to conform to their criteria of relevance. The recent development of a measurement technology, Differential Optical Absorption Spectrometry (DOAS), illustrates the difficulty of this task. An application of spectroscopic technology to the monitoring of the atmosphere's chemical composition, DOAS was developed during the 1980s in several laboratories and firms in Europe. It seemed to offer new solutions for the detection of low-concentration chemical components. Such components, like ozone or sulfur dioxide, are present only at a very low rate in the atmosphere but can have a high degree of toxicity for human health or for the environment. For the researchers and engineers who, in France, Germany, England, Belgium and Sweden, engaged in the construction of DOAS instruments, the technology presented – and it still presents – a great number of interesting advantages when compared to the other measurement technologies then available. But as we will see, it also raised serious metrological questions.

Although the emergence of DOAS technology is a recent feature, it is based on classical principles in the domain of spectroscopy. Absorption spectrometry makes use of the structured absorption of light by chemical components to determine their concentration: When a gas is crossed by a light beam, part of the energy is absorbed at particular wavelengths relative to the chemical components of the gas. An analysis of the light beam with a spectroscopic device enables the identification of the spectral regions of light absorption and, subsequently, the reconstitution of the various concentrations in the gas. The method of *differential* absorption spectrometry is not new

B. Joerges and T. Shinn (eds.), Instrumentation: Between Science, State and Industry, 219–238.
© 2001 *Kluwer Academic Publishers*.

and can be traced back to Dobson (1931). The comparison of the absorption of the solar light at two different wavelengths was then used to evaluate the total atmospheric ozone concentration.[1] But contemporary DOAS systems, devoted to the detection of multiple compounds at very low concentrations, utilize a much more specific technology than did these predecessors, because they do not operate simply by comparison between two absorption rays. Using special detectors and complex mathematical functions, modern DOAS systems reconstitute the spectrum of absorption in the UV and visible domain, smooth it and analyze its fine, differential structure in order to calculate the concentration of several gases.[2] This requires high-performance technical components, such as precise grating spectrometers, sensitive photomultipliers or detectors and powerful computers.

For these reasons, DOAS applications adapted to the tropospheric domain and to the detection of very low concentration chemical components could emerge only in the recent period. It is said that the first instrument comparable to the DOAS systems of today was built by two German researchers in 1978 (see Platt and Perner 1983). But it was mainly after 1985, when diode array detectors[3] appeared on the market of electronic supplies, that the technology could enjoy a new boom and give birth to a generation of reliable and sensitive experimental apparatus. In France, DOAS instruments were developed during this period mainly in a laboratory at the *Centre de Recherche et d'Etudes sur l'Atmosphère Terrestre* (CREAT[4]), as part of research in the field of atmospheric chemistry. In Europe, research on DOAS was structured within a EUREKA program created in 1988 for the purpose of uniting the efforts of several laboratories and firms towards the improvement of the technology – and particularly, its metrology.

From a metrological standpoint, DOAS technology created quite an interesting problem because it generated instruments that proved difficult to calibrate. In spite of this difficulty, or perhaps because of it, scientists and engineers had to devise trials and procedures able to persuade various actors of its reliability as they developed the technology. The diversity of these actors was a problem, because the history of DOAS involved academic scientists as well as small high-tech firms, regulatory institutions and state agencies, industrial test-laboratories, and others. This observation draws attention to the task of adapting metrological practices for specific actors, and it points to a general problem for research-technology, namely the necessity of presenting and promoting an instrument in different social spheres. The history of

---

[1] One of the wavelengths was chosen in the absorption domain of ozone, whereas the second one was not in this domain. Nevertheless, the two wavelengths had to be close enough in order to minimize the influence of large band absorption due to the diffusion of the molecules.

[2] Compared to other spectroscopic analyzes, the DOAS method involves discriminating between the broad and the narrow spectral attenuation features. These narrow attenuations circumscribe the differential spectrum of absorption, which has to be calculated by smoothing the original absorption spectrum. For technical details see Claude Camy-Peyret et al. 1996.

[3] Such components include a large number of detectors (512 or 1024 in 1994, when I investigated the field) on a small array (a few centimeters). When used in conjunction with a grating spectrometer, they allow for the simultaneous detection of many wavelengths, making the reconstitution of a consistent absorption spectrum possible.

[4] Some of the names of persons, products or organizations mentioned in this chapter are pseudonyms. More details on this case study are given in Mallard 1998.

DOAS leads us to question the continuity of metrological practices all along the trajectory of the products of research. From the laboratory to the market, instruments make their way through various *metrological arenas* where their properties are evaluated with specific conditions and criteria. The notion of a metrological arena[5] will be used here in an attempt to grasp the importance of the practices of performance assessment for measuring instruments. It refers both to the composition of an audience and to the constitution of trials that support the evaluation of the validity and precision of the instrument. This chapter will outline some episodes of the history of DOAS in the French and European context, in order to explore the articulation of different metrological arenas that are typical of research-technology.

FROM ATMOSPHERIC CHEMISTRY TO URBAN POLLUTION MONITORING

Since the beginning of the 1980s, Professor Marais, a scientist at CREAT, has been committed to the development and progressive refinement of measurement techniques using UV-visible absorption spectrometry. With a small scientific team of half a dozen technicians, engineers and researchers, he directed the construction of several DOAS instruments and their systematic application to different problems.

Marais' project began with the development of *passive* DOAS systems. In this version of DOAS technology, the instrument's detectors focus on a natural light source (the sun at zenith, or the moon at night). This provides measurements relative to the whole atmospheric column. Ground-based passive DOAS instruments designed in Professor Marais' laboratory were used to observe the variations of atmospheric NO_2, O_3 and NO_3. Balloon-borne instruments were also designed to investigate the stratospheric photochemistry of ozone. These developments took place in a context where the controversies over climatic change and ozone layer depletion reinforced the need for quantitative data concerning the chemical state of the atmosphere. Along with other types of instruments (Lidars, Fourier transform infrared spectrometry, microwave detection, satellite monitoring, etc.), DOAS instruments built at CREAT were used in a series of international research programs, which included campaigns of measurement of the chemical composition of the stratosphere and the establishment of networks of instruments monitoring the Earth's atmosphere. Another kind of application relates to the spatial domain: a variant of the instrument was sent to space in a Russian space probe to measure SO_2 in Venus' atmosphere.

But Marais' commitment to the development of DOAS was not limited to his scientific work. At the end of the 1980s, he helped to create Orion, a small high-tech firm dedicated to the commercialization of his instruments. Through Orion, it be-

---

[5] I use the term 'arena' in a sense close to the interactionists' use. In this tradition, it refers to a material, conceptual or social space in which different participants, representing heterogenous 'social worlds,' negotiate, discuss and confront their point of view concerning a common problem to which they possibly have different approaches; see Clarke 1991. I also take the idea of an arena as a scene of *evaluation*, as it is present in the approach of Nicolas Dodier, in his study of the 'arènes des habilités techniques,' see Nicolas Dodier 1993.

came possible to produce and sell, in limited series, the first generation of passive DOAS instruments created in the laboratory at CREAT.

Another achievement of French promoters of this technology has been their contribution to the application of DOAS to a new context, the monitoring of urban pollution. This application required the use of another version of the technology, based on *active* instruments. Active DOAS systems have an artificial light source and provide measurements at ground level. A special lamp is laid out on a measurement field outdoors, and the measurement relates to the region situated between the lamp and the detector. It corresponds to the average concentration of chemical compounds throughout the path transected by the beam. In order to detect low-concentration compounds, the distance between the light source and the detector has to be quite large. For the various active DOAS systems that were in operation around 1990, this distance ranged between one hundred meters (330 feet) and a few kilometres (3km = 1.9 miles).

Active systems opened the field of urban pollution monitoring for DOAS technology. But in the late 1980s, the leaders of this field were not professional scientists. The only DOAS instrument available in Europe for urban pollution monitoring was built by a Swedish firm, Optim, that had developed the technology at the margin of the academic world. It was also the only instrument available developed for the particular needs of air quality control. CREAT (and the associated firm Orion) was one of the first laboratories to compete with Optim. The team of Professor Marais played a significant role in transferring scientific knowledge and expertise from the domain of atmospheric chemistry to the new field of urban pollution monitoring. In 1988, CREAT initiated the conception of an active DOAS system, and a prototype was built in 1989. The first version of the final commercial apparatus, the *Système d'Observation et de Mesure Atmosphérique* (SOMA) was designed in fall 1990, replication began in 1991. That same year, a run of tests in an official metrological laboratory was launched, and in 1992, Orion sold the first two production units of SOMA. In 1994, Orion was still the only competitor of Optim on the European market for this kind of instrumentation. Whatever the eventual commercial success or failure of SOMA, the very act of bringing an instrument from the laboratory to the market in such a short time illustrates the strong commitment of the French to the development of DOAS, which is also shown by their active participation in the creation and coordination of TOPAS (Tropospheric Optical Absorption Spectroscopy).

A Eurotrac[6] sub-project, TOPAS was created in 1988, with the aim of supporting the efforts of the various European teams engaged in the application of DOAS technology to urban pollution monitoring. In addition to France, research on DOAS was conducted in several laboratories elsewhere in Europe, notably in Belgium, England, Germany and Sweden. At the start, TOPAS involved seventeen academic and industrial partners, including the two competitors, Orion and Optim. Approximately ten teams actually contributed to the project in the period 1988-1994. Although many of these participants had initiated research and development on this topic before

---

[6] Eurotrac is a EUREKA project. It was created in the mid-1980s to support research on the atmosphere of the Earth. Eurotrac was made up of fifteen sub-projects, three of which, including TOPAS, were devoted to the development of measurement technologies.

1988, it appears that the project had a significant effect in the stimulation and coordination of their relatively disparate activities.

For the participants, TOPAS was a possible way of placing DOAS technology in the realm of environmental public policy. According to the promoters of TOPAS, DOAS was potentially a powerful technology poised to supplement, if not replace, the conventional measurement technologies currently in use in the domain of air quality control. Most conventional systems available on the market had a local sampling system: The gas sample was taken from the area surrounding the instrument. DOAS instruments were reputed to have many advantages compared to these competitors. For instance, by their ability to perform average measurement over a large distance, they *de facto* avoided the various difficulties of representativeness, homogeneity and stability of measurement associated with local sampling. DOAS defenders also emphasized its advantages as a "multi-component" technology, meaning that the same instrument could simultaneously measure the concentration of several chemical compounds. In contrast, with conventional commercial instruments, each device could only measure one or two gases, and measurement was based on a specific analysis protocol: UV fluorescence for SO_2, chemiluminescence for nitrogenous components, UV photometry for Ozone, etc. In other words, a single DOAS system was sufficient for measuring a whole range of compounds whereas with conventional techniques, each compound required a separate apparatus.

In order to achieve the full development of a generation of reliable DOAS instruments, it was necessary to foster the exchange of data and expertise among the different scientific teams. This was the main argument for the creation of TOPAS. And, most important, TOPAS provided a framework for obtaining public funding from national governments: Between 1989 and 1992, the French team received more than 120,000 ecus, the Belgian team more than 200,000 ECUs and the Swedish team approximately 450,000 ECUs.[7] Among the concrete non-financial results of the project are the two intercomparison campaigns organized in Brussels in 1992 and Norwich in 1994. During these campaigns, several teams participating in TOPAS brought their instruments to the same experimental field and measured the air pollution, in order to compare results.

These few historical landmarks suggest that the development of DOAS technology operated through its circulation between different spaces: from the laboratory to the field, from the national to the international level, from atmospheric chemistry to urban air-quality control, from academia to the market. The next section introduces the discussion of DOAS as a research-technology, through the perspectives of interstitiality and genericity.

DOAS AS A RESEARCH-TECHNOLOGY

A first point to discuss concerning the relevance of DOAS for research-technology is the characterization of actors. The ability and the desire to "cross boundaries" (Shinn

---

[7] Anyone evaluating or comparing these amounts should take into consideration that they represent the support given to the teams by their national government *through* TOPAS, according to the general funding procedures of EUREKA.

1997). is obvious in the psychology of the French actors concerned. For instance, Marais believes that scientists should not stay isolated in their laboratories and that interaction with non-academic actors should be part of their ordinary activity. His involvement in many programs has led him to view the cooperation between research and industry as normal and necessary for the conduct of experiments and the production of knowledge. He once told me, during my field work, that there were too many "idealists" in academia. He went on to say that he sometime intended to write a paper co-authored by Orion, and he was certain that this would be seen as a provocation by many of his colleagues. In organizing the technology transfer from CREAT to Orion, Marais revealed his strong interest in innovation and proved skillful at rendering public and industrial interests compatible and convergent. Thus, if Marais is a true academic, he belongs to the particular subset of those academics who show interest both in producing knowledge *and* in developing innovation. His academic contributions largely rely on the skills and information accumulated through technological activity; and a brief look at his publication record reveals that almost all his writings are based on results obtained with UV-visible absorption spectrometry instruments developed in his laboratory. Hence, Professor Marais can be characterized as a research-technologist, although he presents the profile of a researcher who tries to draw on the resources of both academia and industry, rather than actually being "in the middle." Is it possible to extend this observation to other DOAS practitioners[8] and to identify an interstitial community? Let us take the example of the actors participating in TO-PAS.

It is worth noting that TOPAS itself was conceived as a project where industrialists and academics could work together. The history of the project in the period 1987–1994[9] reveals the typical tensions of this kind of cooperation. For example, problems occurred in the sharing of information and expertise, and it proved difficult to reconcile the interests of the firms and the academic teams. Consequently, after a short period, the Swedish firm Optim left TOPAS and the French firm Orion remained the only industrial partner in the project. Many researchers working within TOPAS were academics like Marais, interested in constructing and experimenting with new measurement devices. Unlike Marais, they were not necessarily interested in the promotion of innovation as such. The research community of TOPAS seemed to be firmly rooted in academia rather than having an interstitial position between the poles of industry and academia. But this conclusion draws attention to an analytical problem concerning the way in which a community can be granted the attribute of interstitiality. Since interstitiality is a property of groups and not of individuals, the question of how the group is constructed in sociological or historical analysis is very important. It is possible to define the group by the institution (a laboratory, a society) to which ist members belong, or by their commitment to a given political project. But if one considers, as has just been done here, that the technology constitutes the means of defining the community of actors, then there are few reasons to expect that these actors have comparable sociological profiles – for instance, a hybrid career trajectory, or the endeavor of a research-technologist. In this respect, the situation where

---

[8] By this term, I refer to the researchers that are involved in the development of the technology.

[9] I investigated the fieldwork in 1994. All data given in this study relate to the period 1987-1994.

an interstitial research-technology community exactly overlaps a network of researchers engaged in the development of a technology (see for instance Johnston, chapter 7) should be considered a notable singularity. This constitutes an analytic problem to discuss and to solve in further research-technology studies.

To what extent does DOAS technology lead to generic devices? Researchers of the French laboratory at CREAT built several DOAS instruments adapted to different categories of problems. There was a significant technological similarity among the various instruments regarding their optics, electronics and method of analysis. The technical improvements made from one generation to another benefited the whole range of devices. Researchers would sometimes say that "basically, a SOMA was nothing more than a classical passive DOAS instrument equipped with a lamp", notwithstanding the many differences in technical details. On a similar basis, different instruments were built according to their different prospective uses. The conception of the first generation of instruments is intimately linked with an ordinary experimental activity, related to a particular scientific problem; SOMA came from the adaptation of an emerging technology from one domain to another. The firm Orion tried to extend the use of DOAS to other fields of application, for instance the control of polluting emissions in chemical plants. There is clearly a core set of technical practices that characterize the way in which these different topics are tackled by the specialists of DOAS and that make it distinguishable from other measurement technologies. In the experience of the French actors, DOAS is generic in the sense that it circumscribes a basic technology used for the development of measurement systems adapted to different applications.

THE CALIBRATION OF DOAS INSTRUMENTS:
A METROLOGICAL PUZZLE

As we have seen, the development of DOAS technology involved the circulation of instruments through various locations. This kind of trajectory is quite standard for the products of research-technology. Yet from a metrological point of view, such a circulation was far from evident, because each arena featured a particular way of evaluating the instrument's performances. The advantages attributed to the technology by the promoters of DOAS were not always nor immediately recognized by other actors. In this game, specifically metrological issues played a major role for the acceptance or rejection of the technology in different situations.

A typical aspect of the difficulties faced by research-technologists in this situation involves their relations with those whom, for the sake of simplicity, one can call "official metrologists." In general, these actors, in charge of metrological issues for administrative or industrial purposes, operate in state or private agencies and work at the frontier between scientific, industrial and regulatory topics: standardization, certification, legal metrology, etc. Their responsibilities include the preservation of standards, the testing of industrial products, the elaboration of regulations concerning weights and measures and the surveillance of instruments supporting economic transactions in the market.

In the French case, official metrologists entered the DOAS story in 1991, with the intervention of an agency charged with supporting R&D in the environmental domain, namely the ADEME (*Agence de l'Environnement et de la Maîtrise de l'Environnement*). It was part of the mandate of this agency to provide recommendations and requirements for equipment used in local networks of air quality control in the principal cities of France. As the application of DOAS to this field was emerging, it was decided to test the new technology in order to inform the local air-quality control networks on its reliability: Some of these networks had already planned to purchase the Swedish Optim system, which at this time was widely commercialized. ADEME ordered a metrological evaluation of both of the commercial DOAS instruments available. This evaluation was conducted in a public certification laboratory, the INET (*Institut National d'Evaluation des Techniques*). The two instruments concerned were Optim's device and the French SOMA, designed in Marais' laboratory and produced by the firm Orion. The evaluation reports suggest that the main metrological problems faced by DOAS technology came from the absence of a proper *calibration trial* for the instruments.

For many measurement instruments, calibration is performed by means of a particular trial in which the instrument is submitted to a surrogate signal, in the sense given to this word by Collins (1985): a well known and well quantified signal, possibly different from the signal received by the instrument in the situation of normal use, but inducing the same effect. For instance, in the domain of air-pollution monitoring, conventional measuring instruments are calibrated with reference gas bottles, containing a mixture of gases at known concentrations, according to a method that is accepted by most specialists. The instrument is connected to the bottle, and the measurement is adjusted in accordance with these concentrations (see Fig. 11.1). Collins clearly shows that a calibration trial is not convenient *per se* and that the judgement of appropriateness of a particular trial for performing a calibration is historically and socially variable (see Collins 1985: 105). The DOAS case provides us with an interesting example of interpretive flexibility in this type of metrological issue. Indeed, the definition of calibration trial adapted for DOAS instruments was highly problematic. They could not be calibrated like conventional instruments because it would have required building a calibration cell as long as the light path (many hundreds of meters), a very expensive and technically complex operation.

During the official assessment at the certification laboratory of INET in 1991, a substitution method was tried: the 1-kilometer path was replaced by a 2-meter cell, in which the gas could circulate under metrological control. In order to compensate for the reduction of the path, the concentration of the gases in the cell was highly increased. The record shows that this attempt was not very successful. For both instruments, the results of the evaluation were mixed, and the metrologists of INET and ADEME expressed some skepticism about the maturity of the technology. The designers of the instruments contested the evaluation, arguing that the calibration trial was flawed. They were skeptical about the possibility of cooperating with these metrologists. All in all, it proved difficult to reach consensus on the real meaning of the substitutive calibration trial.[10]

This episode, while exemplifying the metrological puzzle of calibration for DOAS technology, also draws attention to an important aspect of the relations be-

---

[10] I have commented on this situation in Mallard 1998.

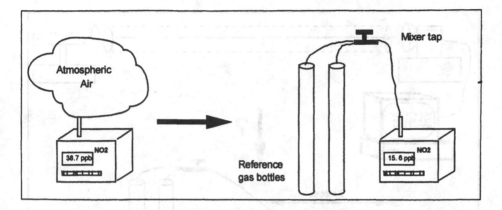

Figure 11.1

tween research-technology and metrology. Research-technologists potentially have to deal with various conceptions of metrological conformity, if the success of their enterprise draws on a capacity to circulate instruments through different contexts, from the laboratory to the market. Indeed, in each arena through which an instrument passes the evaluation of precision and metrological reliability is tied to specific practices and settings. It might be fruitful to explore the conflicts and controversies arising out of divergences on the legitimacy of metrological practices. The trajectory of a DOAS instrument provides us with three different kinds of arenas, to be examined in the following sections: scientific experimentation, official metrology, and the marketing of an instrument.

EXPERIMENTAL METROLOGY IN ACTION

> It is not normal to believe that
> there is a God of calibration.
>
> *A scientist*

Calibration trials are very commonly used to ascertain the quality and precision of instruments. But as the record suggests, this resource cannot be used unproblematically in every situation. Scientists know this very well. For instance, DOAS practitioners claimed that their instruments were precise and sometimes more precise than conventional instruments, although they did not have at their disposal any calibration trial in a strictly metrological sense. And there were very good arguments for believing that this claim was neither insincere nor opportunistic. A closer look at the laboratory work enables one to discern two different kinds of practices that play a major role in the arena of experimental metrology: *familiarity* and *interpolation*.

Familiarity is about the shift between the theory and practice of measurement.[11] When they were asked to account for the good performances of their instruments, the

---

[11] On familiarity as a regime of action, see Thévenot 1994.

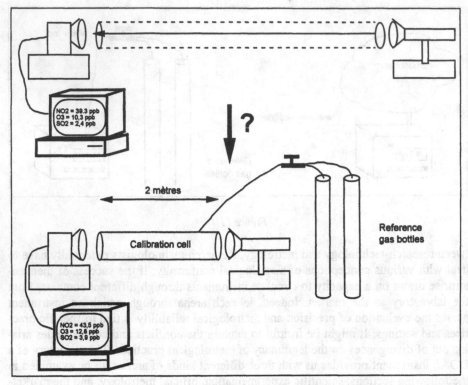

Figure 11.2

promoters of DOAS often stressed the importance of the method of analysis, based on powerful and complex mathematical functions. They exhibited a *theoretical* justification for the good quality of the *practical* result. But DOAS scientists also recognized that this was only part of the story. As has been shown by many scholars in science studies, theory plays a limited role in science – as long as it remains disembodied and . . . theoretical. Beyond the theoretical criteria, the experimenters' familiarity with their instruments intervenes, a familiarity that has been shaped through many years of research. It is largely based on tacit knowledge.[12]

The power of tacit knowledge in the construction of instruments lies partly in the complexity of the technical processes involved and partly in the contextual nature of the knowledge necessary to perform metrological adjustments. The instrument is the inscription of an innumerable series of technical tricks much more than the transposition of a theoretical principle of functioning that could be easily checked by a calibration trial. Why use a 150-watt and not a 75-watt bulb for the lamp of a DOAS instrument? Why choose this supplier for the photodiode array detector, and not another? Is a 1-kilometer path better or worse than a 500-meter path for the sensitivity

---

[12] Drawing on Michael Polanyi (1958), Collins (1985: 56) stresses the importance of tacit knowledge in scientific experiment. On the role of know-how and bodily performance in science, see Otto Sibum 1998.

of detection? What is the best spectral "window" for analyzing the concentration of SO_2? What is the acceptable standard error for a good measurement? What is the estimation of the electronic noise of the system that blurs the real signal? The answers to such questions strongly influence measurement quality, but only the experimenters who took part in the long-term development of the instrument know the exact way this influence is exerted. It is embedded in the history of the instrument, in the many alternative technological solutions that have been explored and eliminated, and in the many local rules that have emerged as conclusions of former unsuccessful experiments. The innumerable micro-choices that have led to the present device are not readable in itself and are only weakly traceable.

As I was told by Professor Marais, "Of course, as a user of the SOMA, you could always open the box and look at the technique or read the software. It would give access to some details of the technology. But not to all the tricks that are inside. There are thousands of tricks inside." Reverse engineering is a risky task when the conception of an instrument relies largely on familiarity. In laboratories, familiarity ensures quality control without always providing for traceability. Familiarity circumscribes a metrological regime in which the performance of the experimenter and the instrument are intimately entwined in the achievement of precision.

Interpolation consists of correcting the equipment performance by comparison with other close indicators. In their daily work, DOAS scientists sometimes used conventional instruments (which were supposed to be absolutely calibrated) in order to check for reasonable agreement with their prototypes. Certain corrective procedures for DOAS apparatus could even require the use of several conventional instruments, measuring the gas concentration at different places along the optical path. Much of the interpolation work was made in measurement campaigns: The instrument was brought to work in the field with other sorts of devices. Measurement campaigns were extremely important for the improvement of the technology, for the gathering of experimental data and for helping to understand how DOAS could be linked to other measurement technologies. Between 1989 and 1994, the instruments of CREAT were used in about ten campaigns at different sites, including Paris, Normandy, Brussels, Geneva and Athens. The two intercomparison campaigns organized under the umbrella of TOPAS, involving mainly DOAS systems, provided a particularly rich context in which to evaluate the possibilities and limits of comparing one instrument with the instrument of another scientific team. Indeed, there is usually no general formula of equivalence between the measurements of the instruments engaged in an interpolation: They do not monitor exactly the same physical properties, they may relate to different samplings, the frequency and place of measurement can be different, and so on.

Interpolation is always a difficult exercise, which asks for specific judgement abilities. It requires elaborating a material and conceptual space for comparison of the outcomes of the different instruments involved. This sort of arena is open to controversy, depending on the way experimenters perceive the reliability of the various instruments involved. Interpolation can also lead to challenging the validity of accepted, authoritative measurement protocols and instruments. For example, results obtained on ozone measurement with passive DOAS systems have led researchers to question the reliability of data given by the satellite TOMS, although TOMS' data is one of the most authoritative sources of information concerning atmospheric chemistry, benefiting from fifteen years of experience in the mapping of atmospheric ozone.

Familiarity and interpolation relate to precision practices that are common and efficient in the arenas of instrument development, an activity that simultaneously requires a strong commitment to technical conception and mastery of the underlying scientific questions. In this sense, these practices are familiar to research-technologists. However, research-technologists also know that these practices are not relevant to the maintenance of an apparatus used in "real context," for instance for an instrument used to produce urban pollution indicators for a state administration.

OFFICIAL METROLOGY AND THE PRACTICES OF PRECISION

Well, so you will probably tell me: "Nevertheless, this instrument maker sold a lot of his DOAS devices!" For sure, but my answer is that he sold them to people who were not reputable metrologists, who do not necessarily know. Because such a device has interesting advantages: it needs very little maintenance, etc. And people see it at night, they think: "It is wonderful, the instrument keeps watch over the city . . ." What is more, it gives data continuously, with an error margin, which is something new for this kind of device. A lot of towns have rushed at it, notably English towns. It seems to be a pure instrument. Now I will tell you what I think: If our decision-makers feel that metrology is an expensive task, if they feel that precise measurements are useless, that 20% error is good enough, then it will close one chapter in the history of metrology. But I don't want to see that. Maybe cities will be monitored by light beams. Maybe these systems will provide data uninterruptedly. Maybe we will get lots of information. But on the basis of the instrument, it will be impossible to say anything. I don't want to see that . . . .

A metrologist

It is precisely the role of official metrology to provide for settings and resources in order for measuring instruments to work properly in end-user contexts, whatever the conditions to which they are submitted. O'Connell (1993) has studied the complex and costly organization of official metrological institutions devoted to the endless maintenance of precision. When metrology is based on artefact standards, the "local creation of universality" is achieved through the circulation of particular objects, with very specific conditions of construction, transportation and use. Since the development of intrinsic standards, the circulation of objects has been replaced by a huge bureaucratic work: NIST is in charge of certifying organizations as to their ability to reproduce correct calibration trials. The origin and constitution of these procedures of metrology are now commonly investigated by historians and sociologists.[13] It appears that today, for many devices on the market, the control by state metrologists concerns the stages of design, manufacture, use and even repair of instruments. At each site associated with an instrument, various checks and calibrations must be activated. The strict application of this kind of bureaucratic scheme is the price to pay in order to ensure that instruments remain precise outside the sphere of instrument development proper.

---

[13] In science studies, the interest for metrology has largely been stimulated by the works of Latour (1987) and Collins (1985). For recent contributions in the historical and sociological debates on the role of metrology in science and society, see Alder 1998, Curtis 1998, Keating and Cambrosio 1997, Mallard 1998, Schaffer 1997, Timmermans and Berg 1997, and Wise 1995.

But one can believe that research-technologists very often have their own say in this process. In so far as they intend to follow their instruments from the laboratory on up to their site of final use, research-technologists may be involved in the definition of the associated metrological schemes. Moreover, when the measurement technology draws on specific metrological features, the point of view of the instrument's designers can conflict with the point of view of official metrologists. This is what happened during the evaluation of the DOAS instruments by INET in 1991. The conflict between engineers/scientists and official metrologists was not limited to the choice of a legitimate trial for calibration; it also extended to the definition of the metrological pattern of verifications that should be applied to DOAS instruments for use in pollution-monitoring networks.

According to the scientists and engineers, DOAS instruments needed to be carefully verified mainly once, when the instrument was about to leave the manufacturer. The drift in the precision of DOAS instruments could be considered insignificant over long periods of measurement. In contrast, the metrologists of INET and ADEME expressed the need to submit DOAS instruments to a relatively classical pattern, comparable to the one used for conventional instruments. This meant that the firms would have to furnish users with a procedure and equipment for regular checks of the instrument. If a "full calibration" of the device was by definition impossible, some partial calibration should be available for end-users. It was possible, for instance, to calibrate the spectrometer with a lamp emitting a standard ray in the UV domain, even if this was not as reliable as a calibration of the full device. The current state of metrological procedures, metrologists said, was not fully satisfying either for the French instrument or for the Swedish one under evaluation. They worried about the calibration of DOAS instruments:

> You cannot do without control because people are suspicious about calibration problems. On these DOAS instruments, calibration is not possible. And this does not work with users, just psychologically: Users need to check periodically, and on conventional instruments, they have their gas bottles and they want control. With DOAS, this impossible because calibration is made once and for all in the laboratory or at the manufacturing plant. So you do not know. Nothing proves that certain electronic components won't drift, that the signal of the spectrometer will not vary with time ....[14]

A thorough examination of the conflict between the metrologists and the research-technologists reveals that it was rooted in diverging conceptions concerning what could make a reliable metrological system for DOAS instruments credible. Particularly, there was an element that did not enter into the metrologists' definition of a credible metrological network, an element that was crucial for the DOAS defenders: absorption cross sections.

The absorption cross section of a chemical compound is the curve representing the propensity of the compound to absorb electromagnetic energy in a given bandwidth. These reference data were used in the analysis of the absorption spectrum calculated by DOAS instruments. According to many of the DOAS scientists and engineers, these data directly representing the properties of natural entities accounted for the temporal stability of the instruments because they made possible a sort of "inher-

---

[14] Interview with M Poulet, INET.

ent calibration." In their discourse, cross sections were equated to a calibration material, comparable to the reference gas bottles used to calibrate conventional instruments. Instead of comparing the value indicated by the instrument with the known concentration of the mixture contained in a gas bottle, DOAS instruments compared the absorption spectrum to absorption cross sections.

In the midst of complex technical debates on the reliability of instruments, this argument that compared gas bottles and cross sections granted the latter an obvious superiority. Cross sections would not drift; they were numerical data insensitive to displacements and changes in the environment. One could put them once and for all in the software, and they would not change. Physical calibration with *real* reference material was just an outmoded procedure, necessary for antique conventional devices. The metrologists did not agree with this position. They criticized a "mode of calibration based on theoretical criteria," potentially generating systematic errors. Particularly, they questioned the origin and quality of cross sections. Were cross sections really universal? Where could one find "good" cross sections? Who was allowed to certify their quality?

Actually, these questions were not irrelevant. Cross sections were produced through sophisticated spectroscopic experiments conducted in specialized laboratories. The current state of knowledge was lacking in this domain: Only the cross sections of some chemical compounds had been published, and these were sometimes calculated over restricted spectral areas. DOAS scientists admitted that bad data sometimes could be found in the literature, and they were careful in choosing one or another set of published cross sections. They paid attention to the reputation of the laboratory, and to the personal contacts they had with spectroscopists. Moreover, the proper use of cross sections sometimes demanded that they "cut and paste" and standardize data coming from different sources and adapt them thoroughly to the instrument.

All these endeavors required a lot of expertise, much more than was required to acquire and use a reference gas bottle. In other words, the reluctance of metrologists largely came from their difficulty in evaluating the quality of cross sections and the relation between this quality and the quality of the final measurements. The associated metrological operations were opaque to official metrologists, notably because there was no conventional, institutional source of authority that could have certified cross sections as a reference material, thus offering a landmark for a non-specialist. This was in complete contrast to calibration procedures and standards for conventional instruments. There, such institutionalized and certified networks existed. The technical and social knowledge permitting one to pick out good cross sections in the scientific literature and to adapt them to a DOAS instrument was part of the metrological arenas of scientists. But it was not in the arena of metrologists.

Confronted by the protestation of metrologists, the two industrial firms reacted in specific ways. The Swedish firm Optim accepted the system of partial calibration. The hybrid character of this solution was readable in the prospectus vaunting the merits of the instrument. One could read: "No gas calibration required" at one place and, in a subsequent footnote this: "Recommended standard calibration to be performed once a month with Optim calibration unit." In contrast, actors of Orion and CREAT radicalized the conflict with the metrologists. Their answer led to a certain extent to the reopening of a general debate about calibration. Drawing on their scientific experience and authority, they tried to convince the metrologists that even if it was not perfect, the metrology of SOMA was not in itself *more* flawed and uncertain than the metrology of conventional measuring instruments. Thus, they testified that in the arena of science, there was no "God of calibration": Any metrological system could potentially be submitted to deconstruction, even if it was rooted in the apparently most commonsensical calibration procedures.

The interaction between official metrologists and research-technologists warrants a further comment. Concerning the cross sections, one may say that it is quite normal that the proper evaluation of their role and quality was possible only for specific actors. Such an evaluation required one to be involved in the whole field of instrumental techniques related to DOAS instruments. This gives research-technologists a special position in the construction of metrological systems or, like here, in a controversy where they have to fight against official metrologists. The reactions of Optim and Orion also deserve attention because they feature two very different responses in such a metrological conflict: on one side, a pragmatic adaptation to the requirements of official metrology in order to secure an access to the market, on the other, an attempt to resolve the conflict on a scientific battlefield. This issue may well exemplify interesting alternatives for the involvement of research-technologists in the construction of new metrological systems.

A MARKET FOR DOAS INSTRUMENTS

> Of course, this problem of calibration is not easy, so a scientist will be able to speculate on it for years and years! When you have a metrological chain, you can always travel up very far, with problems of calibration. When you have a gas bottle, you always wonder how it was calibrated. People tell you that it was calibrated through the precision balance. OK, but the precision balance, how was it calibrated? And so on, you can get back to the standard meter. So of course, this is part of the role and culture of scientists, to ask these kinds of questions. But me, my point of view is that the instrument must sell. If the user wants a cell to calibrate the instrument, even if I am skeptical, maybe I should give him a cell . . . . We will never manage it financially if the instrument is not able to measure the same thing as other devices. Sooner or later, it will be necessary that 1=1 because otherwise, we won't do it. We are not in the position to prove to the rest of the world that measurements with conventional instruments are false!
>
> *A salesman*

Besides their contribution in the elaboration of metrological procedures, research-technologists spend part of their time adapting their instruments to the needs of end-users. The history of SOMA offers many illustrations of this process of adaptation. In its early activities, Orion dealt in passive DOAS instruments to be used by scientists. When it was decided to replicate, manufacture and sell SOMA, things were slightly different: The field of air quality control was less familiar to the actors of Orion and CREAT, and they discovered progressively the necessary adaptations to make the instrument a marketable good in this sector. One of the most visible changes in the instrument was its new design: It received a beautiful (and very functional) box and cover, giving it a more attractive presentation than the scientific prototype it replicated. In the early stages of development, adaptation of the computer interface proved necessary to render it more user-friendly. Progressively, it became clear that the instrument would soon have to shift from a computer technology based on Hewlett Packard hardware, familiar to many scientists, to IBM PC-compatible hardware, more accessible for other users, for instance people working in air-quality control networks. Other questions for the innovators were answered ambiguously, such as the question of price; it remained unclear whether the high price of the device (around 1 million francs) was a handicap or not.[15]

Although design, technology and financial aspects indisputably play a role when one wants to sell an instrument, purely metrological parameters such as sensitivity or precision do not disappear. They remain prevailing parameters in the qualification of this kind of good in a market situation. Thus, the selling of the instrument features another arena, where research-technologists are confronted directly not by scientists or metrologists but by distributors or potential customers whose appreciation of the instrument's reliability might hang on different criteria. One should note that because metrological parameters are largely invisible, difficult to evaluate for the uninitiated and discovered progressively in the course of use, they represent a major qualitative uncertainty in instrument transactions.[16] In these conditions, part of the commercial activity is related to the building of trust and credibility, and to the framing of the customer's space of evaluation in order for the transaction to take place.[17] In this arena, research-technologists have two strategic resources at their disposal to counter the uncertainty about quality: the constitution of a set of *authoritative references* on the one hand and the practices of *demonstration* on the other.

Authoritative references are materialized by certificates of approval, labels or letters of recommendation. Different categories of references can be used for instruments: a certificate obtained from an official metrology laboratory is, of course, a good reference. This was why Orion and Optim were interested in the evaluation of their instrument by the INET laboratory. But both firms also mobilized other kinds of references, notably during the period when the evaluation at INET was not yet finished. Academic scientists in the field were asked to write letters of recommendation that would testify to the good quality of the device and could be shown to potential buyers. The testimony of satisfied customers could also constitute a good authoritative reference.

Another way to prove that the instrument works correctly is to perform *demonstrations*. Demonstration at the customer's site is important for equipment such as measuring instruments, because an innovative – and expensive – product of scientific research will not always be perceived by customers as viable for routine measurements. As I was told by an engineer, "Buying a DOAS system is not like buying a refrigerator. When people consider investing many hundred thousands francs in such a new technology, they like to test it at home before!" Demonstration also occurs in

---

[15] It must be said that only a handful of instruments were really sold, and this was a serious limitation of experience to evaluate the influence of this parameter on the market.

[16] The notion of qualitative uncertainty is used in anthropological economics in order to study the structure of particular markets where, notably, customers do not have the relevant criteria to evaluate product quality before use. See for instance Akerlof 1970, and Karpik 1989.

[17] On the importance of 'framing' in market activities, see Callon 1998.

other places, such as exhibitions. Many of the instruments built by Orion spent a lot of time as demonstration models before anyone decided to buy them. As a resource mobilized to establish trust between actors, demonstration has been much less studied than authoritative references and the underlying economics of credibility and reputation. It seems that demonstration has a rhetoric, a series of specific conditions of felicity. Some anecdotes taken from the experiences of Orion and Optim will illustrate this topic.

Let us consider the role of know-how during demonstrations. At first glance, the existence of specific user know-how could be considered much more problematic in this arena than in a scientific laboratory. If the salesman needs to perform delicate or abstruse operations in order for the instrument to work correctly, the demonstration risks being less convincing. On the other hand, a display of virtuosity can enhance the customer's confidence towards the seller.[18] Thus, the skill of the demonstrator does not constitute a firm criterion solely determining the efficiency of a demonstration. In any case, the demonstrator must be careful to make explicit only the operations that are legitimate for the customer. According to an engineer in charge of setting up DOAS instruments for demonstration, customers sometimes lost part of their enthusiasm for the technology when they discovered that its metrology required *ad hoc* and tricky operations. This suggested to them that the calibration of the instrument was a matter of *bricolage*, whereas in the scientific laboratory these operations were considered as normal or unavoidable. More generally, customers' expectations and cultural criteria concerning reliability can be different, possibly opposite to those of researchers, as this salesman suggests in the case of the meaning of negative concentration values:

> Once, I received a fax from someone who had the instrument in demonstration. He seemed to be very angry. He reported the negative values of chemical concentration measured by the instrument, and he urgently requested our comments on this result. After this episode, I asked the designer of the instrument to set to zero every possible negative value displayed by the device, in order to avoid bad reactions from users.

In contrast, such negative values are not always seen as crippling defects in laboratories, because scientists know and assume that their instruments are calibrated plus or minus a zero value, something is always difficult to determine. Negative values allow for an evaluation of the order of magnitude of this offset.

The study of demonstration practices should consider that in this arena, the demonstrator very often does not totally master the conditions in which the experiment takes place, because he does not operate "at home." This can give rise to painful experiences in negotiations with customers, as another salesman confessed:

> When you install a DOAS instrument, you need a very stable base to lay the lamp and the detector: if they move, even very very little, the beam alignment is changed and measurement is disturbed. And this is very difficult to explain to customers. Because they sometimes believe that they have a stable base while in fact it is not stable enough. I saw something incredible once. The lamp lay on a support made of concrete, apparently stable. But the measurements were systematically bad. It turned out that the concrete was

---

[18] On virtuosity in public demonstration, see Collins 1988.

distorted by dilatation. You look stupid if you tell the customer 'My instrument works, it
is your house that is not stable!' Even if it is true.

A remark finally about the competition of DOAS with other kinds of instruments.
Here, the conservative and consensual strategy seemed to be the most cautious one:
very often, since customers were totally unaware of the details of this new technology, the most compelling demonstration to perform was to show that the DOAS device gave the same results as a conventional analyzer. A quite paradoxical situation,
when one considers the amount of energy spent by DOAS promoters to explain to
metrologists why their instruments gave more stable and more precise results – and
also somehow, different results – than conventional apparatuses.

RESEARCH-TECHNOLOGY AND THE DIVERSITY OF METROLOGICAL ARENAS

There are several conclusions to be drawn from the study of DOAS technology. The
first one relates to the diversity in the modes of production of judgements about the
precision of measuring instruments, depending on the situation. The idea that the
reliability of measuring instruments can be accounted for through the mobilization of
unambiguous calibration trials belongs to a simplistic, pedagogic representation of
metrology. The investigation of metrological arenas all along an instrument's trajectories shows the plurality of configurations of equipment, techniques and methods,
modes of demonstration, criteria of evidence and acceptable social qualifications involved in the production of metrological categories. It should be noted that probably
no stable or fixed hierarchy exists between the different arenas mentioned in this
chapter. One can develop an instrument that is reliable under certain laboratory conditions, but that never satisfies the metrological conditions necessary for use outside
the laboratory. One can succeed in the market without conforming to the metrological trials and criteria required in a scientific experiment. This is what some researchers of TOPAS – and even some metrologists – would say about the Swedish apparatus of Optim, denouncing an instrument beloved by users but not offering enough
metrological guarantees. Similarly, the successful translation of an instrument from
the laboratory to the market does not imply the merging of the different arenas involved, nor the hegemony of the trials and criteria prevailing in a given metrological
arena over others. One might say that the stability of instruments is constructed *in
spite of* the diversity of arenas.

This leads to another important conclusion of this chapter, concerning the specific
involvement of research-technologists in metrological issues. The DOAS story suggests that one does not go continuously from science to the market with a unique
metrological toolbox. Important work of articulation and adjustment has to be done
for an instrument to cross sites where precision is embodied in different sociotechnical settings, and research-technologists are in the best position to accomplish
this task. They help produce metrological correspondences and gateways all along
the trajectory from the laboratory to the end-user. The work of articulation is multiple. For instance, it can bear on the devices: In the present context, this happened
when it was decided to suppress the display of negative values that triggered users'

distrust towards SOMA. In some other situations of conflict, innovators try to leave the object unchanged but to modify the constitution of the arena itself. This happened when the scientists of CREAT tried to convince the metrologists of the certification laboratory that cross sections constituted a reliable argument for metrological stability. The episode also emphasizes the difficulty for research-technologists in imposing metrological procedures, trials and values in an arena that is already highly structured, like the arena of official metrology. Research-technologists cannot always impose new metrological methods and concepts by themselves.

In any case, the specificity of their position in metrological debates could then result from the strategic character of such work of articulation. Because they are engaged in multiple networks and in multiple activities in the realm of science, industry and the state, research-technologists are in a good position to manage the diversity of operations and viewpoints necessary to promote an instrument. Since their enterprise operates through the circulation of their products – sometimes from the early stages of development onwards – through different social and organizational contexts, one can presume that they develop a strategic capacity to anticipate the trials that are prevalent in various metrological arenas. The competition and conflicts that they can have with official metrologists, as in the DOAS case, then seem quite typical; research-technologists and official metrologists are led to occupy the same battlefield, at the interface between science and the market.

In the perspective outlined here, one can suggest that research-technologists have developed a kind of professional competence relative to the task of adapting instruments and metrological practices to diverse situations, endowing them with legitimacy in the eyes of different actors. The specificity of their work as "metrological system builders" would then precisely stem from their ability to take into account, with equal ease, the complex experimental issues implied in measurement, the more or less well-grounded desires of end-users, and the procedural requirements of state metrologists.

France Télécom Recherche & Développement, Paris, France

ACKNOWLEDGMENTS

This chapter has benefited from the suggestions of Sean Johnston and Myles Jackson. I also want to thank Bernward Joerges and Terry Shinn for their stimulative critical comments concerning both the arguments and the narrative.

REFERENCES

Akerlof, George A. (1970). The market for lemons: qualitative uncertainty and the market mechanism. *Quarterly Journal of Economics* 84: 488–500.
Alder, Ken (1998). Making things the same. *Social Studies of Science* 28(4): 499–546.
Callon, Michael (1998). *The laws of the markets.* Oxford: Blackwell.

Camy-Peyret, Claude, Galle, B., Carleer, M., Clerbaux, C., Colin, R., Fayt, C., Goutail, F., Pinharanda, M.N., Pommereau, J., Haussman, M., Platt, U., Pundt, I., Rudolph, T., Hermans, C., Simon, P.C., Vandaele, A.C., Plane, J.M.C., and Smith, N. (1996). Intercomparison of instruments for tropospheric measurements using differential optical absorption spectroscopy. *Journal of Atmospheric Chemistry* 23(1): 51–80.

Clarke, Adele E. (1991). Social worlds/arenas theory as organizational theory, in D.R. Maines (ed.), *Social organization and social process. Essays in honor of Anselm Strauss* (pp. 119–58). New York: Aldine de Gruyter.

Collins, Harry M. (1985). *Changing order. replication and induction in scientific practice.* London: Sage.

Collins, Harry M. (1988). Public experiments and displays of virtuosity: The core-set revisited. *Social Studies of Science* 18: 725–48.

Curtis, Bruce (1998). From the moral thermometer to money. *Social Studies of Science* 28(4): 547–70.

Dobson, G.M.B. (1931). A photoelectric spectrophotometer for measuring atmospheric ozone. *Proceedings of the Physical Society* 43: 324–28.

Dodier, Nicolas (1993). Les arènes des habilités techniques, in B. Conein, N. Dodier and L. Thévenot (eds.), *Raison pratique n° 4. Les objets dans l'action* (pp. 115–39). Paris: Edition de l'Ecole des Hautes Etudes en Sciences Sociales.

Karpik, Lucien (1989). L'économie de la qualité. *Revue française de sociologie* 30(2): 187–210.

Keating, Peter and Cambrosio, Alberto (1997). Interlaboratory life: Regulating flow cytometry, in J.-P. Gaudillière, I. Löwy, and D. Pestre (eds.), *The invisible industrialist: Manufacturers and the construction of scientific knowledge* (pp. 250–95). London: Macmillan.

Latour, Bruno (1987). *Science in action.* Cambridge, MA: Harvard University Press.

Mallard, Alexandre (1996). *Les instruments dans la coordination de l'action. Pratiques techniques, métrologie, instrument scientifique.* Unpublished Ph.D. dissertation. Paris: Ecole des Mines.

Mallard, Alexandre (1998). Compare, standardize, and settle agreement: On some usual metrological problems. *Social Studies of Science* 28(4): 571–601.

Metropolitan measures and their margins of error. Paper presented at the Séminaire du Centre de Recherche en Histoire des Sciences et des Technique, Paris.

Platt, Ulrich and Perner, Dieter (1983). Measurement of atmospheric trace gases by long path differential UV/visible spectroscopy. *Springer Series in Optical Sciences* 39: 95–105.

Polanyi, Michael (1958). *Personal knowledge.* London: Routledge & Kegan Paul.

Schaffer, Simon (1997, 7 January). *Metropolitan measures and their margins of error.* Paper presented at the Séminaire du Centre de Recherche en Histoire des Sciences et des Technique, Cité desSciences et de l'Industrie, Paris.

Shinn, Terry (1997). Crossing boundaries: The emergence of reseach-technology communities, in H. Etzkowitz and L. Leydesdorff (eds.), *Universities and the global knowledge economy: A triple helix of university-industry-government relations* (pp. 85–96). London: Cassell Academic Press.

Sibum, Otto (1998). *Les gestes de la mesure: Joule, les pratiques de la brasserie et la sciences.* Paris: Annales Histoires Sciences Sociales.

Thévenot, Laurent (1994). Le régime de familiarité. Des choses en personne. *Genèses* 17 (September): 72–101.

Timmermans, Stefan and Berg, Marc (1997). Standardization in action: Achieving local universality through medical protocols. *Social Studies of Science* 27: 273–305.

Wise, M. Norton (1995). *The values of precision.* Princeton, NJ: Princeton University Press.

IN CONCLUSION

CHAPTER 12

BERNWARD JOERGES[a] AND TERRY SHINN[b]

RESEARCH-TECHNOLOGY IN HISTORICAL PERSPECTIVE: AN ATTEMPT AT RECONSTRUCTION

The authors of this book have argued that the significance of research-technologies lies in their trans-community positioning or "interstitiality," in the openness or "genericity" of their devices, and in the provision of standardized languages or "metrologies." It remains to be seen where research-technology fits in social studies of science and technology, and how the research-technology perspective contributes to a broader understanding of societal dynamics. In this final chapter we will address these issues in turn.

THE PLACE OF RESEARCH-TECHNOLOGY IN SOCIAL STUDIES OF SCIENCE AND TECHNOLOGY

Research-technology is a long-standing configuration of intellectual and artefact production that has until recently largely gone undiscerned. Research-technology is not another configuration in a possibly novel mode of producing scientific knowledge and devices (see, for instance, Etzkowitz and Leydesdorff 1998 [with an Introduction on 197–201]; Gibbons et al. 1994). The existence and operation of research-technology have been masked by a certain insensitivity, in science and technology studies, to the subtleties of boarder-crossings inside science and between science and other social systems. Indeed, some latter-day students of science and technology have responded to increased boarder-crossings by positing a far-reaching erosion of familiar social forms in science. A better understanding of the process may be acquired by looking at boarder-crossing in terms of the divisions of labor and forms of differentiation in the production of science and technology (Shinn 2000).

Post-seventeenth century science and technology can be depicted as taking three different institutional forms: discipline-related science and technology culture, transitory science and technology culture, and transverse science and technology culture. The study of discipline-related science and technology culture has emphasized institutional and professional elements in the growth of scientific knowledge and distinguished between science and engineering. Analyses of transitory science and technology culture maintain the idea of a demarcation between academia and engineering, but at the same time show how practitioners intermittently pass back and forth

B. Joerges and T. Shinn (eds.), Instrumentation: Between Science, State and Industry, 241–248.

between the two arenas. A study of transverse science and technology culture is congruent with the research-technology perspective. Here, the idea of the institutional boundedness of science and engineering is preserved, but the focus is on situations where back and forth movement is unceasing. It is conceived as a distinct social form in itself, basic both to the maintenance and separateness of disciplinary science and engineering.

Discipline-Related Science and Technology Studies

The history and sociology of science and technology has largely been written in the framework of discipline-related science and technology culture. Innumerable monographs explore the birth, maturity and occasionally the terminal phase of disciplines like astronomy, chemistry, ecology, engineering specialties, phrenology, geology, physics, or micro- and molecular biology (for example, Abir-Am 1993; Gingras 1991; Heilbron and Seidel 1989; Kevles 1978; Lemaine et al. 1976; Mullins 1972; Nye 1993; Rheinberger 1997). The sheer volume of such scholarship is so abundant and omnipresent that inattentive observers of science might erroneously conclude that the history of modern science is principally the history of discipline-related science. In fact, all three science cultures have operated and co-existed for at least two centuries (see, for instance, Pestre 1997). There are sound reasons for the historiographic emphasis on discipline-related science and technology culture. Disciplines are structured around easily identifiable and stable institutions; and disciplines, like most other institutions, produce and leave behind a voluminous paper trail which renders disciplinary analysis more manageable than other forms of analysis. Science disciplines are rooted in the institutions of laboratories, university departments, journals, national and international professional bodies, conferences and congresses, procedures for certifying competence, systems for awarding prizes, formal networks and unofficial connections. Markers like these facilitate the detection and analysis of definite career patterns and categories of scientific production. Moreover, the perceived centrality of institutions in discipline-related science studies has its parallels in the structural analysis of society at large. Connections and congruities between science and society are easily established.

It is in this frame that certain terminologies and notions from non-science realms, such as political and organizational life, have been carried over to probe the world of science. Thus, Richard Whitley's studies of the social and intellectual organization of a large number of scientific disciplines have borrowed crucial vocabularies and insights from the organizational structures of non-science institutions and extended them to the landscape of science's discipline-related culture (Whitley 1984). In a similar vein, general historians have often written the history of science as shaped by world historical tides. An example is Eric Hobsbawm's chapter on "Science" where he elucidates scientific development as the struggle between democratic nations and fascist/communist regimes for supremacy on the world stage (Hobsbawm 1995). Even the classical work of Thomas Kuhn on scientific revolutions may be interpreted as a case in point.

Transitory Science and Technology Studies

Despite their successes, studies of discipline-related cultures have proven deaf to other equally important cultures in science. And yet, an immense amount of science occurs outside the disciplinary matrix. Many careers and much cognition or construction take place in a transitory science and technology culture which is not systematically congruent with orthodox disciplines.

This form of science is not free from the effects of institutional differentiation, but they are dealt with in complex ways which are sometimes overlooked or misunderstood. Intellectual, technical and professional opportunities sometimes arise near the periphery of orthodox fields. In such instances effective research or career-making requires practitioners to step temporarily across the boundaries of their home disciplines, as they seek techniques, data, concepts and colleague cooperation in neighboring specialties. Most of the time, the quest for additional cognitive, material or human resources involves two, or at the most three, disciplines. Practitioner movement consists of a to-and-fro oscillatory pattern. The trajectory remains circumscribed with respect to time and to scale of movement. It is important to note that in transitory science and technology culture practitioners' principal center of identity and action is still disciplinary, even though individuals do traverse fields.

Transitory science and technology cultures subsume two different yet related trajectories. The life and work of Lord Kelvin is emblematic of one pattern. Norton Wise and Crosbie Smith have documented how Kelvin changed from physics to engineering and from engineering back to physics (Wise and Smith 1989; see also the examples of transitory science culture described by Mulkay 1974). As perspectives opened, the man shifted territory. Nevertheless, Kelvin's itinerary remained circumscribed. Moreover, both from the standpoint of the historian and the professional scientist, Kelvin's fundamental allegiance and identity remained discipline-bound, entwined with the orthodox discipline of classical physics.

Alternatively, transitory science and technology culture can lead to the derivation of a new sub-discipline, as in the cases of physical chemistry, biochemistry, biophysics, astrophysics, and geophysics. The list of such creations is long and deeply rooted in the practices of science and technology. In these cases, the oscillatory trajectories of practitioners mentioned above terminate in the establishment of a novel field – a conjunction of two or several established fields. New sub-disciplines are the product of transitory science and technology cultures. In order to understand these cultures and their intellectual/technical achievements, historians and sociologists must concentrate on interfacing and motion.

Yet, to repeat, in these cultures movement and interfacing still tend to be strictly defined and regulated by disciplinary referents. Institutional demarcations and divisions of labor remain paramount, although they are played out in a specific manner. Focus is on career mobility and knowledge fluidity; but both function in a confining and restricted set of institutional coordinates.

Transverse Science and Technology Studies

While in a few respects transitory science and technology cultures resemble transverse cultures, the latter nevertheless represent a distinct mode of production. In transverse science and technology cultures the degrees of freedom and scope of action of practitioners are far greater than in transitory science and technology cultures. For the purpose of our analysis, we will consider research-technology as an exemplar of this last mode of knowledge/artefact production. As documented in chapters 2 and 3 of this book, research-technology reaches back at least one and a half centuries. It then rapidly emerged in Britain, France and the United States. In each of these sites, and during each historical period, it operated alongside the discipline-related and transitory science and technology cultures. The three cultures may in fact be regarded as interdependent and reciprocally enriching one another.

If, as we suggested, transverse science and technology culture, in the guise of research-technology, has been around for a long time, and has often proven important to the growth of scientific knowledge and technology, why then has it been so conspicuously absent from the historiographic palette? Why have historians and sociologists often overlooked its very existence? Part of the answer to this question derives from the fact that participants in transverse science and technology culture are "moving targets."

Research-technology practitioners' association with employers, disciplines and professions is fleeting. The paper trail needed to document their trajectories is thin and fragmentary, making sociological and historical investigation problematic. The difficulty of sound research is exacerbated by the existence of multiple and diverse vehicles for practitioner productions, from conventional scientific publications to patents, confidential reports, exhibits, commercial products, metrological regulations and many others. For scholars whose investigations are rooted in the detection and analysis of stable institutions and sharp divisions of intellectual and material labor, the dealings of transverse culture prove difficult to survey. Similarly, scholars who represent science and technology as seamless technoscience and who ignore gradations of differentiation and divisions of labor, are oblivious to the subtleties and indeed regular structures of research-technology.

Transverse science and technology cultures are characterized by several elements. Practitioners principally draw their identity from projects rather than the disciplines or organizations that they frequent. Yet the perpetuation of well grounded institutions, in the form of academic and technical professions and employers, remains foundational to these cultures. Such defined settings provide necessary inputs for fresh projects in the form of ideas, information and apparatus. They also consume and validate the cognitive/technical products of the participants of transverse cultures. An arena of action in which practitioners are relatively free to shift about constitutes the social and material space wherein novelties may be generated outside the constraints of short-term demands. Two advantages are gained by generalizing research-technology into a separate science and technology culture. First, research-technology's place in the history of science and technology is clarified, as is its historiographical status. Secondly, certain lacunae and contradictions in the operations and representations of discipline-based and transitory science and the relations between them are explained and resolved.

GENERIC INSTRUMENTATION, DIVISIONS OF LABOR
AND DIFFERENTIATION

Attempting to summarize and, to a certain extent, to generalize the lessons to be drawn from the case histories in this volume, it seems to us that the process of an ongoing division of labor and differentiation in research-technology takes place in two interacting spaces: on the one hand in a space of design and dis-embedding of generic instruments by research-technologists, and, on the other hand, in a space of re-embedding of generic equipment, again by research-technologists or practitioners outside research-technology. The first type of practices is located deep inside the interstitial arenas while the second type is performed near their peripheries. Depending on the space research-technologists are operating in at a given time, they shift their stance with respect to differentiation and divisions of labor.

The part of research-technology work which directly involves production of templates strongly relies on divisions of labor. At this point, practitioners require distance and indeed protection from end-user demands and pressures in or outside science. This is necessary for the development of fundamental instrument theory and the design of generic equipment. As long as research-technologists focus on core devices, they remain committed to differentiation and divisions of labor.

However, when they deal with practitioners from industry, the state or disciplinary science, a de-emphasis of divisions of labor and differentiation is required to allow them to enter and re-enter these worlds. Such dealings characteristically occur at two moments in research-technology: when practitioners seek project ideas, concepts, and information from potential local users; and when they engage in demonstrating how generic devices could be tailored to particular local uses. Here, transverse mobility depends on temporarily suspending a commitment to divisions of labor constitutive of the targeted professions and organizations.

The bi-directional boundary crossing by research-technologists in turn induces a partial and temporary relinquishing of customary attachments to differentiation and divisions of labor in the arenas they frequent. As practitioners from industry, metrology, academia, or the military engage in the acquisition and tailoring of generic instrumentation, they too tend to lower the barriers. Once acquisition is completed, members of these professional communities again ground their practices in accustomed divisions of labor.

In sum: gradations of reliance on established divisions of labor are played out in the interstitial arenas where the multi-faceted work of research-technology is performed. Practitioners choose either to maintain their "in-between" positions, which provide the necessary space for generic practices, or to move in and out of neighboring science and industrial cultures. They can move between organizations, research projects, or even paradigms when required; they can also structure practice around a generic instrument-based imperative, appealing to principles of divisions of labor and professional differentiation in order to protect their own project. We see this less as a contradiction or paradox than as a case of flexible institutionalization in a field that requires the wearing of many hats. It is research-technologists' response to the complex set of intellectual, material and social relations that emerged as science, technology and the social order at large have expanded in scale and become progressively differentiated.

GENERIC INSTRUMENTATION, RE-EMBEDDING AND COHESION

We can now advance a stronger thesis, namely, that one impact of generic instrumentation is increased social and intellectual cohesion, which runs counter to the ongoing differentiation and fragmentation of science and society. To the extent that research-technology engenders a form of practice-based universality, it acts as an antidote to the effects of centrifugal forces. How is this possible?

To make a generic instrument effective in end-user audiences' own devices, its adoption entails the incorporation of its protocols. Protocols of generic devices are in turn linked to metrologies. Metrologies contribute to both the constitution of protocols and the circulation of devices. Together with protocols, instrument adopters import implicit working concepts, beliefs about why a generic instrument is effective and ideas about what it can and cannot do; and they import explicit vocabularies, images and notational systems.

Successive re-embeddings in different local material contexts and by different groups yield practitioner assurance that the principles of a template apparatus are solid, and that belief in it is well justified. Belief rooted in local experience and testing gradually gains in objectivity. Practices are independently repeated and are multiplied in numerous environments. This is not the objectivity born of pure reason or the *experimentum crucis*. Objectivization is instead built up through collective practice which is structured around effect-producing materials and procedures. Here, objectivization is practical and cumulative.

As a given generic instrument is tailored differently by various audiences to satisfy their demands, it gives rise to specific niche protocols and vocabularies. If this occurs in many places, ensembles of terminologies and procedures based on the most general principles of the generic device emerge in turn, beyond the local vocabularies. Transverse repertories of protocols, held in common by all users, whatever their local application requirements may be, accumulate. It is important to note that the transverse, quasi-universalizing quality of this process is not only a matter of rhetoric and professional power, but is also very much one of material demonstration and concrete practice in multiple arenas.

The universality born of dis-embedding and endless re-embeddings is a universality of varied experience in countless niches, a universality grounded in informed and legitimate practice. *It is practice-based universality.* The weight of trans-personal conviction, experience and proof stemming from practice-based universality adds to the power of conventional disciplinary tests and procedures habitually employed to buttress sweeping generalizations. It is interesting then to consider the research products of transverse science and technology cultures in terms of materials for "pan-validation." In order for the research outcomes of the transverse science and technology cultures to be seen internally as worthy of being sustained, they must resonate with numerous and diversified outside arenas, whether inside or outside of science.

In sum, cross boundary encounters are grounded in the transverse stock of cognitive and material resources coming out of research-technology programs. Communication between institutionally and cognitively differentiated groups of end-users develops. Seen in terms of its consequences, the dis-embedding/re-embedding cycle of research-technology operates as a cohesive force which counter-balances the dispersive propensities of cognitive and professional specialization – a trait characteristic of contemporary science as well as of much modern social and institutional life. Practitioners from disparate disciplinary horizons and from different walks of life can use the shared vocabularies, techniques, imageries, and notional systems of generic instrumentation to cross their respective boundaries. While dis-embedding/re-embedding is not synonymous with full cognitive and professional integration, research-technology nevertheless figures centrally in the cohesion which ultimately sustains much of science as well as many aspects of social existence. In a more speculative vein, we suggest that the vocabularies and images engendered through research-technology actually help contemporary culture achieve a measure of continuity and stability by furnishing a repertory of shared terminology and common experience to which socio-professional groups from different nations and different cultures can refer, and through which they can address one another.

[a] *Wissenschaftszentrum Berlin für Sozialforschung, Berlin, Germany*
[b] *GEMAS/CNRS, Paris, France*

REFERENCES

Abir-Am, Pnina (1993). From multidisciplinary collaborations to transnational objectivity: International space as constitutive of molecular biology, 1930-1970, in E. Crawford, T. Shinn, and S. Sörlin (eds.), *Denationalizing science, sociology of the sciences yearbook 1992* (pp. 153–86). Dordrecht, The Netherlands: Kluwer Academic Publishers.

Etzkowitz, Henry and Leydesdorff, Loet (eds.) (1998). A triple helix of university-industry-government relations. *Industry & Higher Education* 12(4): 197–258 (with an Introduction on 197-201).

Gibbons, Michael, Limoges, Camille, Nowotny, Helga, Schwartzman, Simon, Scott, Peter and Trow, Martin (1994). *The new production of knowledge: The dynamics of science and research in contemporary societies.* London: Sage.

Gingras, Yves (1991). *Physics and the rise of scientific research in Canada* (transl. by Peter Keating). Montréal: Kingston.

Heilbron, John and Seidel, Robert W. (1989). *Lawrence and his laboratory. A history of the Lawrence Berkeley Laboratory.* Berkeley, CA: University of California Press.

Hobsbawm, Eric (1995). *The age of extremes: The short 20th century, 1914-1991.* London: Abacus Edition, Little, Brown and Co.

Kevles, Daniel (1978). *The physicists.* New York: Alfred Knopf.

Lemaine, Gérard, MacLeod, Roy, Mulkay, Michael, and Weingart, Peter (eds.) (1976). *Perspectives on the emergence of scientific disciplines.* The Hague, Paris: Mouton.

Mulkay, Michael (1974). Conceptual displacement and migration in science: A prefatory paper. *Science Studies* 4(3): 205–34.

Mullins, Nicholas (1972). The development of a scientific specialty: The Phage Group and the origins of molecular biology. *Minerva* 10(1): 51–82.

Nye, Marie Joe (1993). *From chemical philosophy to theoretical chemistry: Dynamics of matter and dynamics of disciplines, 1800-1950.* Berkeley, CA: University of California Press.

Pestre, Dominique (1997). La production des savoirs entre académies et marché. *Revue d'Économie Industrielle* 79: 163–74.

Rheinberger, Hans-Jörg (1997). *Toward a history of epistemic things: Synthesizing proteins in the test tube*. Stanford, CA: Stanford University Press.

Shinn, Terry (2000). Formes de division du travail scientifique et convergence intellectuelle. La recherche technico-instrumentale. *Revue Française de Sociologie* 41(3): 447–73.

Whitley, Richard (1984). *The intellectual and social organization of the sciences*. Oxford: Clarendon Press.

Wise, Norton and Smith, Crosbie (1989). *Energy and empire. A biographical study of Lord Kelvin*. Cambridge: Cambridge University Press.

BIBLIOGRAPHY OF SELECTED REFERENCES

Abbott, Andrew D. (1988). *The system of the professions*. Chicago: University of Chicago Press.

Abir-Am, Pnina (1982). The discourse of physical power and biological knowledge in the 1930s: A reappraisal of the Rockefeller Foundation's policy in molecular biology. *Social Studies of Science* 12: 341–82.

Abir-Am, Pnina (1993). From multidisciplinary collaborations to transnational objectivity: International space as constitutive of molecular biology, 1930-1970, in E. Crawford, T. Shinn, and S. Sörlin (eds.), *Denationalizing science. Sociology of the Sciences Yearbook 1992* (pp. 153–86). Dordrecht, The Netherlands: Kluwer Academic Publishers.

Abir-Am, Pnina (1998). *La mise en mémoire de la science*. Paris: Editions des Archives Contemporaines.

Akerlof, George A. (1970). The market for lemons: Qualitative uncertainty and the market mechanism. *Quarterly Journal of Economics* 84: 488–500.

Alder, Ken (1998). Making things the same. *Social Studies of Science* 28(4): 499–546.

Bachelard, Gaston (1951). *L'activité rationaliste de la physique contemporaine*. Paris: PUF.

Bachelard, Gaston (1993). *Les intuitions atomistiques*. Paris: Boivin.

Badash, Lawrence (1975). Decay of a radioactive halo. *Isis* 66(4): 566–8.

Badash, Lawrence (1979). *Radioactivity in America: Growth and decay of a science*. Baltimore: John Hopkins University Press.

Baird, Davis (1993). Analytical chemistry and the "big" scientific instrumentation revolution. *Annals of Science* 50: 267–90.

Baird, Davis (2000). *Thing knowledge: A philosophy of scientific instruments*. Unpublished manuscript.

Balogh, Brian (1991). *Chain reaction. Expert debate and public participation in American commercial nuclear power, 1945-1975*. Cambridge: Cambridge University Press.

Bayertz, Kurt and Nevers, Patricia (1998). Biology as technology, in R. Porter and K. Bayertz (eds.), *From physico-theology to bio-technology* (pp. 108–32). Amsterdam: Rodopi.

Bechtel, William (1993). Integrating sciences by creating new disciplines: The case of cell biology. *Biology and Philosophy* 8(3): 277–99.

Ben-David, Joseph (1991). *Scientific growth. Essays on the social organization and ethos of science*. Berkeley, CA: University of California Press.

Bennett, Stuart (1993). *A history of control engineering 1930-1955*. London: Peregrinus.

Bensaude-Vincent, Bernadette (1994). Une robe de coton noir. *Cahiers de Science & Vie* 24 : 76–85.

Biervert, Bernd and Monse, Kurt (eds.) (1990). *Wandel durch Technik? Institution, Organisation, Alltag*. Opladen: Westdeutscher Verlag.

Blume, Stuart (1992). *Insight and industry: On the dynamics of technological change in medicine*. Cambridge, MA: MIT Press.

Bordry, Monique and Boudia, Soraya (eds.) (1998). *Les rayons de la vie. Une histoire des applications médicales des rayons X et de la radioactivité en France, 1895-1930*. Paris: Institut Curie.

Boudia, Soraya (1997). The Curie laboratory: Radioactivity and metrology, in S. Boudia and X. Roqué (eds.), *Science, medicine and industry: The Curie and Joliot-Curie laboratories* (special issue) (pp 249–65). *History & Technology* 13(4).

Boudia, Soraya (1997). *Marie Curie et son laboratoire: Science, industrie, instruments et métrologie de la radioactivité en France, 1896-1914*. Ph.D. Dissertation, Université de Paris VII.

Boudia, Soraya and Roqué, Xavier (eds.) (1997). *Science, medicine and industry: The Curie and Joliot-Curie laboratories* (special issue). *History & Technology* 13(4).

Braverman, Harry (1974). *Labor and monopoly capital: The degradation of work in the twentieth century*. New York: Monthly Review Press.

Brown, Frederick L. (1967). *A brief history of the physics department of the University of Virginia, 1922-1961*. Charlottesville, VA: University of Virginia Press.

Bruce, Robert (1973). *A.G. Bell and the conquest of solitude*. Boston: Little Brown.

Bucciarelli, Larry (1994). *Designing engineers*. Cambridge, MA: MIT Press.

Bud, Robert (1978). Strategy in American cancer research after World War II: A case study. *Social Studies of Science* 8: 425–59.

249

Bud, Robert and Cozzens, Susan E. (eds.) (1992). *Invisible connections: Instruments, institutions and science*. Bellingham: SPIE Optical Engineering Press.

Burian, Richard (1992). How the choice of experimental organism matters: Biological practices and discipline boundaries. *Synthese* 92: 151–6.

Burian, Richard (1993).How the choice of experimental organism matters: Epistemological reflections on an aspect of biological practice. *Journal of the History of Biology* 26(2): 362–4.

Busch, Lawrence, Lacey, William B., Burkhardt, Jeffrey and Lacey, Laura L. (eds.) (1991). *Plants, power and profit*. Cambridge, MA: Basil Blackwell.

Cahan, David (1998). *An institute for an empire: The physikalisch-technische Reichsanstalt, 1871-1918*. Cambridge: Cambridge University Press.

Callon, Michel (1998). *The laws of the markets*. Oxford: Blackwell.

Callon, Michel and Law, John (1982). On interests and their transformation: Enrolment and counter-enrolment. *Social Studies of Science* 12: 615–26.

Cattermole, Michael J.G. and Wolf, A.F. (1987). *Horace Darwin's shop: A history of the Cambridge Scientific Instrument Company 1978-1968*. Bristol: Hilger.

Chadarevian de, Soraya and Kamminga, Harmke (eds.) (1998). *Molecularizing biology and medicine: New practices and alliances, 1910s-1970s*. Amsterdam: Harwood Academic Publishers.

Chandler, Alfred D. (1977). *The visible hand: The managerial revolution in American business*. Cambridge, MA: The Belknap Press of Harvard University Press.

Charpak, Georges (1993). *La vie à fil tendu*. Paris: Odile Jacob.

Clarke, Adele E. (1991). Social worlds/arenas theory as organizational theory, in D.R. Maines (ed.), *Social organization and social process. Essays in honor of Anselm Strauss* (pp. 119–58). New York: Aldine de Gruyter.

Collins, Harry M. (1985). *Changing order. Replication and induction in scientific practice*. London: Sage.

Collins, Harry M. (1988). Public experiments and displays of virtuosity: The core-set revisited. *Social Studies of Science* 18: 725–48.

Conein, Bernard, Dodier, Nicolas, and Thévenot, Laurent (eds.) (1993). *Raison pratique n° 4. Les objets dans l'action*. Paris: Edition de l'Ecole des Hautes Etudes en Sciences Sociales.

Connes, Pierre (1992). Pierre Jacquinot and the beginnings of Fourier transform spectrometry. *Journal de Physique II* 2: 565–71.

Connes, Pierre (1995). Fourier transform spectrometry at the Laboratoire Aimé Cotton 1964–1974. *Spectrochimica Acta* 51A: 1097–104.

Connes, Pierre (1984). Early history of Fourier transform spectroscopy. *Infrared Physics* 24: 69–93.

Constant, Edward W. (1983). Scientific theory and technological testability: Science, dynamometers, and water turbines in the 19th century. *Technology and Culture* 24: 183–98.

Crawford, Elisabeth (1980). The prize system of the Academy of Sciences, in R. Fox and G. Weisz (eds.), *The organization of science and technology in France, 1808-1914* (pp. 283–307). Cambridge, UK: Cambridge University Press.

Creager, Angela and Gaudillière, Jean-Paul (2000, in press). Experimental arrangements and technologies of visualization, in J.-P. Gaudillière and I. Löwy (eds.), *Transmission: Human diseases between heredity and infection. Historical approaches*. Amsterdam: Harwood Academic Publishers.

Curie, Eve (1938). *Madame Curie*. Paris: Gallimard.

Curie, Marie (1923). *Pierre Curie*. New York: Macmillan.

Curie, Marie (1935). *Radioactivité*. Paris: Hermann.

Curtis, Bruce (1998). From the moral thermometer to money. *Social Studies of Science* 28(4): 547–70.

Darden, Lindley (1991). *Theory change in science. Strategies from Mendelian genetics*. Oxford: Oxford University Press.

Darden, Lindley and Maull, Nancy (1977). Interfield theories. *Philosophy of Science* 44: 43–64.

Dennis, M.A (1990). *A change of state: The political cultures of technical practice at the M.I.T. Instrumentation Laboratory and the Johns Hopkins University Applied Physics Laboratory, 1930-1945*. PhD Dissertation, Johns Hopkins University.

Dodier, Nicolas (1993). Les arènes des habilités techniques, in B. Conein, N. Dodier and L. Thévenot (eds.), *Raison pratique n° 4. Les objets dans l'action* (pp. 115–39). Paris: Edition de l'Ecole des Hautes Etudes en Sciences Sociales.

Duhem, Pierre (1981). *La théorie physique, son objet, sa structure*. Paris: J. Vrin (1st ed. 1915).

Edge, David and Mulkay, Michael (1976). *Astronomy transformed: The emergence of radio astronomy in Britain*. New York: Wiley.

Elias, Norbert, Martins, Herminio and. Whitley, Richard (eds.) (1982). *Scientific establishment and hierarchies*. Dordrecht, The Netherlands: D. Riedel.

Elzen, Boelie (1986). Two ultracentrifuges: A comparative study of the social construction of artefacts. *Social Studies of Science* 16(4): 621–62.

Engelsberger, Max (1969). *Beitrag zur Geschichte des Theodolits*. Dr.-Ing. Dissertation der Technischen Hochschule München. München: Verlag der Bayerischen Akademie der Wissenschaften.

Etzkowitz, Henry and Leydesdorff, Loet (eds.) (1997). *Universities and the global knowledge economy: A triple helix of university-industry-government relations*. London: Cassell Academic.

Etzkowitz, Henry and Leydesdorff, Loet (eds.) (1998). *A triple helix of university-industry-government relations: The future location of research?* New York: Science Policy Institute.

Etzkowitz, Henry and Leydesdorff, Loet (eds.) (1998). A triple helix of university-industry-government relations. *Industry & Higher Education* 12(4): 197–258.

Etzkowitz, Henry and Leydesdorff, Loet (2000). The dynamics of innovation: From national systems and "mode 2" to a triple helix of university-industry-government relations. *Research Policy* 29(2): 109–23.

Fellgett, Peter (1984). Three concepts make a million points. *Infrared Physics* 24: 95–8.

Fleck, Ludwik (1980). *Entstehung und Entwicklung einer wissenschaftlichen Tatsache*. Frankfurt: Suhrkamp.

Forman, Paul, Heilbron, John, and Weart, L. Spencer (1975). Physics circa 1900. Personnel, funding, and productivity of the academic establishments. *Historical Studies in the Physical Sciences* 5: 1–185.

Fortun, Michael and Mendelsohn, Everett (eds.) (1999). *The practices of human genetics. Sociology of the Sciences Yearbook 1999*. Dordrecht, The Netherlands: Kluwer Academic Publishers.

Fox, Robert and Weisz, George (eds.) (1980). *The organization of science and technology in France, 1808-1914*. Cambridge: Cambridge University Press.

Freedman, Michael I. (1979). Frederick Soddy and the practical significance of radioactive matter. *British Journal for the History of Science* 12(42): 257–60.

Fujimura, Joan (1987). Constructing doable problems in cancer research: Articulating alignment. *Social Studies of Science* 17: 257–93.

Fujimura, Joan (1992). Crafting science: Standardized packages, boundary objects and "translation", in A. Pickering (ed.), *Science as practice and culture* (pp. 168-211). Chicago: The University of Chicago Press.

Galison, Peter (1987). *How experiments end*. Chicago, IL: Chicago University Press.

Galison, Peter (1997). *Image and logic. A material culture of microphysics*. Chicago: The University of Chicago Press.

Galison, Peter and Hebbly, Bruce (eds.) (1992). *Big science, the growth of large-scale research*. Stanford, CA: Stanford University.

Galison, Peter and Thompson, Emily (eds.) (1999). *The architecture of science*. Cambridge;MA: MIT Press.

Gaudillière, Jean-Paul (1998). The molecularization of cancer etiology in the postwar United States: Instruments, politics, and management, in S. de Chadarevian and H. Kamminga (eds.), *Molecularizing biology and medicine: New practices and alliances 1910s – 1970s* (pp. 139–70). Amsterdam: Harwood Academic Publishers.

Gaudillière, Jean-Paul (1999). Circulating mice and viruses: The Jackson Memorial Laboratory, the National Cancer Institute, and the genetics of breast cancer, 1930-1965, in M. Fortun and E. Mendelsohn (eds.), *The practices of human genetics, sociology of science yearbook 1999* (pp. 89–124). Dordrecht, The Netherlands: Kluwer Academic Publishers.

Gaudillière, Jean-Paul (2000, in press). *L'invention de la biomédecine: la reconstruction des sciences du vivant en France après la guerre*. Paris: Editions des Archives Contemporaines.

Gaudillière, Jean-Paul and Löwy, Ilana (eds.) (1998). *The invisible industrialist: Manufactures and the construction of scientific knowledge*. London: Macmillan.

Gaudillière, Jean-Paul and Löwy, Ilana (eds.) (2000). *Transmission: Human diseases between heredity and infection. Historical approaches*: Amsterdam: Harwood Academic Publishers.

Gibbons, Michael, Limoges, Camille, Nowotny, Helga, Schwartzman, Simon, Scott, Peter and Trow, Martin (1994). *The new production of knowledge. The dynamics of science and research in contemporary societies*. London: Sage

Gingras, Yves (1991). *Physics and the rise of scientific research in Canada* (transl. by Peter Keating). Montréal: Kingston.

Goldschmidt, Bertrand (1987). *Pionniers de l'atome*. Paris: Stock.

Gordy, W. (1983). Jesse Wakefield Beams. *Biographical Memoirs* LIV (National Academy of Sciences of the United States of America): 3–49.

Grandy, David A. (1996). *Leo Szilard. Science as a mode of being*. Lanham, MD: University Press of America.

Greenaway, Frank (1978). Instruments, in T.I. Williams (ed.), *A history of technology. Vol. VIII, part 2* (p. 1204). Oxford: Clarendon.

Grinnell, Frederick (1992). *The scientific attitude* (New York and London: Guilford Press.

Gubin, Eliane (1990). Marie Curie et le radium: L'information et la légende en Belgique, in Université Libre de Bruxelles (ed.), *Marie Curie et la Belgique* (pp. 111–29). Brussels: ULB.

Guillot, Marcel (1967). Marie Curie-Sklodowska (1867–1934). *Nuclear Physics* A 103 (October): 1–8.

Hack, Lothar and Hack, Irmgard (1985). "Kritische Massen." Zum akademisch-industriellen Komplex im Bereich der Mikrobiologie/Gentechnologie, in W. Rammert, G. Bechmann, and H. Nowotny (eds.), *Technik und Gesellschaft*, Vol. 3 (pp. 132-58). Frankfurt/Main and New York: Campus.

Hacking, Ian (1983). *Representing and intervening*. Cambridge: Cambridge University Press.

Hacking, Ian (1989). The life of instruments. *Studies in the History and Philosophy of Science* 20: 265–370.

Hacking, Ian (1992). The self-vindication of the laboratory sciences, in A. Pickering (ed.), *Science as practice and culture* (pp. 29–64). Chicago: The University of Chicago Press.

Hacking, Ian (ed.) (1981). *Scientific revolutions*. Oxford: Oxford University Press.

Hagner, Michael and Rheinberger, Hans-Jörg (1998). Experimental systems, objects of investigation, and spaces of representation, in M. Heidelberger and F. Steinle (eds.), *Experimental Essays – Versuche zum Experiment* (pp. 355–73). Baden-Baden: Nomos

Hahn, Otto (1962). *Vom Radiothor zur Uranspaltung. Eine wissenschaftliche Selbstbiographie*. Braunschweig: Vieweg.

Harwood, Jonathan (1986). Ludwik Fleck and the sociology of knowledge. *Social Studies of Science* 16: 173–87.

Hartmann, Dirk and Janich, Peter (eds.) (1996). *Methodischer Kulturalismus: Zwischen Naturalismus und Postmoderne*. Frankfurt/M.: Suhrkamp.

Hasse, Raimund (1996). *Organisierte Forschung. Arbeitsteilung, Wettbeweb und Networking in Wissenschaft und Technik*. Berlin: edition sigma.

Hasse, Raimund and Gill, Bernhard (1994). Biotechnological research in Germany: Problems of political regulation and public acceptance, in U. Schimank and A. Stucke (eds.), *Coping with trouble. How science reacts to political disturbances of research conditions* (pp. 253–92). Frankfurt/Main and New York: Campus.

Heidelberger, Michael (1998). Die Erweiterung der Wirklichkeit, in M. Heidelberger and F. Steinle (eds.), *Experimental essays – Versuche zum Experiment* (pp. 71-92). Baden-Baden: Nomos.

Heidelberger, Michael and Steinle, Friedrich (eds.) (1998). *Experimental essays – Versuche zum Experiment*. Baden-Baden: Nomos.

Heilbron, John and Seidel, Robert W. (1989). *Lawrence and his laboratory. A history of the Lawrence Berkeley Laboratory*. Berkeley, CA: University of California Press.

Heims, Steve J. (1991). *The cybernetics group*. Cambridge, MA: MIT Press.

Helden, Albert van and Hankins, Thomas L. (eds.) (1994). Instruments. *Osiris* 9.

Hessenbruch, Arne (1994). *The commodification of radiations: Radium and X-Ray standards*. PhD dissertation. Cambridge: University of Cambridge.

Hewlett, Richard G. and Anderson, Oscar E. Jr., (1962). *A history of the United States atomic energy commission*. Volume I, *The New World 1939/1946* and Volume II, *Atomic shield 1947/1952*. University Park, PA: The Pennsylvania State University Press.

Hewlett Richard G. and Holl, Jack M. (1989). *Atoms for peace and war, 1953-1961*. Berkeley, CA: University of California Press.

Hippel, Eric von (1988). *The sources of innovation*. Oxford: Oxford University Press.

Hobsbawm, Eric (1995). *The age of extremes: The short 20th century, 1914-1991*. London: Abacus Edition, Little, Brown and Co.

Hobsbawm, Eric and Ranger, Terence (1983). *The invention of tradition*. Cambridge: Cambridge University Press.

Hohlfeld, Rainer (1988). Biologie als Ingenieurskunst. Zur Dialektik von Naturbeherrschung und synthetischer Biologie. *Ästhetik und Kommunikation* 69: 61–7.

Hohn, Hans-Willy and Schimank, Uwe (1990). *Konflikte und Gleichgewichte im Forschungssystem*. Frankfurt/Main und New York: Campus.

Holl, Jack M. (1997). *Argonne National Laboratory, 1946-96*. Urbana and Chicago: University of Illinois Press.

Holstein, Jean (1979). *The first fifty years at the Jackson Laboratory*. Bar Harbor: The Jackson Laboratory.

Hounshell, David (1992). Du Pont and the management of large-scale research and development, in P. Galison and B. Hebbly (eds.), *Big science, the growth of large-scale research*. Stanford, CA: Stanford University Press.

Hoyningen-Huene, Paul and Hirsch, Gertrude (eds.) (1988). *Wozu Wissenschaftsphilosophie?* Berlin: Walter de Gruyter.

Hughes, J.A. (1993). *The radioactivists: Community, controversy, and the rise of nuclear physics*. PhD Thesis, University of Cambridge.

Hughes, Jeff A. (1998). Plasticine and valves: Industry, instrumentation and the emergence of nuclear physics, in J.-P. Gaudillière and I. Löwy (eds.), *The invisible industrialist. Manufactures and the production of scientific knowledge* (pp. 58–1101). London: Macmillan Press.

Hughes, Thomas (1976). *Thomas Edison, professional inventor*. London: HMSO.

Hurwic, Anna (1995). *Pierre Curie*. Paris: Flammarion.

Jackson, Myles W. (2000). *Spectrum of belief: Joseph von Fraunhofer and the craft of precision optics*. Cambridge, MA: MIT Press.

Jackson, Myles W. (1996). Buying the dark lines of the solar spectrum, in J. Buchwald (ed.), *Archimedes: New studies in the history and philosophy of science and technology I* (pp. 1–21). Dordrecht, The Netherlands: Kluwer Academic Publishers.

Jackson, Myles W. (1999). Illuminating the opacity of achromatic lens production: Joseph von Fraunhofer and his monastic laboratory, in P. Galison and E. Thompson (eds.), *The architecture of science* (chapter 6). Cambridge, MA: MIT Press

Jacob, Margaret C. (ed.) (1994). *The politics of Western science, 1640-1990*. New Jersey: Humanities Press.

Joerges, Bernward (1996). *Technik: Körper der Gesellschaft*. Frankfurt/M.: Suhrkamp.

Joerges, Bernward (1999). High variability discourse in the history and sociology of technology, in O. Coutard (ed.), *The governance of large technical systems* (pp. 259–290). London: Routledge.

Johnston, Sean F. (1991). *Fourier Transform infrared: A constantly evolving technology*. London: Ellis Horwood.

Johnston, Sean F. (1994). *A notion or a measure: The quantification of light to 1939*. PhD Dissertation. Leeds, UK: University.

Johnston, Sean F. (1996a). Making light work: Practices and practitioners of photometry. *History of Science* 34: 273–302.

Johnston, Sean F. (1996b). The construction of colorimetry by committee. *Science in Context* 9: 387–420.

Karpik, Lucien (1989). L'économie de la qualité. *Revue française de sociologie* 30(2): 187–210.

Kay, Lily (1993). *The molecular vision of life*. Oxford: Oxford University Press.

Kay, Lily (1986). W. M. Stanley's Crystillization of the Tobacco Mosaic Virus, 1930-1940. *Isis* 77: 450–72.

Kay, Lily (1988). The Tiselius electrophoresis apparatus and the life sciences, 1930-1945. *History and Philosophy of the Life Sciences* 10: 51–72.

Keating, Peter and Cambrosio, Alberto (1997). Interlaboratory life: Regulating flow cytometry, in J.-P. Gaudillière, I. Löwy, and D. Pestre (eds.), *The invisible industrialist: Manufacturers and the construction of scientific knowledge* (pp. 250–74). London: Macmillan.

Kevles, Daniel (1971). *The physicists: The history of a scientific community in modern America*. New York: A. Knopf.

Kleinman, Daniel L. (1995). *Politics on the endless frontier*. Durham: Duke University Press.

Kohler, Robert (1991). *Partners in science: Foundations and natural scientists, 1900-1945*. Chicago: University of Chicago Press.

Kohler, Robert (1994). *The Lords of the Fly: Drosophila genetics and the experimental life*. Chicago: University of Chicago Press,

König, Wolfgang (1980). Technical education and industrial performance in Germany: A triumph of heterogeneity, in R. Fox and A. Guanini (eds.), *Education, technology and industrial performance, 1850–1939* (pp. 65–87). Cambridge: Cambridge University Press, 1980.

Krige, John (1996). The ppbar project, in J. Krige (ed.), *History of CERN*. Vol. 3 (pp. 207–74). Amsterdam: North-Holland.

Krige, John (ed.) (1996). *History of CERN*. Vol. 3. Amsterdam: 1996.

Krige, John (2000). Crossing the interface from R&D to operational use: The case of the European meteorological satellite. *Technology and Culture* 41(1): 27–50.

Krimsky, Sheldon, Ennis, James G., and Weissman, Robert (1991). Academic corporate ties in biotechnology: A quantitative study. *Science, Technology and Human Values* 16(3): 275–87.

Krohn, Wolfgang, Layton, Edwin T., and Weingart, Peter (eds.) (1978). *The dynamics of science and technology*. Dordrecht: D. Riedel Publishing Co., 1978)

Kuhn, Thomas S. (1962). *The structure of scientific revolution*. Chicago: University of Chicago Press.

Landa, Edward R. (1981). The first nuclear industry. *Scientific American* 247: 154–63.

Lanouette, William (1992). *Genius in the shadows: A biography of Leo Szilard; The man behind the bomb*. New York: Scribner's.

Latour, Bruno (1987). *Science in action*. Cambridge, MA: Harvard University Press.

Latour, Bruno (1992). *Aramis ou l'amour des techniques*. Paris: La Découverte.

Latour, Bruno (1993). *We have never been modern*. Cambridge: Harvard University Press.

Latour, Bruno and Woolgar, Steve (1986). *Laboratory life: The social construction of scientific facts*. Princeton, NJ: Princeton University Press.

Laudan, Larry (1981). A problem-solving approach to scientific progress, in I. Hacking (ed.), *Scientific revolutions* (pp. 144–55). Oxford, New York, etc.: Oxford University Press.

Layton, Edwin T. (1971b). *The revolt of the engineers: Social responsibility and the American engineering profession*. Cleveland, OH: Press of Case Western Reserve University.

Layton, Edwin T. (1971a). Mirror-image twins: The communities of science and technology in 19th century America. *Technology and Culture* 12, 562–80.

Lemaine, Gérard, MacLeod, Roy, Mulkay, Michael, and Weingart, Peter (eds.) (1976). *Perspectives on the emergence of scientific disciplines*. The Hague, Paris: Mouton.

Lenoir, Timothy (1997). *Instituting science. The cultural production of scientific disciplines*. Stanford, CA: Stanford University Press.

Lenoir, Timothy (1986). Models and instruments in the development of electrophotography, 1845-1912. *Historical Studies in the Physical and Biological Sciences* 17: 1–54.

Lenoir, Timothy and Hays, Marguerite (2000). The Manhattan Project for biomedicine, in P.R. Sloan (ed.), *Controlling our destinies: Historical, philosophical, social and ethical perspectives on the human genome project* (pp. 29–62). South Bend, IN: University of Notre Dame Press.

Loewenstein, Ernest V. (1966). The history and current status of Fourier transform spectroscopy. *Applied Optics* 5: 845–54.

Löwy, Ilana (1995). *Between bench and bedside*. Cambridge, CA: Harvard University Press.

Löwy, Ilana and Gaudillière, Jean-Paul (1998). Disciplining cancer: Mice and the practice of genetic purity, in J.-P. Gaudillière I. Löwy (eds.), *The invisible industrialist: Manufactures and the construction of scientific knowledge* (pp. 209–49). London: Macmillan.

Lundgreen, Peter, Horn, Bernd, Krohn, Wolfgang, Kueppers, Günter, and Paslack, Rainer (1986). *Staatliche Forschung in Deutschland, 1870–1980*. Frankfurt, New York: Campus.

MacDonald, Keith M. (1995). *The sociology of the professions*. London: Sage Publications.

Magner, Lois N. (1994). *A history of the life sciences*. New York, Basel, Hong Kong: Marcell Dekker, Inc.

Maienschein, Jane (1991). Epistemic styles in German and American embryology. *Science in Context* 4(2): 407–27.

Maines, David R. (ed.) (1991). *Social organization and social process. Essays in honor of Anselm Strauss*. New York: Aldine de Gruyter.

Mallard, Alexandre (1996). *Les instruments dans la coordination de l'action. Pratiques techniques, métrologie, instrument scientifique*. Unpublished Ph.D. dissertation. Paris: Ecole des Mines.

Mallard, Alexandre (1998). Compare, standardize, and settle agreement: On some usual metrological problems. *Social Studies of Science* 28(4): 571–601.

Malley, Marjorie (1979). The discovery of atomic transformation: Scientific styles and philosophies in France and Britain. *Isis* 70(252): 213–23.

Mannheim, Karl (1986). *Conservatism: A contribution to the sociology of knowledge*. London and New York: Routledge and Kegan Paul.

Mayr, Ernst (1982). *The growth of biological thought*. Cambridge, MA: Cambridge University Press.

McNeill, William H. (1991). *Hutchins' University. A memoir of the University of Chicago 1929-1950*. Chicago: The University of Chicago Press.

Mertz, Lawrence (1971). Fourier spectroscopy, past, present, and future. *Applied Optics* 10: 386–9.

Mindell, David (1996). *"Datum for its own annihilation": Feedback, control and computing, 1916–1945*. Ph.D. dissertation. Cambridge, MA: Massachusetts Institute of Technology.

Morris, Peter and Travis, Anthony (1997). The role of physical instrumentation in structural organic chemistry, in J. Krige and D. Pestre (eds.), *Science in the twentieth century* (pp. 715–40). Amsterdam: Harwood Academic Publishers.

Mulkay, Michael(1974). Conceptual displacement and migration in science: A prefatory paper. *Science Studies* 4(3): 205–34.

Mullins, Nicholas (1972). The development of a scientific specialty : The Phage Group and the origins of molecular biology. *Minerva* 10(1): 51–82.

Nelson, Rodney D. (1992). The sociology of styles of thought. *British Journal of Sociology* 43(1): 25–54.

Noble, David (1977). *America by design: Science, technology and the rise of corporate capitalism*. New York: A. Knopf.

Noble, David (1984). *Forces of Production: a Social History of Industrial Automation*. New York: Knopf.

Nye, Mary J. (1986). *Science in the provinces. Scientific communities and provincial leadership in France, 1860-1930*. Berkeley, CA: University of California Press.

Nye, Marie Joe (1993). *From chemical philosophy to theoretical chemistry: Dynamics of matter and dynamics of disciplines, 1800-1950*. Berkeley, CA: University of California Press.

Olesko, Kathy M. (1991). *Physics as a calling. Discipline and practice in the Königsberg seminar for physics*. Ithaca, NY: Cornell University Press.

Oster, Gerald (1966). A young physicist at seventy: Hartmut Kallmann. *Physics Today* 19: 51–4.

Osteryoung, Janet (1982). Developments in electrochemical instrumentation. *Science* 218: 261–65.

Pasachoff, Naomi (1996). *A.G. Bell*. New York: Oxford University Press.

Patterson, James T. (1987). *The dread disease, Cancer and modern American culture*. Cambridge, MA: Harvard University Press.

Paul, Harry W. (1985). *From knowledge to power: The rise of the science empire in France, 1860-1939*. Cambridge, UK: Cambridge University Press.

Pestre, Dominique (1984). *Physique et physiciens en France, 1918-1940*. Paris: Éditions des Archives Contemporaines.

Pestre, Dominique (1997). La production des savoirs entre académies et marché. *Revue d'Économie Industrielle* 79: 163–74.

Pestre, Dominique (1997). The moral and political economy of French scientists in the first half of the 20th century, in S. Boudia and X. Roqué (eds.), *Science, medicine and industry: The Curie and Joliot-Curie laboratories. History & Technology* (special issue), 13(4): 241–8.

Pfetsch, Frank (1970). Scientific organisation and science policy in imperial Germany, 1871–1914: The Foundation of the Imperial Institute of Physics and Technology. *Minerva* 8(4), 557–80.

Pflaum, Rosalynd (1989). *Grand obsession. Madame Curie and her world*. New York: Doubleday.

Pickering, Anthony (1984). *Constructing quarks: A sociological history of particle physics*. Chicago: University of Chicago Press.

Pickering, Andrew (ed.) (1992). *Science as practice and culture*. Chicago: University of Chicago Press.

Pickering, Andrew (1995). *The mangle of practice*. Chicago: University of Chicago Press.

Pinault, Michel (1997). The Joliot-Curies: Science, politics, networks, in S. Boudia and X. Roqué (eds.), *Science, medicine and industry: The Curie and Joliot-Curie laboratories. History & Technology* (special issue), 13(4): 307–24.

Pinch, Trevor J. (1986). *Confronting nature: The sociology of solar-neutrino detection*. Dordrecht, The Netherlands: D. Reidel Publishers.

Pinell, Patrice (1992). *Naissance d'un fléau: Histoire de la lutte contre le cancer en France (1890-1940)*. Paris: Métailié.

Polanyi, Michael (1958). *Personal knowledge*. London: Routledge & Kegan Paul.

Porter, Roy and Bayertz, Kurt (eds.) (1998). *From physico-theology to bio-technology*. Amsterdam: Rodopi.

Pyatt, Edward (1983). *The national physical laboratory: A history*. Bristol: Adam Hilger, 1983.

Quine, Willard V.O. (1969). *Ontological Relativity and other Essays*. New York: Columbia University Press.

Quine, Willard V.O. (1972). *Methods of logic*. New York: Holt, Rinehart and Winston.

Quine, Willard V.O. (1986). *Philosophy of logic*, 2nd ed. Cambridge, MA: Harvard University Press.

Rabinow, Paul (1996). *Making PCR. A story of biotechnology*. Chicago: The University of Chicago Press.

Rabkin, Yacob (1987). Technological innovation in science: The adoption of infrared spectroscopy by chemists. *ISIS* 78(291): 31–54.

Rader, Karen (1995). *Making mice: C.C. Little, The Jackson Laboratory and the standardization of* mus musculus *for research*. PhD thesis. Bloomington, IN: Indiana University.

Rammert, Werner, Bechmann, Gotthard, and Nowotny, Helga (eds.) (1985). *Technik und Gesellschaft*, Vol. 3. Frankfurt: Campus.

Rasmussen, Nicolas (1997). *Picture control. The electron microscope and the transformation of biology in America, 1940-1960*. Stanford, CA: Stanford University Press.

Rasmussen, Nicolas (1997). The midcentury biophysics bubble: Hiroshima and the biological revolution in America, revisited. *History of Science* 35: 245–99.

Rasmussen, Nicolas (1998). Instruments, scientists, industrialists and the specificity of 'influence: The case of RCA and biological electron microscopy, in J.-P.Gaudillière and I. Löwy (eds.), *The invisible industrialist: Manufactures and the construction of scientific knowledge* (pp. 173–208). London: Macmillan.

Reid, Robert (1974). *Marie Curie*. London: Collins.

Rheinberger, Hans-Jörg (1992). *Experiment, Differenz, Schrift*. Marburg: Basilisken Presse.

Rheinberger, Hans-Jörg (1997). *Toward a history of epistemic things: Synthesizing proteins in the test tube*. Stanford, CA: Stanford University Press.

Rheingans, Friedrich G. (1988). *Hans Geiger und die elektrischen Zählmethoden, 1908-1928*. Berlin: D.A.V.I.D. Verlagsgesellschaft.

Roqué, Xavier (1994). La stratégie de l'isolement. *Cahiers de Science & Vie* 24: 46–67.

Roqué, Xavier (1997). Marie Curie and the radium industry: A preliminary sketch, in S. Boudia and X. Roqué (eds.), *Science, medicine and industry: The Curie and Joliot-Curie laboratories. History & Technology* (special issue), 13(4): 267–81.

Rosenberg, Nathan (1994). *Exploring the black box: Technology, economics, and history*. Cambridge: Cambridge University Press.

Schaffer, Simon (1991). Late Victorian metrology and its instrumentation: A manufactory of ohms, in R. Budd and S.E. Cozzens (eds.), *Invisible connections: Instruments, institutions, and science* (pp. 23–56). London: SPIE Optical Engineering Press.

Scheibe, Arnold (1987). *Bedeutung der wissenschaftlichen Institute für die private Pflanzenzüchtung*. Hamburg und Berlin: Verlag Paul Parey.

Schimank, Uwe and Stucke, Andreas (eds.) (1994). *Coping with trouble. How science reacts to political disturbances of research conditions*. Frankfurt: Campus.

Servos, John (1980). The industrial relations of science: Chemical engineering at MIT, 1900–1939. *ISIS* 71: 531–49.

Shapin, Steven and Schaffer, Simon (1993). *Leviathan et la pompe à air: Hobbes, Boyle, entre science et politique*. Paris: La Découverte.

Shimshoni, Daniel (1970). The mobile scientist on the American instrument industry. *Minerva* 8(1): 58–89.

Shinn, Terry (1980). The genesis of French industrial research, 1880–1940. *Social Information* 19: 607–40.

Shinn, Terry (1985). How French universities became what they are. *Minerva* 23: 159–65.

Shinn, Terry (1988). Hiérarchie des chercheurs et formes des recherches. *Actes de la Recherche en Sciences Sociales* 74: 2–22.

Shinn, Terry (1993). The Bellevue grand electroaimant, 1900-1940: Birth of a research-technology community. *Historical Studies in the Physical Sciences* 24, 157–87.

Shinn, Terry (1994). Science, Tocqueville, and the state: The organization of knowledge in modern France, in M.C. Jacob (ed.), *The politics of western science, 1640-1990* (pp. 47–80). New Jersey: Humanities Press.

Shinn, Terry (1997). Crossing boundaries: The emergence of reseach-technology communities, in H. Etzkowitz and L. Leydesdorff (eds.), *Universities and the global knowledge economy: A triple helix of university-industry-government relations* (pp. 85–96). London: Cassell Academic Press.

Shinn, Terry (1998). Instrument hierarchies: Laboratories, industry and divisions of labour, in J.-P. Gaudillière I. Löwy (eds.), *the invisible industrialist: Manufactures and the production of scientific knowledge* (pp. 102–21). London: Macmillan Press.

Shinn, Terry (1998). L'effet pervers des commémorations en science, in P. Abir-Am (ed.), *La mise en mémoire de la science* (pp. 225–47). Paris: Editions des Archives Contemporaines.

Shinn, Terry (1999). Change or mutation? Reflections on the foundations of contemporary science. *Social Science Information* 38(1, March), 149–76.

Shinn, Terry (2000). Formes de division du travail scientifique et convergence intellectuelle. La recherche technico-instrumentale. *Revue Française de Sociologie* 41(3): 447–73.

Sibum, Otto (1998). *Les gestes de la mesure: Joule, les pratiques de la brasserie et la sciences*. Paris: Annales Histoires Sciences Sociales

Sloan, Phillip R. (ed.) (2000). *Controlling our destinies: Historical, philosophical, social and ethical perspectives on the human genome project*. South Bend: University of Notre Dame Press.

Star, Susan Leigh and Griesemer, James R. (1989). Institutional ecology, "translations," and boundary objects: Amateurs and professionals in Berkeley's Museum of Vertebrate Zoology, 1907-39. *Social Studies of Science* 19: 387–420.

Starr, Paul (1982). *The social transformation of American medicine*. New York: Basic Books.

Straub, Joseph (1986). Aus der Geschichte des Kaiser-Wilhelm-/Max-Planck-Instituts für Züchtungsforschung. *Berichte und Mitteilungen der Max-Planck-Gesellschaft* 2: 11–36.

Stubbe, Hans (1959). Gedächtnisrede auf Erwin Baur, gehalten am 25. Todestag (2. Dezember 1958) in Müncheberg/Mark. *Der Züchter* 29(1): 1–6.

Suppe, Frederick (1974). *The structure of scientific theories*. Chicago: University of Illinois Press.

Swenson, Loyd S. (1972). *Ethereal aether: A history of the Michelson-Morley-Miller aether-drift experiments, 1880-1930*, Austin, TX: University of Texas Press.

Thackray, Arnold and Myers, Minor (2000). *Arnold O. Beckman: One hundred years of excellence*. Philadelphia: Chemical Heritage Foundation.

Thévenot, Laurent (1994). Le régime de familiarité. Des choses en personne. *Genèses* 17 (September): 72–101.

Timmermans, Stefan and Berg, Marc (1997). Standardization in action: Achieving local universality through medical protocols. *Social Studies of Science* 27: 273–305.

Université Libre de Bruxelles (ed.) (1990). *Marie Curie et la Belgique*. Brussels: ULB.

Vanderlinden, Jacques (1990). Marie Curie et le radium "belge", in Université Libre de Bruxelles (ed.), *Marie Curie et la Belgique* (pp. 91–109). Brussels: ULB.

Van Helden, Albert and Hankins, Thomas (eds.) (1994). Instruments (special issue). *Osiris* 9.

Varian, Dorothy (1983). *The inventor and the pilot: Russel and Sigurd Varian*. Palo Alto, CA: Pacific Books.

Vincent, Bénédicte (1997). Genesis of the pavillon Pasteur of the Institut du Radium, in S. Boudia and X. Roqué (eds.), *Science, medicine and industry: The Curie and Joliot-Curie laboratories. History & Technology* (special issue), 13(4): 293-305.

Weart, Spencer R. (1979). *Scientists in power*. Cambridge, MA: Harvard University Press.

Weart, Spencer R. (1988). *Nuclear fear. A history of images*. Cambridge, MA: Harvard University Press.

Weingart, Peter (1978). The relationship between science and technology – A sociological explanation, in W. Krohn, E.T. Layton, and P. Weingart (eds.), *The dynamics of science and technology* (pp. 251–86). Dordrecht, The Netherlands, and Boston: D. Riedel.

Whitley, Richard (1984). *The intellectual and social organization of the sciences*. Oxford: Clarendon Press.

Whitley, Richard (1982). The establishment and structure of the sciences as reputational organizations, in N. Elias, H. Martins, and R. Whitley (eds.), *Scientific establishment and hierarchies* (pp. 313–53). Dordrecht, Boston and London: D. Riedel.

Williams, Mari E.W. (1994). *The precision makers: A history of the instrument industry in Britain and France 1870-1939*. London: Routledge.

Wise, Norton (ed.), *The Values of Precision* (Princeton: Princeton University Press, 1995)

Wise, Norton and Smith, Crosbie (1989). *Energy and empire. A biographical study of Lord Kelvin*. Cambridge: Cambridge University Press.

Yoxen, Edward (1982). Giving life a new meaning: The rise of the molecular biology establishment, in N. Elias, H. Martins, and R. Whitley (eds.), *Scientific establishment and hierarchies* (pp. 313-57). Dordrecht, Boston, London: D. Riedel.

Zimmerli, Walther Ch. (1988). Ethik der Wissenschaft als Ethik der Technologie. Zur wachsenden Bedeutsamkeit der Ethik in der gegenwärtgen Wissenschaftsforschung, in P. Hoyningen-Huene and G. Hirsch (eds.), *Wozu Wissenschaftsphilosophie?* (pp. 391–418). Berlin: Walter de Gruyter.

Zimmerli, Walther Ch. (1990). Handelt es sich bei der gegenwärtigen Technik noch um Technik? Elf Thesen zu "langen Wellen" der Technikentwicklung, in B. Bievert and K. Monse (eds.), *Wandel durch Technik? Institution, Organisation, Alltag* (pp. 289–310). Opladen: Westdeutscher Verlag.

Zubord, C. Gordon, Schepartz, Saul A., Leiter, John, Endicott, Kenneth M., Carrese, Martin L., and Baker, Christopher G. (1966). History of the cancer chemotherapy program. *Cancer Chemotherapy Reports* 50: 349–81.

Zubord, C. Gordon, Schepartz, Saul A., and Carter, Stephen C. (1977). Historical background of the National Cancer Institute's drug development trust. *National Cancer Institute Monographs* 45: 7–11.

LIST OF CONTRIBUTORS

Gaudillière, Jean Paul, INSERM, Centre de Recherches Medecine, Sciences, Société, 182 boulevard de la Villette, F-75019 Paris, France.

Hasse, Raimund, Rheinisch-Westfälische Technische Hochschule Aachen, Karman-Forum Institut für Soziologie, D-52056 Aachen, Germany.

Hohlfeld, Rainer, Berlin-Brandenburgische Akademie der Wissenschaften, Arbeitsgruppe Wissenschaft und Wiedervereinigung, Jägerstrasse 22-23, D-10117 Berlin, Germany.

Jackson, Myles W., Willamette University, Humanities Center/History of Science, Eaton Hall 314, 900 State Street, Salem, Oregon 97301, USA.

Joerges, Bernward, Wissenschaftszentrum Berlin für Sozialforschung, Forschungsgruppe "Metropolenforschung", Reichpietschufer 50, D-10785 Berlin, Germany.

Johnson, Ann, Fordham University, Department of History, Dealy Hall, Bronx, NY 10458, USA.

Johnston, Sean F., University of Glasgow, Crichton Campus, Science Studies, Rutherford Building, Dumfries DG1 4ZL, Scotland.

Mallard, Alexandre, Laboratoire "Usages Créativité Ergonomie", France Télécom Recherche & Développement, FTRD/DIH/UCE, 3840 rue du Général Leclerc, F-92794 Issy Moulineux Cedex 9, France.

Nevers, Patricia, Universität Hamburg, Fachbereich Erziehungswissenschaft, Institut 9, Von-Melle-Park 8, D-20146 Hamburg, Germany.

Rheinberger, Hans-Jörg, Dir. Max-Planck-Institut für Wissenschaftsgeschichte, Wilhelmstrasse 44, D-10117 Berlin, Germany.

Roqué, Xavier, Universitat Autònoma de Barcelona, Centre d'Estudis d'Història de les Ciències (CEHIC), Edifici Cc, E-08193 Bellaterra, Spain.

Shinn, Terry, GEMAS/CNRS, Maison des Sciences de l'Homme, 54, boulevard Raspail, F-75270 Paris Cedex 06, France.

Zimmerli, Walther, President, Universität Witten/Herdecke, Alfred-Herrhausen-Strasse 50, D-58455 Witten, Germany.

BIOGRAPHICAL NOTES ON CONTRIBUTORS

Jean-Paul Gaudillière is historian of biology and medicine at INSERM. He is the author of articles on the history of biology, cancer research and medical genetics. With Ilana Löwy he is the editor of *The invisible industrialist: Manufactures and the construction of scientific knowledge* (Macmillan 1998). He is currently working on a book comparing the development of biomedicine in postwar France and United States.

Raimund Hasse is a sociologist at the University of Aachen. In 1998/9 he was a research fellow at the University of Wisconsin, Madison. His main research fields are science and technology studies and organization research. Publications include: *Organisierte Forschung* (Berlin 1996); *Neo-Institutionalismus* (with G. Krücken, Bielefeld 1999). Currently he is studying the transnational diffusion of political and economic changes.

Rainer Hohlfeld, after receiving his Ph.D. in bacterial genetics at the University of Cologne, worked until 1980 at the Max-Planck-Institute for Research on the Conditions of Life in the Scientific-Technical World, and from 1981 to 1993 at the Institute for Science and Society in Erlangen, an institute devoted to the study of science and technology in East Germany. His work has focused on the sociological analysis of biological and biomedical research. He is currently at the Berlin-Brandenburg Academy of Sciences in Berlin.

Myles W. Jackson is currently an Alexander von Humboldt Research Fellow at the Max-Planck-Institut für Wissenschaftsgeschichte in Berlin. He has published numerous articles on Goethe and science and on the history of German optics. He has just completed a book entitled, *Spectrum of Belief: Joseph von Fraunhofer and the craft of precision optics* (MIT Press 2000). He is currently working on a project dealing with the triangular exchange among composers and musicians, musical instrument makers, and physicists in nineteenth-century Germany.

Bernward Joerges is a researcher at the Wissenschaftszentrum Berlin für Sozialforschung (WZB) where he heads the Metropolitan Studies Group. He is also professor of sociology at the *Technische Universität Berlin*. He has published widely in German, English and French in the fields of science, technology and urban studies. Book publications include *Technik ohne Grenzen* (Suhrkamp, Frankfurt a.M. 1994, with Ingo Braun), *Technik - Körper der Gesellschaft* (Suhrkamp, Frankfurt a. M. 1996) and *Körper-Technik: Aufsätze zur Organtransplantation* (edition sigma, Berlin 1997).

Ann Johnson is an assistant professor of the history of science and technology at Fordham University in Bronx, NY. She works on the culture and epistemology of engineering, especially the ways new ideas and products are generated between the public and private sectors. She wishes to thank Norton Wise for his support in attending the 1997 Sociology of the Sciences Workshop.

Sean F. Johnston is a lecturer in Science Studies at Crichton Campus of the University of Glasgow in Dumfries, UK. His research has dealt with the emergence of technical communities and intellectual bases for spectroscopy and light measurement in the nineteenth and twentieth centuries, and he currently is writing a history of chemical engineering in Britain with Colin Divall. He was an International Scholar of the Society for the History of Technology (USA) in 1996 and 1997.

Alexandre Mallard received his PhD in the sociology of innovation from the Ecole des Mines de Paris in 1996, and is now a researcher at the Centre National d'Etudes des Télécommunications. His dissertation focused on metrology and standardization in the automobile industry, and air quality control. He is currently working on the professional use of communication and information technologies, and on the role played by consumer associations in information technologies markets.

Patricia Nevers is a Professor of Biology Education at the University of Hamburg. She studied biology and chemistry at the University of Freiburg and obtained her Ph.D. in genetics from the same university. She has worked in research in bacterial and plant genetics, taught school, developed teaching materials dealing with gene technology for teachers and schools, and participated in a project in the philosophy and sociology of science examining styles of thought in plant genetics, which is the topic of her contribution to this book. Currently she is involved in investigating children's attitudes towards nature and forms of knowledge employed by children in relating to nature.

Hans-Jörg Rheinberger (born 1946). Studied Philosophy and Biology in Tübingen and Berlin. Molecular biologist and historian of science. 1978–1990 research scientist at the Max Planck Institute for Molecular Genetics in Berlin, 1990-1994 Assistant Professor at the Institute for the History of Medicine and Science, University of Lübeck, 1994–1996 Associate Professor at the Institute for Genetics and General Biology, University of Salzburg, since 1997 Director at the Max Planck Institute for the History of Science in Berlin. Numerous research papers in molecular biology and history of science. Books among others: *Experiment, Differenz, Schrift* (Basiliskenpresse, Marburg 1992); *Toward a history of epistemic things. Synthesizing proteins in the test tube* (Stanford University Press, Stanford 1997).

Xavier Roqué (born 1965) is a lecturer in the history of science at the Universitat Autònoma de Barcelona. Following a degree in physics, he took his PhD in the history of science at the Universitat Autònoma de Barcelona in 1993. He held postdoctoral positions in the Department of History and Philosophy of Science of the University of Cambridge and the Centre de Recherche en Histoire des Sciences et des Techniques/CNRS, at the Cité des Sciences et de l'Industrie. He has a number of publications on the history of twentieth-century physics.

Terry Shinn is Director of Research in sociology at the Centre National de la Recherche Scientifique in Paris. He has written extensively on the history and sociology of science and engineering education as well as on the organization of industrial research, university and industry relations, the organization of scientific disciplines and the roles of science popularization. The author is currently completing a book on the history of French research-technology.

Walther Zimmerli, born 1945 in Zurich, studied philosophy, German and American literature, and different other subjects at Yale, Goettingen, and Zurich; PhD 1971, habilitation 1978 (both Zurich); 1978–1988 full professor of philosophy at Braun-schweig and acting professor at Goettingen, 1988–1996 chaired professor in a joint appointment at Bamberg and Erlangen, 1996–1999 chaired professor at Marburg. Since 1999 president of Witten/Herdecke private university (UWH). Visiting appointments in the U.S., Australia, Japan, South Africa, and different European countries. Main areas of research: philosophy of science and technology, applied ethics, aesthetics, social and political philosophy. Numerous publications, including 30 books and some hundred articles or chapters in books and journals.

AUTHOR INDEX

---

*Numbers in italics refer to reference pages.

Sociology of the Sciences

1. E. Mendelsohn, P. Weingart and R. Whitley (eds.): *The Social Production of Scientific Knowledge*. 1977   ISBN Hb 90-277-0775-8; Pb 90-277-0776-6
2. W. Krohn, E.T. Layton, Jr. and P. Weingart (eds.): *The Dynamics of Science and Technology*. Social Values, Technical Norms and Scientific Criteria in the Development of Knowledge. 1978   ISBN Hb 90-277-0880-0; Pb 90-277-0881-9
3. H. Nowotny and H. Rose (eds.): *Counter-Movements in the Sciences*. The Sociology of the Alternatives to Big Science. 1979
        ISBN Hb 90-277-0971-8; Pb 90-277-0972-6
4. K.D. Knorr, R. Krohn and R. Whitley (eds.): *The Social Process of Scientific Investigation*. 1980 (1981)   ISBN Hb 90-277-1174-7; Pb 90-277-1175-5
5. E. Mendelsohn and Y. Elkana (eds.): *Sciences and Cultures*. Anthropological and Historical Studies of the Sciences. 1981
        ISBN Hb 90-277-1234-4; Pb 90-277-1235-2
6. N. Elias, H. Martins and R. Whitley (eds.): *Scientific Establishments and Hierarchies*. 1982   ISBN Hb 90-277-1322-7; Pb 90-277-1323-5
7. L. Graham, W. Lepenies and P. Weingart (eds.): *Functions and Uses of Disciplinary Histories*. 1983   ISBN Hb 90-277-1520-3; Pb 90-277-1521-1
8. E. Mendelsohn and H. Nowotny (eds.): *Nineteen Eighty Four: Science between Utopia and Dystopia*. 1984   ISBN Hb 90-277-1719-2; Pb 90-277-1721-4
9. T. Shinn and R. Whitley (eds.): *Expository Science*. Forms and Functions of Popularisation. 1985   ISBN Hb 90-277-1831-8; Pb 90-277-1832-6
10. G. Böhme and N. Stehr (eds.): *The Knowledge Society*. The Growing Impact of Scientific Knowledge on Social Relations. 1986
        ISBN Hb 90-277-2305-2; Pb 90-277-2306-0
11. S. Blume, J. Bunders, L. Leydesdorff and R. Whitley (eds.): *The Social Direction of the Public Sciences*. Causes and Consequences of Co-operation between Scientists and Non-scientific Groups. 1987   ISBN Hb 90-277-2381-8; Pb 90-277-2382-6
12. E. Mendelsohn, M.R. Smith and P. Weingart (eds.): *Science, Technology and the Military*. 2 vols. 1988   ISBN Vol, 12/1 90-277-2780-5; Vol. 12/2 90-277-2783-X
13. S. Fuller, M. de Mey, T. Shinn and S. Woolgar (eds.): *The Cognitive Turn*. Sociological and Psychological Perspectives on Science. 1989   ISBN 0-7923-0306-7
14. W. Krohn, G. Küppers and H. Nowotny (eds.): *Selforganization*. Portrait of a Scientific Revolution. 1990   ISBN 0-7923-0830-1
15. P. Wagner, B. Wittrock and R. Whitley (eds.): *Discourses on Society*. The Shaping on the Social Science Disciplines. 1991   ISBN 0-7923-1001-2
16. E. Crawford, T. Shinn and S. Sörlin (eds.): *Denationalizing Science*. The Contexts of International Scientific Practice. 1992 (1993)   ISBN 0-7923-1855-2
17. Y. Ezrahi, E. Mendelsohn and H. Segal (eds.): *Technology, Pessimism, and Postmodernism*. 1993 (1994)   ISBN 0-7923-2630-X
18. S. Maasen, E. Mendelsohn and P. Weingart (eds.): *Biology as Society? Society as Biology: Metaphors*. 1994 (1995)   ISBN 0-7923-3174-5
19. T. Shinn, J. Spaapen and V. Krishna (eds.): *Science and Technology in a Developing World*. 1995 (1997)   ISBN 0-7923-4419-7

Sociology of the Sciences

20. J.Heilbron, L. Magnusson and B.Wittrock (eds.): *The Rise of the Social Sciences and the Formation of Modernity*. Conceptual Change in Context, 17501850. 1996 (1998)
ISBN 0-7923-4589-4

21. M. Fortun and E. Mendelsohn (eds.): *The Practices of Human Genetics*. 1998
ISBN 0-7923-5333-1

22. B. Joerges and T. Shinn (eds.): *Instrumentation: Between Science, State and Industry*. 2000
ISBN 0-7923-6736-7

KLUWER ACADEMIC PUBLISHERS – BOSTON / DORDRECHT / LONDON